Aquaculture, Resource Use, and the Environment

Aquaculture, Resource Use, and the Environment

Claude E. Boyd
Professor, Water Quality
School of Fisheries, Aquaculture and Aquatic Sciences
College of Agriculture
Auburn University
Alabama, USA

Aaron A. McNevin
Director, Aquaculture
World Wildlife Fund
Washington, DC, USA

WILEY Blackwell

Published by John Wiley & Sons, Inc., Hoboken, New Jersey.
Published simultaneously in Canada.

For general information on our other products and services or for technical support, please contact our Customer Care Department within the United States at (800) 762-2974, outside the United States at (317) 572-3993 or fax (317) 572-4002.

Wiley also publishes its books in a variety of electronic formats. Some content that appears in print may not be available in electronic formats. For more information about Wiley products, visit our web site at www.wiley.com.

Library of Congress Cataloging-in-Publication Data

Boyd, Claude E.
  Aquaculture : resource use, and the environment / Claude E. Boyd and Aaron McNevin.
    pages cm
  Includes index.
  ISBN 978-0-470-95919-0 (cloth)
  1. Aquaculture–Environmental aspects.   I. McNevin, Aaron.   II. Title.
  SH135.B69 2015
  639.8–dc23

                                                                          2014038157

A catalogue record for this book is available from the British Library.

Cover images: Fish farm ©iStock/ imagedepotpro, The Shrimp farming ©iStock/ pinyoj
Printed and bound in Malaysia by Vivar Printing Sdn Bhd

1   2015

Contents

Foreword

Aquaculture is the fastest growing food production system in the world. And, it is not because it plays such a minor role in global food production. In the past five years, the production of seafood from all aquaculture that is consumed directly by people exceeded that of all global fisheries. Two years ago, production from aquaculture exceeded that of all beef on the planet. For the past 30 years, aquaculture production has grown on average, 7–10% per year globally.

Put another way, aquaculture production has surpassed both fisheries and beef production in a span of about 30 years compared with ten thousand years or more that it took the latter to achieve such levels. The learning curve of aquaculture has been steep, and mistakes have been made. But if you think about the amount of production, the amount of time, and the overall effort involved, aquaculture producers have learned very quickly how to become more efficient and produce more with less—both fewer inputs and impacts.

The math of the planet is simple—population × consumption = the planet's resources. This can be mitigated by technology and improved practices, but some parameters are hard and fast limits to the planet's carrying capacity. Today we have 7 billion people and they consume an average of 7 billion units of resources. By 2050 we will have at least 9 billion people and they are expected to have 2.9 times as much income per capita and consume about twice as much per capita. So by 2050, the projections of the World Bank and others suggest that we will have 18 billion units of consumption. If nothing is done differently in 37 years, the global footprint of consumption will double what it is today as each person will consume twice as much as they do now.

No one actually believes that the business as usual case will continue. People are ingenious. We have invented new technologies and found solutions to problems like this when Malthus first raised this issue. However, the question is how much and how quickly can we reduce the footprint of each unit of consumption, for example, how much land, water, feed, soil erosion, N, P, K, other nutrients, and pesticides will be used to produce each unit of consumption. If we indeed have 18 billion units of consumption by 2050, we will need to reduce the impact of each unit of consumption by more than 60% of what it is today just to be within the resource limitations of the planet. Yet, World Wildlife Fund's Living Planet Report suggests that we are living at 1.5 planets worth of resources today, that is, we are living beyond the carrying capacity. For farmers, that is the equivalent of eating our seed. To actually return to a balance with the planet's resources we will need to reduce the impacts of each unit of consumption by 75–80%.

Which food production systems can improve this dramatically? To date, only aquaculture has achieved such performance. Recent history has shown that aquaculture, when most efficient, can support the levels of performance required to live within the planet's resource boundaries. But it will not be easy.

And, the journey has not been easy to date. In the 1990s, NGOs became concerned by the rapid growth of the global aquaculture industry. Inevitably it was local NGOs (both social and environmental) that saw the impacts more quickly because they were living with them. This was true of salmon, shrimp, tilapia, catfish, *Pangasius*, trout, shellfish, and even seaweed. Later, global NGOs became aware of the issues and began to engage the sector beginning in the early to mid-1990s.

The initial interaction was not pretty. In fact it was very contentious and in at least some cases involved not only confrontations at various meetings but also lawsuits and countersuits. No "side" felt that it was being listened to or accurately portrayed by the other. The hot issues early on in the debate were mangroves, disease, access to resources, escapes, chemicals, predator control, zoning and carrying capacity, and a range of social issues from worker rights to cumulative impacts on local communities. Without a credible baseline, much less science-based consensus about what was happening within specific production areas, the arguments went on without resolution.

At this point, the World Bank, NACA, UN FAO, and WWF agreed to work together to build awareness about the issues, and consensus about a way forward, for a single species produced by aquaculture—shrimp. This work cost $1 million (coming mostly from the World Bank, and the MacArthur and Avina Foundations), spanned three years (1999–2002), explored issues in 30 countries, produced 44 studies, involved 140 different researchers, and involved more than 7000 experts in local, regional, and national meetings. In the end, these efforts produced the most up-to-date analyses about the impacts of shrimp aquaculture production, what was being done to address them, and the economics of moving forward. This work took the steam out of the debate, built awareness about the actual reality of shrimp aquaculture, and built consensus about how to move forward. There were still detractors—extremes at both ends, the NGO and producer level—but a middle ground had been found and there was agreement about how to move forward. That is not to say that there were no ongoing battles over real issues—Ecuador and shrimp and British Colombia and salmon are good examples—but these were the exceptions whereas before they had been the rule.

In 2004, the first species-specific aquaculture dialogues were launched beginning with the Salmon Aquaculture Dialogue (SAD) on Valentine's Day in Washington, DC. And, while three of the eight founding members of the SAD were actually suing each other at the time, it did not turn into another Valentine's Day Massacre as had been predicted. In fact, even before the meeting, the 130 participants had agreed on six of the seven key impacts that would consume the work of the SAD for 8 years. Science-based presentations at the meeting kicked off the discussions around each of the issues and where there was conflicting science, two presenters were asked to make the cases.

The aquaculture dialogues focused the attention of producers, NGOs, and researchers on the real issues. There was a lot of education and exchange going both ways. But if anything, I would argue that the NGOs gained the most from the exchanges (with some notable exceptions regarding issues like feed in/out ratios!) because the information about the aquaculture industry that was in the public domain was not up-to-date. Most published information was at least 10 years old. For an industry with a steep learning curve, 10 years is a lifetime.

Depending upon the species, the dialogues were started from 2004 to 2007. It took each dialogue 5–8 years to develop credible standards. All told, it cost about $10 million (there were additional in-kind contributions for travel, accommodation, etc.) and 10 years to run the dialogues, generate the standards, develop guidance documents, create the Aquaculture Stewardship Council (ASC) and hand the standards off to the ASC.

Specifically, it took about $3.5 million to launch the SAD in 2004 and support its work to its conclusion in 2012 when the standards were handed off to the ASC. This is not an insignificant amount of time or money. At the producer level, salmon aquaculture is a $5.4 billion per year industry, however. In that context, the investment of time and money does not seem so large if it actually improves the industry's performance.

Unbeknownst to the NGOs, over the past 2 years, salmon aquaculture producers created the Global Salmon Initiative (GSI). In August, the GSI was launched at AquaNor in Trondheim, and 15 CEOs announced that as a group representing some 70% of global production they were committed to be 100% certified according to ASC standards by 2020. For them, the ASC was the gold standard.

This is the first time that any sector in the food business has made such a commitment. That in itself is groundbreaking. However, what is even more important is the fact that the companies have all agreed to share information about how to improve performance more quickly. In short, these companies see sustainability as a precompetitive issue. They recognize that the poor performance of one company can affect all the others—and not just their access to markets but also their license to operate as well. In short, the products these companies sell are still competitive, but how they are produced and what their impact is on the environment is precompetitive—they need to work on it together not only to achieve improved performance, but to achieve it faster and cheaper as well.

To do this, the companies will need an open source database to allow them to share their information about better management practices as well as their costs and payback periods. In addition, they need to share lessons about practices and paths to avoid as well as lessons learned the hard way. This interaction has also made it easier for the companies to share concerns about the industry and to identify trends and issues much more quickly than they have in the past.

The impact of the GSI does not stop there, however. Within 24 hours of their announcement, the world's largest chocolate company contacted me to see if the same thing could be done for cocoa. The next day the head of a global food brand asked if something similar could be done for palm oil. And the third day, three people asked if it would be possible to start a similar group for shrimp aquaculture producers interested in the ASC. The GSI is not readily transferable to other industries—salmon aquaculture production is highly concentrated at the producer level. Other producer groups are far larger in number and more geographically diverse. Still, the GSI has shown that working together is not only possible, it is practical.

The global aquaculture industry has come a long way. Thirty years ago the industry began to grow at phenomenal rates without many outsiders even paying attention. Mistakes were made. Twenty years ago, NGOs began to call on the industry to address them and then condemned the industry when this did not appear to happen quickly enough.

In the late 1990s, the global work on shrimp aquaculture undertaken jointly by the World Bank, NACA, the UN FAO, and WWF marked a major turning point. This was the first time that NGOs had worked with other institutions to understand the global impacts of a single aquaculture industry and sort out which concerns were more science based and which, perhaps, less so. The research on shrimp pivoted the discourse from confrontation to science-based analysis, awareness, and consensus about key impacts. This gave way to a period of 10 years during which globally credible standards were developed for the 12 species produced by aquaculture that have the highest value and volume globally. These science-based, multi-stakeholder aquaculture dialogues produced globally credible standards that were then handed off to the independent ASC.

But as the GSI demonstrates, the work is not done, it is just beginning. Now is when we need to find ways for producers to begin a stepwise approach to continuous improvement. It is unrealistic to expect all producers to be able to meet credible standards without changing the way they culture different species. A stepwise approach is essential. It should start by ensuring that producers are operating legally, that is, they have the legal right to the resources they use, only use legal inputs, do not exceed legally required pollution levels, obey all relevant laws, and keep records, and report as required in the country of production.

The next step would be to identify the low hanging fruit—the best management practices that would allow producers to improve performance, increase efficiency, and have higher net earnings. Every producer has some ability to do this, though some certainly have more room for improvement than others. Ideally, savings or earnings from these early efforts would allow producers to invest in other technologies or practices that will have a longer payback period. If these are sequenced right, considerable improvements can be made that will improve performance and increase income even in the short term.

For some producers, production may be for domestic markets. For them certification may not be beneficial as it will not provide additional market access. However, the improved performance associated with meeting credible global standards can make any producer more profitable whether certification is a good option or not. No producer can focus on the thousands of impacts they see in their operations every day. The standards represent global consensus about the most important 6–10 impacts and then the performance level indicators that suggest better performance.

With each species, documenting and sharing how the better producers have improved their performance and reduced their impacts while increasing their net income is important. We need systems for sharing this information with other producers but also to show governments, so that enabling conditions can be created that encourage producers to become better. This generates income for producers, economic development, taxes, and reduced impacts that would otherwise have to be addressed by society at large.

The journey to date of global aquaculture has been amazingly rapid. Producers, researchers, and those providing technical assistance have done much to improve the performance of this nascent industry during a period of very steep learning curves, probably steeper than any ever before experienced by any other food sector. They should be proud of what they have accomplished. NGOs have also helped focus producers and the industry more broadly on being strategic and focusing on a few impacts rather than trying to focus on all of them at once. And, together, all of these groups have shifted the focus from practices to performance, to results. Practices are a means to an end, but the end, the performance, is what needs to be measured.

Aquaculture is the future of seafood. But, make no mistake, there is still much to be done if we are going to double production by 2050 without using more resources. No one producer or institution can do everything. Everyone can do something. Together, we can make a difference. Think about it.

Jason Clay
Senior Vice President, Market Transformation
World Wildlife Fund
Rome, Italy

Foreword

If you happen to travel frequently, as I do, perhaps you too marvel at how commonplace, safe, and comfortable air travel has become. Such achievements, of course, did not happen all at once. It took many tries before Orville and Wilbur Wright's motorized glider took flight to become the world's first viable airplane. Humanity's remarkable progress in science and engineering has been achieved in countless small steps that have built upon each other in an age-old process of continuous improvement. Even major scientific breakthroughs have been the product of preceding advances. As Isaac Newton said, "If I have seen further than others, it is by standing upon the shoulders of giants."

In a corresponding way, humanity has also added an increasing environmental burden to our planet over the ages. As our population continues to grow, we recognize the impending limits of our consumption, and we strive for sustainability. The journey of aquaculture toward sustainability has also been one of continuous improvement that has come in steps—sometimes small steps and even missteps.

In the case of shrimp farming, Dr. Motosaku Fujinaga achieved the first breakthroughs in closing the life cycle of shrimp during the 1930s. Early shrimp farmers made mistakes such as choosing mangrove sites for shrimp ponds and relying on antibiotics to manage diseases. As with pioneers in aviation, they learned and improved. Today's operators use specific pathogen-free, genetically improved animals to produce high yields in biosecure ponds with zero water exchange and vegetable-protein feeds. They continue to err, but their mistakes are fewer and farther between.

Urging aquaculture forward is the rising global demand for seafood. This is driven not only by increasing population, but also by the rising middle class in China and elsewhere, which is increasing per capita consumption. Marine fisheries cannot supply this increasing demand, because landings have been stagnant for over a decade and most of the valuable species are either fully exploited or over exploited. Aquaculture is the only means of meeting rising seafood demand, and it has become the fastest growing sector of global food production.

While global aquaculture production doubled during the 1990s, its growth has slowed since then, due to increasing constraints such as environmental limits, disease outbreaks, and availability of feed ingredients. The way forward is to produce more efficiently and responsibly. This is gradually being achieved through advances in genetic improvement, recycling of wastes, zone management, reduced reliance on fishmeal in feeds, and other innovations.

In today's age of instant access to information, consumers are keen to know more about the environmental, social, and food safety attributes of farmed seafood.

Market acceptance relies more and more on compliance with international standards of best practice as indicated by certified eco-labels.

From this author's perspective, as President of the Global Aquaculture Alliance (GAA), certification standards are a unifying force in guiding the aquaculture industry forward in its journey of continuous improvement toward sustainability. We are still in the early stages of this journey, and we have much to learn. It is tempting to immediately set aspirational standards in hopes of stimulating a quantum leap toward a future goal, but stakeholders may disengage if the bar is set too high. Imagine challenging medieval man to develop a system of air travel where one can enjoy a hot meal while seated comfortably in a plane flying between continents! The enormity of an unrealistic challenge may paralyze further progress.

GAA's Best Aquaculture Practices (BAP) certification standards for aquaculture facilities seek to effect immediate improvements by engaging as many facilities as possible through stringent but pragmatic standards. As combined efforts, great and small, of researchers and producers around the world continually raise the bar for what aquaculture can be, certification standards are raised in step. It is a dynamic ever-advancing process. As Heraclites said over 2000 years ago, "No man ever steps in the same river twice, for it's not the same river and he's not the same man."

The work contained within this volume helps explain these complex and evolving issues, which are so important to the future of our seafood supply and our planet. Dr. Claude Boyd is eminently qualified to guide this discussion, because he has been an active player in the development of the aquaculture sustainability movement. Dr. Aaron A. McNevin also has much experience with aquaculture certification through his work with the World Wildlife Fund Aquaculture Dialogues. He provides insight about the reasons that environmental NGOs have taken certain positions on aquaculture.

Dr. Boyd assisted GAA in the development of its initial BAP standards for shrimp farms—released in 2003 as the seafood industry's first such certification standards. Since then, BAP certification has come to encompass farms for salmon, tilapia, *Pangasius* and other farmed species, as well as hatcheries, feed mills, and processing plants. Standards for mussel farms and revised standards for finfish and crustacean farms were released in early 2013. The annual volume of BAP certified products now exceeds 2.1 million tonne.

The Global Aquaculture Alliance appreciates the extensive knowledge, research, and insights that Drs. Boyd and McNevin share in this new publication. They address both a historical perspective and an excellent overview of some of the challenges in land and water use, energy consumption, protein conversion, and conservation of biodiversity. Fittingly, the final chapters describe best management practices and certification programs that help guide aquaculture on its journey to responsibly feed the world.

George W. Chamberlain
President, Global Aquaculture Alliance
Saint Louis, MO, USA

Foreword

Aquaculture is the only form of agriculture I can think of that evolved from mainly subsistence food production into an important part of international economy within one human generation. Coincident with—or perhaps caused by—its rapid expansion came dramatic changes in the way people viewed the relationship between aquaculture and the environment. All this provided a unique opportunity for individual scientists of a certain age to personally witness the arc of aquaculture's development and environmental performance.

In the spring of 2000, a colleague and I were flying from Jackson, Mississippi, to Washington, DC, for the first of many meetings we would have over the next 4 years with the United States Environmental Protection Agency's Office of Water. Earlier that year, the Agency had announced its intent to implement the Clean Water Act for aquaculture by developing national effluent limitations. Aquaculture was one of the nine industries identified for federal rulemaking as part of a 1992 court-ordered consent decree (Natural Resources Defense Council et al. vs. Browner). To many of us, including my airborne colleague and me, the fact that *our* way of growing food had, by inference, been identified as one of the nine most notorious polluters in the country was nothing less than shocking. Looking out the plane's window as we flew over the southeastern states on that bright, clear day, it was difficult to understand how anyone could take an unbiased look at the scope of human activities and conclude that aquaculture was one of the country's greatest water pollution threats. After all, we had spent a lifetime working to increasing aquaculture production in the United States. What had happened?

When I entered the aquaculture community as an Auburn University student 40 years ago, most of my acquaintances believed that aquaculture's role in world agriculture was to produce low-cost, protein-rich food for peoples in underdeveloped countries. Many of my fellow students were freshly out of the Peace Corps, having spent a couple of years helping people in Africa and Asia build ponds, fertilize them with agricultural byproducts, and grow carps, tilapias, clarid catfishes, and other hardy fishes for local consumption. Examples of commercial aquaculture certainly existed at the time, although some of my friends looked upon "growing fish for money" as a bourgeois corruption of a noble cause. Environmental impacts? Most people viewed aquaculture as, at worst, a benign endeavor and, at its best, the soundest imaginable way to grow food. If you can find a copy, take a look at the 1972 book *Aquaculture* by Bardach, Ryther, and McLarney. That was our bible in those days but you will struggle to find the phrase "environment impact" or the word "sustainable" anywhere in it. About 10 years after that book's publication, things really started to change.

John Ryther, coauthor of the book *Aquaculture* and my mentor when I worked for Woods Hole Oceanographic Institution in the mid-1970s, wrote a famous paper in the October 1972 issue of *Science* wherein he estimated the potential sustained yield of marine capture fisheries based on calculations of the ocean's primary productivity. He concluded that fish landings would level off in a decade at somewhere near 100 million metric tons annually. He slightly overestimated landings but he was eerily prescient in his predicted timing: By the mid-1980s, capture fish landings peaked at about 85 million metric tons. Meanwhile, global seafood demand continued to increase as populations grew, incomes rose, and seafood was appreciated as part of a healthy diet. The difference between non-expanding supply from capture fisheries and rapidly expanding demand had to be derived from aquaculture. A milestone of sorts was reached in 2013, when the world's seafood supply from aquaculture exceeded, for the first time, seafood obtained from capture fisheries.

During the 1980s and 1990s, goals and culture practices quickly evolved in response to new profit opportunities and (often overwrought) encouragement from aquaculture development institutions. New aquaculture sectors developed with the goal of producing higher value products, often for export. Production practices changed in efforts to squeeze more food—or more money—out of less water. These new practices had higher rates of resource use and greater (or at least different) environmental impacts than traditional aquaculture. All too often, greed overwhelmed good judgment and farms were poorly designed, poorly operated, or simply built in the wrong place.

All this occurred during a time of heightened environmental awareness and advocacy. Highly visible problems in marine shrimp and salmon aquaculture led to closer scrutiny of aquaculture in general. An explosion of publications and books ensued, wherein the environmental impacts of aquaculture were cussed, discussed, or rebutted. With some important exceptions (the role of netpen salmon farming in sealice dispersal comes to mind), interest in identifying problems seems to have crested and emphasis has shifted to finding solutions and developing environmental certification programs (the two are not the same). So, it seems to be a good time to take stock of where aquaculture has been and how far it has come in relation to resource use and environmental impacts. This new book contains Claude Boyd's assessment of environmental and resource use in aquaculture along with an environmental NGO viewpoint given by Aaron A. McNevin. The book is, to my knowledge, the first to attempt a synoptic summary of the topic.

In his 2001 book *The Skeptical Environmentalist*, Bjørn Lomborg—a Danish political scientist and statistician—critically reviewed data on a variety of global environmental issues and tried to assess what the data say about mankind's future. Predictions about the future are, by their nature, uncertain and Lomborg believes that many environmental scientists often support only the gloomiest scenarios—sometimes for self-serving reasons. The book's critics—who included nearly all environmentalists and many, if not most, environmental scientists—responded that Lomborg was equally guilty by dancing only with the most optimistic end of confidence intervals. This was a difficult argument for either side to win; after all, whose uncertainty is most uncertain? But the harshest criticism of the book was leveled at Lomborg personally. In a sort of reverse fallacy of "arguing from authority," some scientists labeled Lomborg as an outsider lacking credentials to write such a book. That is, he must be wrong because he is not one of us (a working environmental scientist).

Some people may see this new book as a narrowly focused version of *The Skeptical Environmentalist*. The senior author, Claude Boyd, is certainly skeptical—if not downright cynical—about science. I have known him for 40 years and he has never met a data point he did not question. But this book differs from Lomborg's book in important ways, past merely the book's scope. First, no one can criticize the book based on the authors' credentials—Professor Boyd has had a career of great breadth and productivity. He is an aquatic scientist of the first order and has worked throughout the world with private industry, environmental groups, and certification organizations. But being the world's expert on a topic does not necessarily make that person's opinions correct. What sets this book apart is that while Lomborg's book angered only one side of the aisle, this new book has something to provoke everyone. While taking the environmental community to task for overstating many of aquaculture's impacts and taking those impacts out of logical context, it also points out aquaculture's many blemishes and past scars, and argues that improved environmental performance is essential if aquaculture is going to responsibly fulfill its future role as the world's major seafood supplier.

A truly unique feature of this book is the inclusion of perspectives from Aaron A. McNevin. He worked for the World Wildlife Fund's aquaculture program during the aquaculture dialogues, joined academia for 3 years, and presently directs the aquaculture work at WWF. The WWF was one of the first nongovernmental organizations to productively engage the aquaculture community on environmental issues and McNevin's comments on each chapter provide fascinating insight from another point of view.

So read this book and I hope it makes you think. As Boyd points out in the concluding chapter, humans have many shortcomings, but that one unique trait—the ability to think—is our only hope for the future.

Craig S. Tucker
Project Leader
USDA-ARS Warmwater Aquaculture Research Unit
Stoneville, MS, USA

Abbreviations

Units of measurement

Billion tonne	GT
Calorie	cal
Capita	cap
Centigrade degree	°C
Centimeter	cm
Day	d
Gram	g
Hectare	ha
Hectare-centimeter	ha-cm
Hectare-meter	ha-m
Horsepower	hp
Hour	h
Joule	j
Kilogram	kg
Kilometer	km
Liter	L
Meter	m
Microgram	μg
Micrometer	μm
Milliliter	mL
Million tonne	Mt
Minute	min
Parts per billion	ppb
Parts per million	ppm
Parts per trillion	ppt
Second	sec
Tonne	t
Watt	W

Organizations

AAES	Alabama Agricultural Experiment Station
ACC	Aquaculture Certification Council
ACP	Alabama Catfish Producers

ADB	Asian Development Bank
ADEM	Alabama Department of Environmental Management
APHIS	Animal and Plant Health Inspection Service
ASC	Aquaculture Stewardship Council
ASEM	Asia-Europe Meeting
ASI	Accreditation Services International
EDF	Environmental Defense Fund
eNGO	Environmental non-governmental organization
EU	European Union
FAO	Food and Agriculture Organization (of the United Nations)
FDC	Fish Diseases Commission
FOS	Friend of the Sea
FSC	Forest Stewardship Council
GAA	Global Aquaculture Alliance
IDH	Sustainable Trade Initiative
IEA	International Energy Agency
IFAD	International Fund for Agricultural Development
IFDC	International Fertilizer Development Center
IFOAM	International Federation of Organic Agriculture Movements
IPCC	Intergovernmental Panel on Climate Change
ISEAL	International Social and Environmental Labeling Alliance
ISO	International Standards Organization
MSC	Marine Stewardship Council
NACA	Network of Aquaculture Centres in Asia-Pacific
NASS	National Agricultural Statistics Service
NGO	Non-governmental organization
NRCS	Natural Resource Conservation Service
NSF	National Sanitation Foundation
OECD	Organization for Economic Cooperation and Development
OIE	World Organization for Animal Health (in French: Office International des Epizooties)
SGS	Société Générale de Surveillance
UN	United Nations
UNCSD	United Nations Commission on Sustainable Development
UNEP	United Nations Environmental Program
UNFCCC	United Nations Framework Convention on Climate Change
US	United States
USAID	United States Agency for International Development
USDA	United States Department of Agriculture
USDHHS	United States Department of Health and Human Services
USEIA	United States Energy Information Agency
USEPA	United States Environmental Protection Agency
USFDA	United States Food and Drug Administration
USGS	United States Geological Survey
WB	World Bank
WFP	World Food Program
WHO	World Health Organization

WRI   World Resources Institute
WWF   World Wildlife Fund

Other terms

AFR   Aquaculture to freshwater ratio
BAP   Best aquaculture practice
BMP   Best management practice
CAAP   Concentrated aquatic animal production
EIA   Environmental impact assessment
FCE   Feed conversion efficiency
FCR   Feed conversion ratio
FCR_d Dry matter-based feed conversion ratio
FIFO   Fish in–fish out ratio
FMR   Fish meal ratio
GDP   Gross domestic production
GIS   Global information systems
GMO   Genetically modified organism
HACCP   Hazard analysis and critical control points
HRT   Hydraulic retention time
IPM   Integrated pest management
LCA   Life cycle assessment
MSD   Material safety data
NPDES   National Pollutant Discharge Elimination System
PA   Protected area
PCR   Protein conversion ratio
PER   Protein efficiency ratio
REDD   Reduced emission from forest deforestation and degradation
Thai CoC   Thailand Code of Conduct
Thai GAP   Thailand good aquaculture practices
TMDL   Total maximum daily load
WPR   Waste production ratio

Preface

The world's ecosystems are increasingly altered by human activities but they are resilient and have provided sufficient resources and ecological services for the growing population. But there are serious doubts that they can continue to do so. Because irreparable damage to the structure and function of the world's major ecosystems could threaten even the existence of humans, there is growing concern that mankind is on track for unprecedented ecological disaster.

A rational person has difficulty accepting predictions about impending ecological disaster. The human race seems to have always been faced with insurmountable impediments. Although there have been temporary setbacks, humans have been able to overcome and thrive. The doomsday prophets have always been wrong, so why should we believe them now? Moreover, logical individuals are confused by experts who often disagree on the issues; glib talkers on both sides of the environmental divide who seldom have their facts straight, but receive the most media attention; and politicians who favor positions that are supported by the majority of voters or give in to special-interest groups. Actions to solve resource use and environmental disputes are controversial and political because winners in these quandaries often are determined by legislation and by policies and regulations developed by governmental agencies. Add to this mix of confusion the competitiveness of individuals and special-interest groups, greed, and an often poorly informed populace, and it is not surprising that a logical, science-based way of dealing with resource use, waste disposal, and environmental stewardship has not been forthcoming.

Today—at least in most developed countries—the environmental movement is gaining ground. There is no doubt that the public in the western world is more aware of environmental issues (but possibly no more knowledgeable), governments worldwide are more concerned with environmental protection, and most producers of goods and services are more attuned to avoiding negative environmental impacts than in the past. There also is evidence that the environmental movement is spreading to the developing world, but in countries with widespread poverty, political instability, or armed conflicts, it is difficult to successfully promote environmental values.

Production of food through agriculture is necessary to support the world's population. Nevertheless agriculture causes negative environmental impacts because it alters land use, requires water, nutrients, energy, and other resources, and it contributes to air and water pollution. The rise of mankind as the dominant species in world ecosystems coincided with the beginning of agriculture that lessened the reliance of humans on hunting, fishing, and gathering. The capture fishery has provided a sizeable fraction of the animal protein supply throughout human history because the sea contained a seemingly inexhaustible supply of aquatic animals. Although aquaculture has been practiced on a small scale in Asia and Europe for at least 2500 years, only recently has the capture fishery failed to meet the demand for fisheries products, allowing aquaculture to thrive. Aquaculture should continue to grow in importance because capture fisheries have reached their sustainable limit.

Modern agriculture is still being refined; but most current crop and livestock species were adopted, their culture methods developed, and vast expanses of natural ecosystems converted to cropland and pastureland before the onset of environmentalism. Aquaculture, on the other hand, is developing in an era in which there is much more concern about the environment and this endeavor is being held to much higher environmental standards than those that were imposed on agriculture. There is much less understanding of aquaculture production techniques by consumers, politicians, government regulators, and environmentalists than there is about methods of agriculture.

In addition to criticism of aquaculture by environmental advocates, the capture fishing industry tends to consider aquaculture as a competitor. There is much advertising that extols—without supporting evidence—the superiority of wild fisheries products over aquaculture products. In reality, capture fisheries are equivalent to early settlers in the United States depending upon deer, squirrels, bison, and other wild animals for meat. Capture fisheries cause much more environmental damage than aquaculture, but they receive less criticism from environmentalists than is leveled at agriculture or aquaculture. There is, however, a nexus between capture fisheries and aquaculture because fish meal and fish oil made from wild fish are included as an ingredient in feed for many aquaculture animals.

The major environmental issues upon which environmentalists tend to focus, land and water use, energy consumption, use of fertilizers, grains, and fish meal, water pollution, carbon emissions, introduction of exotic species, and resulting negative impacts on biodiversity, all apply to aquaculture. However, in comparison with many other endeavors, in most instances, aquaculture is a rather minor contributor to wasteful resource use and negative environmental impacts. Of course resources should be conserved and environmental perturbations should be minimized or prevented whenever possible, and aquaculture should be practiced in an environmentally responsible manner. Fisheries products, however, are an essential component of world food supply, and aquaculture should not be hampered by unnecessary regulations or have its image tarnished by embellished statements.

The senior author has been involved in aquaculture for nearly 50 years and owes a great debt to the industry for providing him a livelihood, interesting career, and the opportunity to travel almost everywhere. He feels compelled to attempt to accurately depict resource use and the negative effects of aquaculture on the environment and to place them into perspective with those resulting from the production of other human necessities. Moreover, he feels that it is important to emphasize efforts underway to lessen the negative impacts of aquaculture.

Contentious issues have more than one side, and the views of an aquaculture insider are no doubt skewed. To compensate for this inevitability, at the end of every chapter of this book is a section on the most prevalent views of the environmental NGOs (eNGOs) prepared by the junior author who was employed 7 years by the

World Wildlife Fund and after a 3-year hiatus in academia, returned to WWF to direct their aquaculture work. These sections will assist readers to understand stakeholder views that resulted in much of the attention paid to natural resource use and environment impacts by aquaculture over the past two decades.

The eNGO perspective is not conveyed appropriately through outsider examination of the goals and activities of these organizations. The nature of eNGOs and the means by which they operate and choose to prioritize issues of natural resource use is fundamentally different than that of the scientific and academic sectors. It is only possible to understand this perspective with intimate knowledge of operational and priority setting mechanisms of each eNGO. Moreover, the challenges that are inherent in the efficient and responsible use of natural resources in aquaculture and other human activities have both a scientific and humanistic perspective. The scientific perspective is more straightforward than the humanistic perspective because the latter is relative to the specific stakeholders who are examining resource use. Further specific stakeholder groups are not uniform in their views, and definitions of terms such as "sustainable," "responsible," or "appropriate" are relative.

The WWF is the largest privately funded environmental conservation organization in the world. The WWF initiated and coordinated the largest stakeholder-driven discussion forums (Aquaculture Dialogues) on aquaculture to date. Aquaculture Dialogues were convened for the most important aquaculture species or species groups in international trade (tilapia, *Pangasius*, bivalve mollusks, abalone, freshwater trout, marine shrimp, *Seriola* and cobia, and salmon) with the objective of arriving at consensuses on tolerable and measurable performance targets to give better context to the term "responsible aquaculture." It is true that the perspective of one individual who has worked for only one of the eNGOs actively engaged in the aquaculture sector is far from comprehensive. However WWF played a pivotal role in the development of standards for third party certification of commercial aquaculture businesses. This work was conducted through a multi-stakeholder engagement, which requires not only an understanding of the views of various stakeholders but a recognition that these views must be addressed and reconciled with the views of other stakeholders to move toward a consensus on aquaculture natural resource use. The standards developed in the Aquaculture Dialogues have been more widely supported by the environmental community than other aquaculture standards because of greater eNGO engagement in the standard-making process.

Our thoughts about environmental issues expressed in this book have been influenced by discussions with many commercial aquaculturists, aquaculture scientists, and environmentalists—especially E. W. Shell, Craig S. Tucker, George Chamberlain, Jason Clay, Julio Queiroz, John Hargreaves, and Richard Hulcher. We are particularly appreciative for June Burns' help, for without it, the manuscript for this book would have never been finalized. Financial support from the Butler/Cunningham Endowment to Auburn University to foster efforts on environment issues in agriculture was greatly beneficial to the preparation of this book.

Claude E. Boyd and **Aaron A. McNevin**

Chapter 1

An overview of aquaculture

Aquaculture is an old pursuit that only became common during the last 75 years. Today nearly everyone has heard of aquaculture and realizes that one can purchase either farm-reared or wild-caught fisheries products. The dictionary definition of aquaculture is "the cultivation or rearing of aquatic animals and plants." But there is no consensus—even among aquaculture experts—on the best definition of aquaculture.

Despite most people having heard of aquaculture, very few, including most professional and lay environmentalists, have much knowledge of the important aquaculture species/species groups and of the various culture systems and methodologies used to produce aquatic organisms. This chapter provides a simple discussion of aquaculture species, production methods, and associated environmental issues. Land and water requirements, nutrient sources, energy use, and water management techniques will be featured in particular, because many of the negative impacts of aquaculture result from these factors.

History

The first writings about aquaculture are from China about 2,500 years ago; the writings were about carp culture that had originated several centuries before (Stickney 2000). The Egyptians may have been involved in fish culture before the Chinese, and the Romans cultured oysters and possibly other species. Shrimp culture dates back to around 800 AD in Asia and freshwater aquaculture has been practiced in several Asian countries for many centuries (Stickney 2000). Aquaculture was fairly common in Europe—especially in central Europe—during most of the second millennium AD. For example, there were 75 000 ha of carp ponds in Bohemia alone by the end of the fourteenth century—more than exist in that region today (Berka 1986). By the sixteenth century, the pond area in Bohemia reached a maximum of 180 000 ha, but the area declined considerably soon afterward.

Aquaculture, Resource Use, and the Environment, First Edition. Claude E. Boyd and Aaron A. McNevin.
© 2015 John Wiley & Sons, Inc. Published 2015 by John Wiley & Sons, Inc.

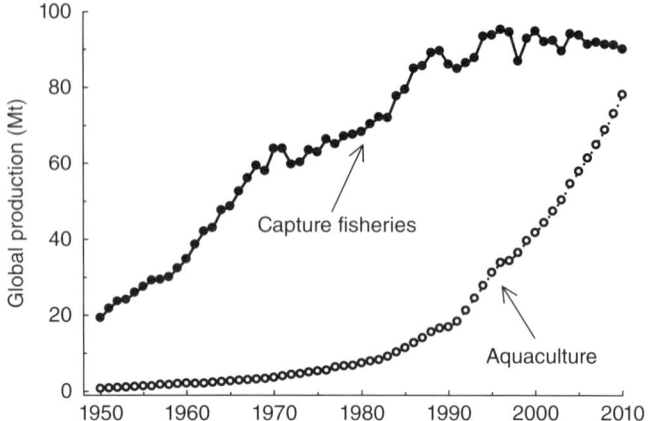

Figure 1.1 Total world fisheries and aquaculture production since 1950. *Source:* FAO (2011).

The real "boom" in aquaculture began in the 1950s and 1960s in many countries including the United States. Growth of aquaculture was relatively slow at first; in the early 1950s, it produced less than 1 Mt/year and reached only about 2.5 Mt/year by 1970. Since 1970 aquaculture has grown at an average rate of about 8% annually reaching about 63.6 Mt in 2011 (Fig. 1.1). This rapid growth in aquaculture has occurred because the capture of fish and other aquatic organisms from natural waters (Fig. 1.1) has apparently reached or exceeded its sustainable limit, and the difference in demand and wild catch must be supplied by aquaculture.

Culture species

The species for aquaculture include both plants and animals. In 2010, there were about 19 Mt of aquatic plant production and 59.9 Mt of aquatic animal production by aquaculture. The aquatic plant production was nearly all in marine water, but aquatic animal production was further separated into freshwater, brackishwater, and marine species (Table 1.1).

Aquaculture animals that will be the focus of this book consist mainly of molluscs, crustaceans, and fish. These groups also are further subdivided; for example, freshwater fish may be listed as salmonids, tilapia and other cichlids, carps and other cyprinids, catfish, etc. Nontropical aquaculture species often are classified according to water temperature optima for growth: coldwater (<10°C); coolwater (10–20°C); warmwater (>20°C). Tropical species cannot survive when water temperature declines below about 20°C for several hours or days.

A total of 527 species of aquatic organisms are reported as aquaculture species by the Statistics Unit of the Fisheries and Aquaculture Department of the Food and Agriculture Organization (FAO) of the United Nations. Finfish dominate freshwater aquaculture, and several species of carp and related fishes comprise about two-thirds of total finfish production. Although marine animal aquaculture consists mainly of bivalve mollusc production (Table 1.2), the culture of marine fish is expected to

Table 1.1 World fisheries and aquaculture production (aquatic plants excluded) and utilization for 2011.

Sector	Production (Mt)
Inland	
Capture	11.5
Aquaculture	44.3
Total inland	55.8
Marine	
Capture	78.9
Aquaculture	19.3
Total marine	98.2
Total capture	90.4
Total aquaculture	63.6
Total world fisheries	154.0
Utilization	
Human consumption	130.8
Nonfood uses	23.2
Food fish supply (kg/capita)	18.8

Source: Modified from FAO (2012).

increase in the future. Brackishwater aquaculture is mostly Penaeid shrimp culture. Although freshwater animal aquaculture production exceeds marine animal aquaculture production, in 2010 there was about 18.4 Mt of marine seaweed cultured. Add this to marine animal production and the total marine aquaculture was approximately equal to freshwater aquaculture production in 2010.

Water sources and culture systems

Water is a primary requirement for aquaculture, and features of the water supply are priority considerations in selecting species and production systems suitable for

Table 1.2 World aquaculture production by culture environment—2010.

Type and culture environment	Quantity (Mt)
Fish, crustaceans, and molluscs	
Freshwater	36.9
Brackishwater	4.7
Marine	18.3
Subtotal	59.9
Aquatic plants	
Freshwater	0.1
Brackishwater	0.5
Marine	18.4
Subtotal	19.0
Total aquatic organisms	**78.9**

Source: FAO. Fishery Statistics. Yearbook of Fishery Statistics. Accessed: 10/15/13. URI: ftp://ftp.fao.org/FI/STAT/summary/default.htm

a particular location. Water temperature determines whether a site is suitable for coldwater, coolwater, warmwater, or tropical species. Of course water temperature varies seasonably, and at some warmwater sites, it is possible to culture coldwater or coolwater species in winter and tropical species in summer. For example, tilapia can be cultured in summer and rainbow trout can be cultured in winter—although it has only been done in research—in central Alabama (32° north latitude) in the United States.

The range of species that can be cultured at a particular location also will be restricted by salinity, that is, freshwater, brackishwater, or marine species. Freshwater has salinity less than 1 g/L while ocean water has salinity around 35 g/L. Brackishwater is intermediate in salinity between freshwater and ocean water. Inland waters—especially in arid regions—may be brackish with salinities as high as 5–10 g/L. In extremely arid regions, some inland waters and waters in estuaries with low exchange rates with the ocean may be hypersaline with salinities above 40 g/L. At some sites, there are seasonal differences in salinity that influence selection of culture species. More often than not the kind of production system selected for the particular species at a given site results from the climate and the type of water source and its seasonal availability.

Water quality and water quantity also strongly influence selection of culture species and production systems. Moreover, water quality often changes in production systems as a result of management inputs to enhance production (Boyd and Tucker 1998).

Ponds

Ponds are used widely for culture of fish and crustaceans. A recent estimate places the global water surface area of aquaculture ponds at about 11 000 000 ha (Verdegem and Bosma 2009). Watershed ponds are constructed by installing a dam to impound runoff. Many watershed ponds (Fig. 1.2) receive only overland flow following rains. Water levels in most watershed ponds fluctuate with changes in rainfall and they may

Figure 1.2 Watershed pond showing the dam and a portion of the watershed (left); a complex of watershed ponds on the E. W. Shell Fisheries Center at Auburn University in Alabama (United States) (right).

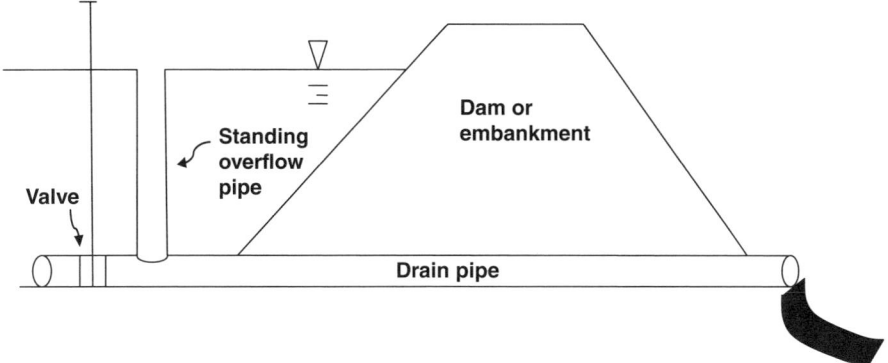

Figure 1.3 Illustration of overflow structure and drain pipe in a pond.

fall drastically during droughts. Ponds may be arranged in series to form a water-harvesting scheme (Boyd et al. 2009), and water that naturally overflows and seeps from ponds at higher elevation maintains water levels in lower ponds during dry weather. Where there is abundant groundwater, wells may be developed to supplement water supply for watershed ponds as necessary (Yoo and Boyd 1994).

Ponds usually have a water control structure to allow excess water to overflow and to drain ponds (Fig. 1.3). A grass-lined spillway should be provided to bypass large inflows and prevent overtopping of dams following unusually heavy rainfall events.

Ponds may be constructed by damming streams. These ponds have stable water levels but they also may have high flushing rates. Short water retention time may reduce the effectiveness of liming and fertilization in augmenting productivity. Overflow structures with intakes near the bottom in the deepest area of ponds sometimes are installed to avoid discharge after heavy rainfall events of plankton that inhabits the upper layer of illuminated water (Fig. 1.4).

Ponds can be constructed by excavating a basin in which to store water (Fig. 1.5). Excavated ponds usually are small because of the large volume of earth that must be removed to form them. Water for filling excavated ponds may come from wells, streams or lakes, irrigation systems, or overland flow. Where the water table is

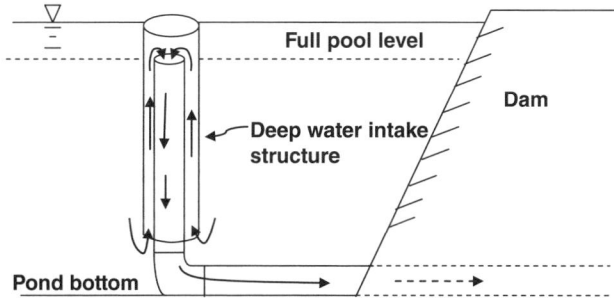

Figure 1.4 Deep-water intake structure.

Figure 1.5 An excavated fish pond in Thailand. Courtesy of David Cline.

shallow, excavated ponds may fill by groundwater seepage. Excavated ponds cannot be drained, and their water levels may fall drastically in the dry season as the water table declines. Nevertheless, small, excavated ponds are important for providing fish for poor, rural farm families in many Asian nations.

The most suitable type of pond for aquaculture is the embankment pond constructed by building an earthen embankment around an area in which to store water (Fig. 1.6). Ponds may be fitted with water control structures such as shown in Fig. 1.3, or water gates with dam boards for controlling water level may be installed. Little overland flow enters embankment ponds because the watershed is limited to the sides and tops of embankments. There are few regions in the world where rainwater and overflow will maintain water levels in embankment ponds year round. Water must be supplied from external sources such as wells, streams, lakes, irrigation systems,

Figure 1.6 Embankment ponds used for channel catfish farming in the United States.

Figure 1.7 Plastic-lined ponds on the Claude Peteet Mariculture Center, Gulf Shores, Alabama (United States). Courtesy of David Cline.

lakes, estuaries or the sea. For example, the US channel catfish industry in western Mississippi and southeastern Arkansas uses embankment ponds filled by well water. Marine shrimp farming also is conducted almost exclusively in embankment ponds. The great advantages of embankment ponds are that water levels can be controlled, ponds can be drained and refilled according to management schedules, and water exchange may be implemented to improve water quality.

Where soils are highly permeable, clay blankets or impermeable, plastic liners may be installed in ponds to reduce seepage loss (Fig. 1.7). In addition, plastic liners in heavily aerated ponds prevent erosion by aerator-generated water currents (Avnimelech 2012).

The intensity of aquaculture in ponds varies greatly (Avault 1996; Boyd and Tucker 1998). Yields of culture species based on natural productivity will be low— seldom more than 50–500 kg/ha/year but nutrient inputs are made in aquaculture to allow more production. Manures and fertilizers can be used to increase production— depending upon the species, production may reach 500–2000 kg/ha. Much greater production can be achieved using feed, and the combination of feed and mechanical aeration provides the highest production. Fish and shrimp production in ponds with feeding alone normally ranges from 1000 to 3000 kg/ha, but with mechanical aeration, production often exceeds 5000 kg/ha and may reach 15 000–20 000 kg/ha. Ponds usually average 1–1.5 m in depth, and on a volume basis the culture density is 0.033 to 2 kg/m^3.

In most kinds of pond culture, only a small fraction of the total water area and volume is necessary to support the culture species. In extensive production, the remainder of the pond space is necessary to produce food organisms for the culture species. In more intensive production with feed inputs the rest of the pond space serves as an internal waste treatment area (Boyd et al. 2007). Where water exchange is used a portion of the waste is flushed from the pond, that is, the waste load is externalized for treatment by natural waters.

Figure 1.8 Round ponds at a shrimp farm in Belize.

Shapes, water surface areas, and depths of ponds vary greatly. Topography of the original land surface strongly influences the morphometry of watershed ponds, but as a general rule, watershed ponds take the shape of an irregular semicircle, and average depth is about 0.4 times maximum depth (Boyd and Boyd 2012). Water surface area may vary from a few hundred square meters to several hectares. Embankment ponds often are square or rectangular—a shape approaching the 2:1 rectangle likely is the most common (Yoo and Boyd 1994). There has been some use of round ponds (Fig. 1.8) because some feel that this shape enhances aeration-induced water circulation. These ponds are more expensive to construct than rectangular or square ones, and as discussed by Boyd and Tucker (1998), from the aspect of water circulation, a square pond is not greatly different from a round one.

Bottoms of embankment ponds normally are constructed with gentle slopes and cross slopes to facilitate draining. Water depth quickly increases from edges to a depth of 0.75 m or more; maximum depths seldom exceed 2–3 m and average depths usually are 1–2 m. Excavated ponds normally are rectangular or square with depths of 1–3 m. Table 1.3 gives categories of typical water surface areas for the three major hydrologic types of ponds.

Flow-through systems

Rainbow trout often are cultured in raceways supplied by gravity flow with water from springs, streams, or lakes. Raceways constructed of concrete and often located

Table 1.3 Categories of ponds based on water surface area.

	Hydrologic pond type		
Size category (ha)	Watershed	Embankment	Excavated
Small	<1	<0.25	<0.1
Medium	1–5	0.25–2.5	0.1–0.5
Large	>5	>2.5	>0.5

Figure 1.9 A trout raceway in the United States.

in series are probably the most common flow-through culture units (Fig. 1.9), but small, earthen ponds, tanks, or other units also may be used. Water typically is exchanged at a rate of two or three times the volume of culture units per hour by gravity flow and discharged from the lowermost culture unit into natural water bodies. Re-aeration occurs where water falls from the end of one culture unit into the beginning of the next, and mechanical aerators or pure oxygen contact systems can be installed to supplement the dissolved oxygen supply.

Raceways contain much higher densities of culture animals than ponds. For example, rainbow trout may be reared at densities of 80–160 kg/m^3 (Soderberg 1994).

Cages and net pens

Fish often are produced in enclosures placed in natural water bodies, reservoirs, and ponds. The most common types of enclosures are cages and net pens. Cages range in size from about 1 m^3 to more than 1000 m^3 (Fig. 1.10), and fish density may range from <20 to >200 kg/m^3 (Schmittou 1993). Cages typically float in the water and they are moored to the bottom. Water flows through cages to exchange waste-laden water and replenish dissolved oxygen used in respiration. Uneaten feed and feces fall through cages and settle to the bottom in the vicinity. Cages are periodically moved to new locations to allow benthic communities affected by sediment to recover—a process called fallowing.

Net pens are similar to cages but they are made by placing netting around posts inserted in the bottom (Fig. 1.11). Net pens may cover areas of a few to a thousand square meters and extend into water up to 2 or 3 m in depth. Stocking density in net pens typically is much less than in cages.

Figure 1.10 Large cages in a lake (left); a small cage in a pond (right). Courtesy of David Cline.

Water reuse systems

Water reuse systems allow greater production per unit of water volume and improve the efficiency of water use. They are particularly useful in arid regions with scant supplies of water and in areas where land for ponds is unavailable or overly expensive. Fish and other organisms grown in small volumes of water can be fed more efficiently and other treatments applied more easily than in less intensive culture. Water reuse systems also reduce the volume of effluents to lessen the pollution potential of aquaculture production facilities.

The simplest method of water reuse is to produce more than one crop of aquatic animals in the same water. Channel catfish farming in the United States is a good example of this practice. Ponds are not drained for harvest. Marketable-sized fish are removed with a grading seine and additional fingerlings are stocked to replace them. Ponds typically are drained at intervals of 6–10 years (Boyd et al. 2000).

The water passing through raceways and other culture systems can be pumped back to the grow-out units and reused. An example of this methodology is illustrated in Fig. 1.12.

Figure 1.11 Example of a net pen culture system.

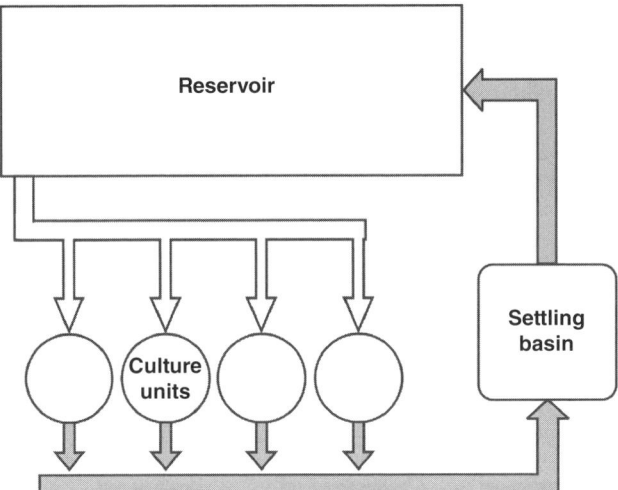

Figure 1.12 Schematic of an outdoor, water reuse system for tilapia culture.

The water from culture units is passed through a sedimentation pond to remove the coarse solids and then held in a pond for natural water purification before reuse. In some systems, one or more additional species are cultured in the treatment pond.

Water recirculation systems of much greater technological complexity are promoted by some innovators (Timmons et al. 2001). These systems rely upon physical, chemical, and biological wastewater treatment technology to purify water for reuse (Fig. 1.13). The entire system may be constructed in a green house or other heated structure to allow year-round production in temperate climates.

Biofloc technology is increasingly used for intensive culture of shrimp and a few fish species (Serfling 2000; McIntosh 1999; Avnimelech 2012). These systems usually consist of completely lined ponds with a large amount of mechanical aeration—often more than 50 hp/ha. Because of high feed input, the phytoplankton bloom diminishes and is replaced by a bacterial suspension or floc. Feed input usually consists of regular aquaculture feed plus crushed grain, molasses, or other source of organic matter. Bacteria decompose the organic carbon source using ammonia from metabolic wastes of the culture species. The bacteria floc is rich in protein and serves as food for the culture species. Thus by combining feed and organic matter and using nitrogenous wastes from the feed to stimulate production of microbial protein, the crude protein input for production of the aquaculture species can be lessened considerably in comparison with normal feed-based aquaculture. Moreover, water exchange normally is not employed, and at harvest, the water containing the floc can be transferred to another pond for later reuse.

New production technologies that are functionally hybrids among pond, raceway, cage, and water reuse systems include the in-pond raceway system (Brown et al. 2011) and the partitioned aquaculture system (Brune et al. 2003). Fish are held at high density in floating raceways through which water is flushed by paddlewheel aerators or air-lift pumps. In the simplest form of the partitioned aquaculture system, a small area of a pond is partitioned off for confining the culture species, and water is

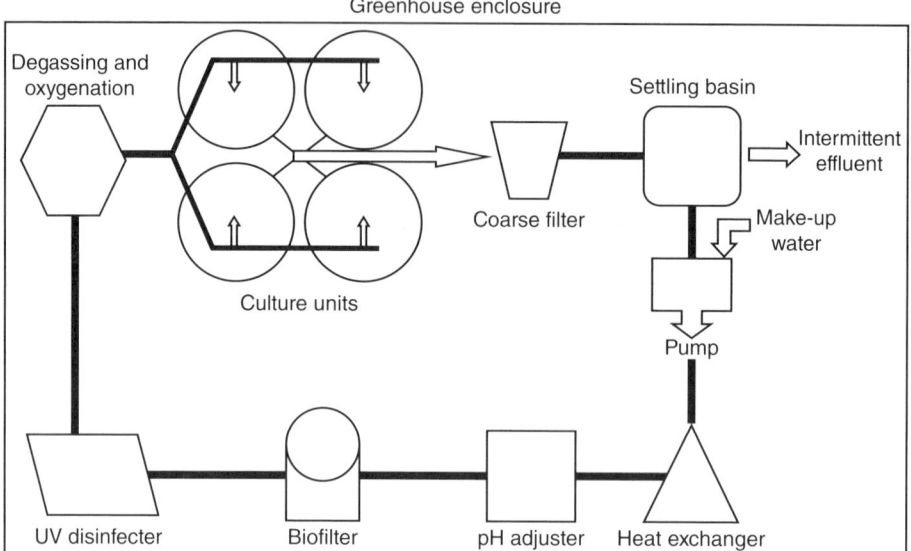

Figure 1.13 Schematic of a water reuse system with water purification equipment and enclosed in a greenhouse.

exchanged between the culture area and the rest of the pond which serves as a waste treatment area (Fig. 1.14).

Mollusc and seaweed culture

Bivalve molluscs are produced by bottom culture methods in which spat are laid on sediment, rocks, or other solid surfaces for grow-out (Fig. 1.15). However,

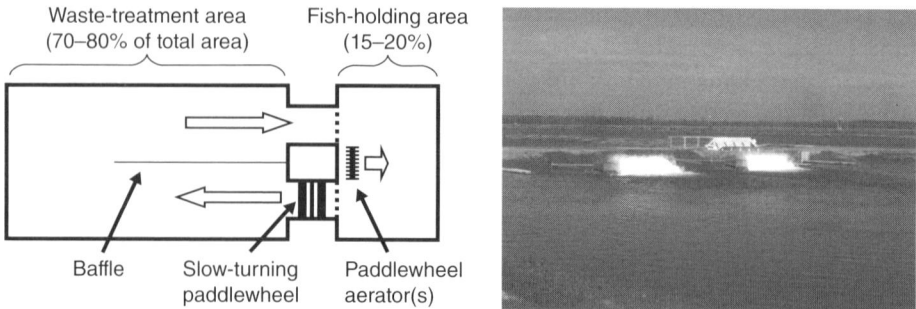

Figure 1.14 A simple, partitioned aquaculture system at the Delta Research and Extension Center, Stoneville, Mississippi (United States). Fish are held in the smaller part of the divided pond and water is circulated between the waste-treatment area and the fish-holding area by a slow-turning paddlewheel. The standard, paddlewheel aerators prevent low dissolved oxygen concentration in the fish-holding area at night.

Figure 1.15 Oyster culture in off-bottom cages in the intertidal zone. Courtesy of David Cline.

off-bottom culture for bivalves other than clams is more efficient for it prevents ben-thic predators, eliminates impaired sediment quality as a limiting factor, and allows three-dimensional use of the water column. Spat may be transferred to longlines attached to rafts, stakes, or racks for grow-out (Boyd et al. 2005).

Floating or suspended culture of many species of seaweed is achieved by fixing seaweed propagules on ropes or nets and attaching them to rafts, nets, or longlines (Boyd et al. 2005). A few species of aquatic plants such as *Gracilaria* and *Caulerpa* normally are cultured in ponds.

Environmental issues

Many entities to include primary, secondary, and higher educational institutions, governmental agencies, international development organizations, and nongovern-mental organizations (NGOs) are interested in promoting wise resource use, pol-lution abatement, and environmental sustainability. There has been growing con-cern for several decades over excessive resource use, pollution, and overpopulation leading to unsustainability of the earth's ecosystems. The large environmental NGOs (eNGOs)—Sierra Club, World Wildlife Fund, Greenpeace, The Nature Conservancy, Environmental Defense Fund, and The Audubon Society to name a few—have lead the charge for environmental responsibility.

The eNGOs initially had little or no interest in aquaculture. The growth of aqua-culture during the mid-twentieth century was mainly in Asia and it consisted mainly of production of filter-feeding fish in manured ponds for family use or sale in local markets. Environmentalists saw such aquaculture as a way of producing food in rural areas of poor countries that required little input of resources and allowed use of agricultural wastes for a beneficial purpose. In fact environmentalists were ini-tially enamored with aquaculture and dubbed it the "blue revolution"—this name obviously was inspired by the term "green revolution" given to greater grain yields resulting from use of improved varieties of wheat, rice, and other grain species and large inputs of fertilizer nutrients and water for irrigation.

In the 1960s and 1970s, it became apparent that many fisheries products popular with consumers in wealthy countries could be profitably produced by aquaculture. This revelation led to the emergence of various kinds of feed-based aquaculture, for example, production of trout, salmon, marine shrimp, and channel catfish, for the markets in the United States and Canada, Europe, and Japan. The increasing demand for fisheries products in wealthy countries opened up the possibility for a lucrative export market for certain aquaculture species—particularly marine shrimp and salmon. Large areas of coastal land in certain South American and Asian countries were converted to shrimp ponds, and cage culture of salmon in coastal waters became common in a few countries—Norway and Chile in particular. The aquaculture industry suddenly was no longer so appealing to eNGOs; it had taken on many of the features of large-scale agribusiness (Bailey and Skladany 1991; Khor 1995).

The complaints against aquaculture by environmentalists have focused on feed-based aquaculture for export markets (Naylor et al. 1998, 2000). Relatively little criticism has been directed at aquaculture for domestic markets—other than in developed countries where high value products are produced using feeds, for example, trout or salmon production in the United States or Europe. Tilapias, although often cultured in developing countries for domestic consumption, have become important export products that are increasingly produced with feeds. Thus tilapia production also has drawn the ire of eNGOs. The eNGOs, however, do not appear to be greatly concerned about the culture of seaweed and molluscs.

The main issues in the aquaculture–environment controversy are listed in Table 1.4. These general areas of concern are not unlike the list of concerns that eNGOs would have for terrestrial agriculture, or for that matter, most other industries. Of course, the concerns listed in Table 1.4 do not apply equally across all aquaculture species or all production methods.

Table 1.4 Major issues that concern environmentalists about aquaculture.

Land use modifications
  Wetland alterations, cropland conversion
Excessive use of freshwater
Water pollution
  Eutrophication, turbidity and sedimentation, and salinization
Excessive energy use and greenhouse gas emissions
Wasteful use of fish meal
Negative effects on biodiversity
  Pollution, capture of wild aquatic animals for broodstock, exotic species, genetically modified
    species, bird and other predator control, entrainment of small aquatic organisms in pumps,
    spread of diseases to wild populations
Antibiotics and other chemicals
  Antibiotic resistance in organisms, toxicity of certain compounds, residues in aquaculture
    products
Socioeconomic issues
  Conflicts with other resource users, worker safety and other rights, etc.

In later chapters, considerable attention will be given to the relative impacts of different culture species and methods. However, for now a few examples will suffice. Channel catfish *Ictalurus punctatus* and Pacific white shrimp or whiteleg shrimp *Litopenaeus vannamei* are both produced in earthen ponds to which feed is applied and that may be intensively aerated. However, there is much more environmental concern over *L. vannamei* culture than *I. punctatus* culture for the following reasons:

- Shrimp ponds tend to be sited in more ecologically sensitive areas than are catfish ponds;
- Wild-caught postlarvae and broodstock have been used in shrimp culture, but broodstock and fingerling catfish have been farm-reared;
- Compared to shrimp feed, catfish feed is very low in fish meal content;
- Water exchange is not commonly used in catfish farming as it is in shrimp farming;
- Shrimp ponds are drained for harvest of every crop (1–2 times per year); catfish ponds are drained for repairs at an interval of 6–10 years.

Pond culture is often considered more environmentally responsible than cage culture because:

- All wastes from cages enter directly into the water body containing the cages, but in ponds, a large proportion of the wastes is assimilated by natural processes before discharge;
- Cages often are placed in public waters, while ponds tend to be located mainly on privately owned land;
- Fish escapes from cages are more difficult to avoid than are those from ponds.

It already has been mentioned that culture of aquatic animals without the use of feeds is usually considered by environmentalists to be more environmentally desirable than feed-based aquaculture. However, this assumption should not be accepted *a priori*. There are many "trade-offs" that should be considered when evaluating the environmental effects of aquaculture (Boyd et al. 2007). For example, low-input aquaculture in ponds does not require feed and aeration or produce highly polluted effluent, but production is low. More land area must be converted to water surface area in ponds for extensive culture than in those for intensive culture for the same amount of production. The reduction in energy, feed ingredient use, and waste discharge resulting from low-input aquaculture must be weighed against the greater amount of land and water needed per unit of production.

Governments are increasingly interested in environmental protection and most governments have imposed environmental regulations upon aquaculture. The nature of these regulations varies from country to country as does the level of enforcement. Of course eNGOs are well aware of the role of national priorities and "special interests" in governmental regulations, and they tend to feel that governments cannot be trusted to put environmental sustainability ahead of these interests. Therefore, the eNGO lobby works hard to influence legislation related to the environment. For example, the US Environmental Protection Agency (USEPA) recently made an effluent rule for US aquaculture (Federal Register 2004). This rule-making process was the result of a report on US aquaculture that was published by the Environmental Defense Fund (EDF) (Goldburg and Triplett 1997) and subsequent lobbying by this organization.

There is increasing public awareness of environmental issues, and this environmental awakening can do more to promote wise resource use and environmental sustainability than possibly any other factor. If consumers want products produced by environmentally responsible methods, it places demands on producers, governments, importers, wholesalers, and retailers to supply these products. Aquaculturists are beginning to use better production practices, governments are becoming more serious about environmental regulations, importers are seeking products resulting from environmentally responsible methods, and retailers want to cater to the wishes of their clientele who are increasingly concerned about environmental, social, and food safety issues. The upshot is that there is a growing demand for product labels that tell where and how products were produced (Boyd and McNevin 2011).

Conclusions

Aquaculture is essential in meeting future needs for aquatic animal protein because most capture fisheries have reached or exceeded their sustainable limit. The majority of freshwater aquaculture and a considerable amount of coastal aquaculture production come from ponds of which there are about 11 000 000 ha worldwide. In addition, raceways, cages and net pens, and water recirculating systems are important in freshwater, and cage culture is important in the marine environment. Much marine aquaculture is for molluscs and employs open-water systems. The intensity of aquaculture is being increased through use of fertilizers, feeds, and mechanical aeration.

Aquaculture can result in wasteful resource use and negative environmental impacts; thus, it has recently been subjected to much criticism by environmental advocate groups. Much effort is currently being made to improve resource use efficiency and reduce the negative environmental impacts of aquaculture. The collaborative efforts between eNGOs and some large-scale seafood buyers in sourcing aquaculture products is leading to a greater understanding of environmental and resource use issues by both parties.

The eNGO perspective

Most of the eNGOs engaged in issues related to aquaculture are driven by a broader ocean conservation mission. Most eNGOs have taken stances on aquaculture which are driven by the desire to conserve the world's ocean.

Although most eNGOs started to raise the issue of the effects of aquaculture on the oceans only peripherally, their efforts have increased dramatically over the past two decades. The eNGOs' concerns with conservation of the ocean have also provided a strategic framework for this group of stakeholders to prioritize certain aquaculture species and production systems. The prioritization is rooted in the impacts that could be realized in the marine environment.

With this background, it is easier to understand why the eNGOs first started their engagement in aquaculture with marine shrimp and salmon farming. Marine shrimp farming, up until the late 1980s, relied on wild postlarvae collected from estuaries. Further the earlier practices of constructing shrimp ponds in coastal zones gave rise to concerns because of the importance of these areas for nursery and breeding areas for a variety of aquatic and terrestrial organisms.

There are few places left in the world where significant natural runs of Atlantic salmon can be found, most notably in Russia and Iceland. Atlantic salmon farming which increased rapidly in the 1990s was another obvious concern for the eNGOs as cages were being placed throughout the coastal waters of Europe, Canada, and the United States. Later in the 1990s, Chile became a major player in farmed salmon industry. The salmon industry relies primarily on the production of Atlantic salmon. This species is sparsely found in the natural environment and placement of Atlantic salmon farms in regions that have existing and healthy populations of Pacific salmon (Canada and the Pacific Northwest of the United States) has raised concerns over the impact of these nonnative species on the natural salmon populations.

Shrimp and salmon aquaculture also tend to have two other common characteristics that raise concern for eNGOs. The first is the release of effluents into the natural environment, thus posing a pollution threat to the marine environment. The second impact is the utilization of fish meal and oil made from wild fish as a component in manufactured aquafeeds.

While shrimp and salmon farming were among the highest priorities of eNGOs, much of the rest of aquaculture was tainted by these two sectors for many years. However, in recent years, there has been a growth in interactions between the eNGOs and the aquaculture industry and there is a greater understanding that aquaculture is a varied industry with some forms of production posing greater environmental risk than others. Part of the drive for understanding aquaculture to a greater degree resulted from partnership agreements that some eNGOs have with large seafood buyers where the eNGOs are to comment on, review, or suggest purchasing strategies that are more environmentally benign. Examples of specific aquaculture agreements include Monterey Bay Aquarium's Seafood Watch Program and Bon Appetite Management Company; New England Aquarium and Darden Restaurants; World Wildlife Fund and Costco Wholesale Corporation; Environmental Defense Fund and Wegman's; Conservation International and Wal-Mart; Sustainable Fisheries Partnership and High Liner Foods.

Some of these agreements involve the exchange of funds for services or resources provided by the particular eNGO. Some of the eNGOs listed above also have multiple retailer or buyer partner agreements.

Part of the challenge for the eNGO community working on aquaculture issues is the lack of funding available to them. Most eNGOs obtain funding from grants to work on aquaculture issues. The grants of any sizeable nature have come mainly from the David and Lucile Packard Foundation. The Packard Foundation has a long history of funding many of the abovementioned organizations for ocean conservation work, but they were probably the only large foundation that was willing to fund mainstream eNGOs to address threats from aquaculture.

It is important to point out that the eNGOs in the United States have the luxury of large private foundations such as Gates, Packard, Moore, Hewlett, MacArthur, etc. The magnitude of financial support by private foundations in other parts of the world to fund eNGOs is miniscule. Thus, the traditional mechanism for fund-raising in other parts of the world is through membership. Interestingly membership compels an eNGO to act or carry out projects in a manner that would please the bulk of the members. Most of the large eNGOs in the United States have members but do not feel the same pressures as those outside the United States because of the private foundation cushion.

As the aquaculture industry has matured so have the eNGOs working toward a more sustainable vision of aquaculture. There is a prevalent recognition within the eNGO community that much of aquaculture is better than its wild fishery counter-part. However, the perception of aquaculture by the eNGOs will likely be driven by key impacts identified through the examination of the shrimp and salmon farm-ing industry. These main impacts include water pollution, utilization of wild fish as a feed ingredient, introduction of exotic species or escapes and habitat conversion (particularly in coastal environments) and chemical use. An ideal aquaculture facility would be one that is closed from the environment (addressing escapes, introductions and exotic species and water pollution) and one that cultures a species that does not require a high protein diet (to address wild fish utilization). Additionally a system that is low enough in intensity (often "organic") tends to be the eNGO solution for reductions of chemical or therapeutic inputs into a particular system.

There are few examples of these ideal systems that supply a significant amount of aquaculture to global markets or to domestic markets. This complicates the position of many eNGOs that have mainstream and large-scale private sector partnerships with the large seafood buyers of the world. These eNGOs are effectively pigeon-holed into making some level of concession to aquaculture operations that are viewed as "sub-optimal" but not "bad" players in their eyes. Some of the more extreme factions of the eNGOs such as Mangrove Action Project, Sea Shepherd, Greenpeace, the Coastal Alliance for Aquaculture Reform and other grassroots organizations see the large eNGOs as "selling out to" or "green-washing" industry by making these concessions, and in a few cases some of these concessions have been egregious and have been nothing more than a means to generate publicity. Nevertheless with the growing cooperation of the aquaculture industry and the eNGOs, there is a greater need to understand the utilization of natural resources in the aquaculture industry relative to other large food production activities such to prioritize and coordinate activities and targets.

References

Avault Jr., J. W. 1996. *Fundamentals of Aquaculture*. Baton Rouge: AVA Publishing Company, Inc.

Avnimelech, Y. 2012. *Biofloc Technology—A Practical Guide Book*, 2nd ed. Baton Rouge: The World Aquaculture Society.

Bailey, C. and M. Skladany. 1991. Aquaculture development in tropical Asia. A re-evaluation. *Natural Resources Forum* 15:66–72.

Berka, R. 1986. A brief insight into the history of Bohemian carp pond management. In R. Billard and J. Marcel, editors, *Aquaculture of Cyprinids*, pp. 35–40. Paris: Institut National de la Rescherche Agronomique.

Boyd, C. E. and C. A. Boyd. 2012. Physiochemical characteristics of ponds. In J. W. Neal and D. W. Willis, editors, *Small Impoundment Management in North America*, pp. 49–82. Bethesda: American Fisheries Society.

Boyd, C. E. and A. A. McNevin. 2011. An early assessment of the effectiveness of aquaculture certification and standards. Report to RESOLVE, Washington, DC.

Boyd, C. E. and C. S. Tucker. 1998. *Pond Aquaculture Water Quality Management*. Boston: Kluwer Academic Publishers.

Boyd, C. E., A. A. McNevin, J. Clay, and H. M. Johnson. 2005. Certification issues for some common aquaculture species. *Reviews in Fisheries Science* 13:231–279.

Boyd, C. E., S. Soongsawang, E. W. Shell, and S. Fowler. 2009. Small impoundment complexes as a possible method to increase water supply in Alabama. Proceedings 2009 Georgia Water Resources Conference, University of Georgia, Athens, Georgia.

Boyd, C. E., C. S. Tucker, A. McNevin, K. Bostic, and J. Clay. 2007. Indicators of resource use efficiency and environmental performance in fish and crustacean aquaculture. *Reviews in Fisheries Science* 15:327–360.

Boyd, C. E. J. Queiroz, J. Lee, M. Rowan, G. N. Whitis, and A. Gross. 2000. Environmental assessment of channel catfish, *Ictalurus punctatus*, farming in Alabama. *Journal of the World Aquaculture Society* 31:511–544.

Brown, T. W., J. A. Chappell, and C. E. Boyd. 2011. A commercial-scale, in-pond raceway system for *Ictalurid* catfish production. *Aquacultural Engineering* 44:72–79.

Brune, D. E., G. Schwartz, A. G. Eversole, J. A. Collier, and T. E. Schwedler. 2003. Intensification of pond aquaculture and high rate photosynthetic systems. *Aquacultural Engineering* 28:65–86.

FAO (Food and Agriculture Organization). 2012. *The State of World Fisheries and Aquaculture*. Rome: FAO Fisheries and Aquaculture Department.

Federal Register. 2004. Effluent limitation guidelines and new source performance standards for the concentrated aquatic animal production point source category: final rule. *Federal Register* 69(162): 51892–51930. Office of the Federal Register, National Archives and Records Administration, Washington, DC.

Goldburg, R. J. and T. Triplett. 1997. *Murky Waters: Environmental Effects of Aquaculture in the United States*. Washington: Environmental Defense Fund.

Khor, M. 1995. The aquaculture disaster, third world communities fight the "Blue Revolution". *Third World Resurgence* 59:8–10.

McIntosh, R. P. 1999. Changing paradigms in shrimp farming: I. general description. *Global Aquaculture Advocate* 2(4/5):42–45.

Naylor, R. L., R. J. Goldburg, H. Mooney, M. Beveridge, J. Clay, C. Folke, N. Kautsky, J. Lubchenco, J. Primavera, and M. Williams. 1998. Nature's subsidies to shrimp and salmon farming. *Science* 282:883–884.

Naylor, R. L., R. J. Goldburg, J. H. Primavera, N. Kautsky, M. C. M. Beveridge, J. Clay, C. Folke, J. Lubchenco, H. Mooney, and M. Troell. 2000. Effects of aquaculture on world fish supplies. *Nature* 405:1017–1024.

Schmittou, H. R. 1993. High density fish culture in low volume cages. American Soybean Association, Singapore.

Serfling, S. A. 2000. Closed-cycle, controlled environment systems: the Solar Aquafarms story. *Global Aquaculture Advocate* 3(3):48–51.

Soderberg, R. W. 1994. *Flowing Water Fish Culture*. Boca Raton: CRC Press.

Stickney, R. R. 2000. History of aquaculture. In R. R. Stickney, editor, *Encyclopedia of Aquaculture*, pp. 436–446. New York: John Wiley and Sons.

Timmons, M. B., J. M. Ebeling, F. W. Wheaton, S. T. Summerfelt, and B. J. Vinci. 2001. *Recirculating Aquaculture Systems*. Ithaca: Cayuga Aquaculture Ventures.

Verdegem, M. C. J. and R. H. Bosma. 2009. Water withdrawal for brackish and inland aquaculture, and options to produce more fish in ponds with present water use. *Water Policy* 11(Suppl 1):52–68.

Yoo, K. H. and C. E. Boyd. 1994. *Hydrology and Water Supply for Aquaculture*. New York: Chapman and Hall.

Chapter 2

World population

Assessments of global resource use, pollution, and sustainability of world ecosystems should consider the effects on these variables of human population distribution and growth and other demographic variables. As human population increases, there is greater demand for resources and more wastes are generated. The human population is not uniformly distributed across continents and countries, and neither are resource use and waste production. Effects may differ from place to place but the world's ecosystems—including marine ecosystems—tend to be interconnected and global effects must be considered.

Countries with a higher per capita income tend to use more resources per unit of population than do countries with a poorer populace. As countries become more affluent, they tend to consume resources and produce waste at a rate greater than expected from population increase alone. Today it is economically possible to transport large volumes of inexpensive raw materials, manufactured items, and perishable foods from one continent to another, reducing the importance for certain primary industries in importing countries and transferring negative impacts associated with these industries to exporting countries. Food production is a good example. Many countries import a large part of their food supply from other countries. This reduces the use of water, land, and other resources for food production in importing countries but increases the use of these resources in the exporting country. In assessments of environmental issues at country or regional levels, it is important to consider the influence of such factors as population density and growth, import–export balances, and level of development.

A historical perspective on human population growth as it relates to technological advancement and improvement in the human condition is instructive in understanding the present population status. It is also necessary for predicting how populations of countries, continents, and the world will change in the future and influence resource use and the environment.

Aquaculture, Resource Use, and the Environment, First Edition. Claude E. Boyd and Aaron A. McNevin.
© 2015 John Wiley & Sons, Inc. Published 2015 by John Wiley & Sons, Inc.

The purpose of this chapter is to provide an overview of human population—past, present, and future.

Historical demographics

Demographics are the statistical data of a population to include birth rates, death rates, estimated number of people in the population, rate of population growth or decline, etc. The ability of governments to collect such data accurately and share the information with those wanting to study local, regional, continental, or global demographics has traditionally been quite limited. In fact many of the statistics such as birth and death rates were not defined until the nineteenth century. Thus, population statistics have improved in reliability over time, and of course, much more is known about the history of human population in some regions or countries than in others. The most reliable demographic data are from about 1850 to present, and certainly the reliability of these data is much better since the early 1950s when international organizations such as the United Nations (UN) began to work with member countries to obtain population data at frequent intervals and to maintain these data in accessible archives. Of course the most reliable population data are at the local level, and the reliability of estimates becomes less accurate as the census area increases in complexity. In addition, some countries have better census programs than do other countries.

The human race apparently had a global population of around 1 million during many millennia before the advent of agriculture about 10 000 BC (Mazoyer and Roudart 2006). During this era, humans depended on hunting, fishing, and gathering for food; they also had little control over their living conditions. Human population numbers were regulated basically by the same factors controlling other animal populations. The development of agriculture radically changed the relationship of humans to their environment by providing them a degree of control over their food supply. This lessened famine as a bottleneck to population growth and allowed the number of humans occupying the earth simultaneously to increase. By 5000 BC, there were around 15 million people, and by 1 AD, the global population is thought to have stood at around 150–200 million—mostly concentrated in what is now eastern China and in the Mediterranean region (Tanton 1994). The population on average required more than a millennium to double during the period 10 000 BC to 1 AD (Fig. 2.1).

The population continued to grow at a slow rate during the first millennium AD reaching about 310 million people by the end of this period (Fig. 2.1). In the second millennium AD, the population increased at a greater rate, and a global population of 1 billion was attained soon after 1800. It had taken mankind tens of thousands of years to attain a population of 1 billion. Beginning about 1800, the growth of the global population exploded; it doubled again by the late 1920s—a period of around 120 years to gain the second billion. After the 1920s, the time necessary to add 1 billion people continued to decline: 3 billion by 1959 (32 years); 4 billion by 1974 (15 years); 5 billion by 1987 (13 years); 6 billion by 1999 (12 years). By midyear 2012, the 7 billion mark was reached about 14 years after the population reached 6 billion. The number of years for the population to add 1 billion appears to have

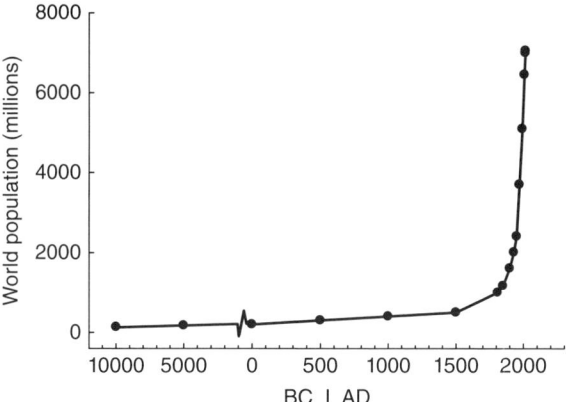

Figure 2.1 World population growth over time.

reached a minimum and is expected to increase suggesting that the rate of increase is slowing (http://donsnotes.com/reference/population-world.html).

Population growth rate on a global basis is determined by two factors: the birth rate and the death rate. These two variables usually are reported in births and deaths per annum for each 1000 inhabitants. The birth rate minus the death rate equals the natural population increase rate. When the two rates are equal, population growth lessens to replacement rate—zero population growth. Of course for an individual country, the natural population increase rate must be adjusted for the net migration rate.

For most of human history, society was highly agrarian and the death rate was high. Parents needed to have many children so that some would survive to help with agricultural chores and support the parents in their old age. A narrow difference between birth and death rates occurred worldwide for most of human history (http://www.globalchange.umich.edu/globalchange2/current/lectures/pop_socio/pop_socio.html). This situation allowed only a slow increase in population and famines, wars, and disease epidemics could cause population growth to slow, and at times, global population decreases occurred. However, around 1750 scientific and technological advances in nearly all human pursuits, and especially in sanitation and medicine, lead to a decline in death rates in the more developed countries. There was a great increase in population growth but the birth rate soon began to decline in these countries. It was no longer necessary for parents to have so many children because the non-agrarian population was increasing and more children were surviving until adulthood and population growth rate declined. In less developed countries, the death rate began to decline in the early 1900s, but the overall birth rate did not begin to fall until after 1950 (Reher 2004) and it still is much greater in some countries than others. This has resulted in a huge population increase in less developed countries as opposed to more developed countries. The birth rate in 2000 was about 30 per 1000 people in less developed countries and only 12 per 1000 people in more developed countries but the death rate was about the same in countries at both levels of development (Reher 2004).

A change in attitude about having children that probably is related to affluence has resulted in a negative growth rate in some more developed countries. Add to this the decrease in birth rate in less developed countries, and we have the reason for the slowing of world population growth rate. Of course in some of the world's poorer countries, and especially in some African countries, the population growth rate remains quite high.

Despite famines, epidemics, wars, and natural catastrophes some of which have claimed tens of millions of people (Janssen et al. 2012)—except for brief periods of declines—global population has been increasing for about 12 millennia. This increase became exponential about 1750 and this pattern of increase has continued until present.

A larger population requires more food, and in the eighteenth and nineteenth centuries, a large amount of land already was required to produce the world's food supply. This fact led to many discourses on impending global famine and societal collapse because of excessive population. Probably the most famous doomsday prophet was Thomas Malthus who in 1778 argued that population increased exponentially while food production increased arithmetically (Gilbert 1993)—a hypothesis often referred to as the "Malthusian Check." Malthus proclaimed that population would soon outgrow the capacity for food production and population would be controlled by famines, wars, and diseases. Others have continued to make similar predictions; one of the more well known of the modern doomsday prophets is Paul Ehrlich who wrote the book "Population Bomb" (Ehrlich 1971). However, technological advances have been at a greater level than envisioned by Malthus, Ehrlich, and others, and the world population is continuing to increase. Of course there have been many wars, famines, diseases, and natural catastrophes since Malthus' time but the human race has overcome and flourished.

Advances in medicine and sanitation lessened mortality from infections and common diseases that historically had been a great bottleneck to population growth. Improvements in industrial output, transportation, communications, housing, and other infrastructure necessary to support the growing population also occurred. More land also was devoted to agricultural production, because the growing population required more food. However, intensification of agriculture was a major factor necessary for the population increase since the mid-1950s. Without the increase in agricultural yields per unit area (Table 2.1), there would not have

Table 2.1 Global agricultural production, 1961–2010.

Year	World population (billions)	Global agricultural production (Mt)	Global agricultural area (billions ha)	Global average yield (kg/ha)	Food available (kg/cap)
1961	3.09	2567	4.46	576	831
1971	3.77	3360	4.58	734	891
1981	4.51	4070	4.67	865	902
1991	5.39	5233	4.86	1077	971
2001	6.20	6219	4.97	1251	1003
2010	6.90	7670	4.93	1550	1112

Source: FAO. FAOSTAT. Production. 2013. Accessed: 10/15/2013. http://faostat3.fao.org/faostat-gateway/go/to/download/Q/*/E.

been enough agricultural land to provide sufficient food for the population boom. The amount of agricultural land increased by only about one-tenth between 1961 and 2010. But food production tripled, and there was a modest amount of more food per capita in 2010 than in 1961. Based on 1961 agricultural yields about 147 million km^2 of agricultural land would be needed to produce the amount of food produced in 2010—more area than the earth's land surface excluding Antarctica.

Failure of past prophecies of doom to materialize is no deterrent to modern-day doomsday prophets—there are many in all types of news media. The predictions of these prognosticators will likely to be no more accurate than those of their predecessors because the future often is shaped by unexpected, unprecedented, or unpredictable events. As the famous physicist Neils Bohr once stated "prediction is very difficult especially if it is about the future." Nevertheless the dire predictions by environmentalists may benefit society by leading to actions to minimize real but possibly overstated impacts. For example, in the late 1970s and 1980s, the furor over the effects of acidic rain lead to significant reductions in sulfur emissions, and the current obsession with climate change likely will evoke a large reduction in greenhouse gas emissions.

The early environmental movement focused on wilderness preservation, and a growing population obviously favored conversion of wilderness area to land for forestry, agriculture, and other exploitations. Therefore, the environmentalists advocated population control. There has been little success, however, in convincing people that they should be less procreative—especially in less developed countries where children are needed for family farm labor and for the parents' welfare in their later years. There also are conflicts between population control and religion and human rights. Most countries apparently have greatly tempered efforts at birth control and focus more on planned parenthood. With the notable exception of the one child policy in China, birth control programs probably have not been highly effective in slowing of population growth in most countries. The major factors leading to lower population growth apparently are greater affluence and better education.

World population and its distribution

As of September 11, 2013 at 14:15 hour, the world population stood at an estimated 7 110 692 393 people (http://www.census.gov/main/www/popclock.html). It is interesting to note that this number is about 6.5% of the estimated 107.6 billion human births that have occurred during the history of the species (http://www.prb.org/Articles/2002/HowManyPeopleHaveEverLivedonEarth.aspx).

The populations of the different continents are listed in Table 2.2. Asia contains about 60% of the current global population. Africa has the second greatest population making up 15% of the global population. Europe follows with around 8% of the world's people. The population of the two Americas combined is not a lot greater than the population of Europe. Oceania (Australia and South Pacific Islands) contains only about 0.5% of the planet's inhabitants.

Table 2.2 Population, land area, and population density for the world's continents.

Continent	Population (millions)	Land area (×1,000 km²)	Density (individuals/km²)
Asia	4265	31 119	137
Africa	1099	29 642	37.1
Europe	741	22 168	33.4
North America (including the Caribbean)	552	24 393	22.6
South America	402	17 824	22.6
Oceanic (Australia and South Pacific Islands— not to include Indonesia and Timor)	36	8515	4.2
Total	7095	133 661 (excluding Antarctica)	53.1

Source: Made with data from Veregin (2010).

The population spread uniformly over the earth's land surface would have a density of 53.1 individuals/km²—about 2 ha/person. The continental population densities range from 4.2 individuals/km² (about 25 ha/person) in Oceania to 137 individuals/km² in Asia (slightly less than 1 ha/person). Of course the population density is a misleading statistic because much land in all continents is unoccupied by people because of severe climate, unfavorable terrain, etc.

The 20 most populous countries in the world are listed in Table 2.3. The combined population of the top 10 countries in the list is 4144 million (58% of global population); adding the next 10 most populous countries brings the total to 5021 million (71% of global population). The land area of the 10 most populated nations is 50 516 000 km², 38% of the world's total land area of 133 661 000 km² (Antarctica excluded)—58% of the world's population is in countries occupying only 38% of the global surface area. Adding the next 10 countries increases the amount of land surface area to 59 590 km² (45% of total)—58% of the global population occupying 45% of the land surface area. Much of Russia is essentially uninhabited because of its harsh, northern climate. If Russia is removed from the list in Table 2.3, the other 19 countries have 68% of the world population but occupy only 32% of the land area.

Population density ranges from 8.7 individuals/km² in Russia to 1262 individuals/km² in Bangladesh. Of course Russia is an exception; the next lowest population density among the top 20 countries is 33 individuals/km² in the Democratic Republic of the Congo. The average population density for the 20 countries is 84.3 individuals/km² (112.9 individuals/km² excluding Russia) compared to the global population density of 53.1 individuals/km².

Data in Table 2.3 reveal that countries in parts of the world with a long history of civilization, for example, countries in Asia, Africa, and Europe, tend to have greater population densities than countries such as the United States, Brazil, and Mexico in the "new" world. This suggests that countries with old civilizations have had more time to develop dense populations than have countries in the new world. Also the relatively small proportion of global land area contained in the more populous countries suggests that natural resources and climate for population development have been more favorable than in less populous countries—again Russia is an exception.

Table 2.3 Populations, land areas, and population densities for the 20 most populous countries in the world in 2013.

Country	Population (millions)	Land area (×1000 km²)	Population density (individuals/km²)
China	1350	9570	141
India	1221	2973	411
United States	317	9148	35
Indonesia	251	1812	139
Brazil	204	8459	24
Pakistan	193	771	250
Nigeria	175	911	192
Bangladesh	164	130	1262
Russia	142	16 378	8.7
Japan	127	364	350
Mexico	116	1943	60
Philippines	106	298	356
Vietnam	92	310	297
Ethiopia	94	1000	94
Egypt	85	995	85
Germany	81	349	234
Turkey	81	770	105
Iran	80	1531	52
Congo (Democratic Republic of Congo)	75	2267	33
Thailand	67	511	131

Source: Population data taken from United Nations Department of Economic and Social Affairs (2013); land area taken from http://data.worldbank.org/indicator/AG.LND.TOTL.K2.

The future

The doomsday prophets have been wrong up until the present. Of course there have been famines, natural catastrophes, pandemics and regional epidemics, and world wars as forecasted, but the population has continued to grow on a global basis. Moreover, the present world situation leaves much to be desired. Some countries have governments that are essentially nonfunctional and many of these are involved in civil warfare. There also are frequent wars between countries or groups of countries.

Of 215 countries, 107 are listed in a UN database have annual per capita incomes less than US$5000 and 43 of these countries have annual per capita incomes less than US$1000 (http://unstats.un.org/unsd/demographic/products/socind/inc-eco.htm).

Despite the extent of poverty in the world, Kharas and Gertz (2010) predicted that the world middle class is poised to expand exponentially in the coming decades as a result of improving economies in China and several other Asian nations. Kharas and Gertz defined the middle class as households with daily expenditures between US$10 and US$100 per person in purchasing power parity. Based on this definition, the middle class numbered 1.845 billion in 2009, and it was predicted to increase to 3.249 billion by 2020 and to 4.884 billion by 2030. Thus, the middle class could increase from 30% of global population in 2009 to more than 60% by 2030.

The middle class has more money to spend on food and other goods and services. Food production will therefore have to be increased at a greater rate than population growth. This will place a huge demand on resources and ecosystems in coming decades.

In addition to the growth of the middle class, the world population—especially in low-income countries—will become increasingly urban. According to the City Mayors Society (http://www.citymayors.com/society/urban_growth.html), the expected increase of 2.12 billion will occur almost entirely in low-income countries. The world's annual urban growth rate is projected at 1.8% as compared to 0.1% for rural areas. This shift in population means that there will be a greater demand for goods and services in cities. Not only must food production increase, there must be improvement in methods for moving food from farms in rural areas to cities.

The number of malnourished people in the world probably exceeds 1 billion and even more lack adequate housing, sufficient water supply, and proper sanitation. Famines still occur in some countries but not for lack of world food supply. Today most famines result from inability of governments even with the assistance of international donor organizations to distribute food because of poor transportation infrastructure, corruption, armed conflicts, or anarchy in the regions of famine.

It is instructive to consider the 16 dimensions of the human population problem presented by Brown et al. (1998). According to these authors, the main factors affecting population growth are grain production, freshwater supply and availability, biodiversity, climate change, ocean fish catch, availability of jobs, cropland area, forest area, housing, energy, urbanization, natural recreational areas, education, waste, meat production, and income. Most of these factors are directly or indirectly related to food production. Moreover, there are relationships among the factors. For example, to produce more grain requires more water for irrigation, results in more carbon emissions that may influence climate change, and may even require more cropland leading to clearing of forests and loss of biodiversity. There also is much interest in biofuels, a renewable resource, in order to reduce dependence on petroleum, coal, and other nonrenewable energy sources. Use of corn, soybeans, and other crops to make fuels competes with the use of these crops for human and animal food causing food prices to rise. Therefore, assuming present day efficiencies in resource use, improving the status of one of the factors considered essential for population growth may worsen the status of one or more of the other factors.

An in-depth discussion of the status of factors affecting population growth is beyond the scope of this chapter but it is noteworthy that aquaculture can be a contributor to several of the factors. From a positive standpoint, aquaculture supplements the oceanic fish catch to provide enough fisheries products to meet the world demand. It also lessens fishing pressure on some fish populations—although it increases the demand for small, pelagic marine fish for making fish meal and oil to use in aquafeeds. Aquaculture also creates jobs and produces income. On the negative side, aquaculture can change land use patterns (replace traditional cropland and wetlands), consume energy, contribute to carbon emissions and climate change, produce wastes, negatively impact biodiversity, and compete with other users of freshwater.

Brown et al. (1998) emphasized the growing dangers of continued population increase and they suggested that predicted estimates of future population growth

Table 2.4 Population in 2010 and estimates for the next four decades by continent, by level of country development, and for the world.

Continent or other category	Population in millions				
	2010	2020	2030	2040	2050
Africa	1015	1261	1532	1827	2138
Asia	4133	4531	4841	5049	5167
Europe	734	731	718	698	671
North America	539	595	648	695	739
South America	396	440	477	504	520
World	6853	7597	8259	8820	9284
Less developed countries	5622	6338	6984	7539	8005
More developed countries	1231	1259	1275	2181	1279

Source: http://www.census.gov/compendia/statab/2011.

may not be realized because of limitations imposed by one or more of the 16 factors listed earlier. It has been over a decade since Brown et al. (1998) published their 16 factors affecting population growth. The world population was nearing 6 billion and it reached 7 billion by mid-2012. The factors obviously have not yet caused a bottleneck in population growth.

The importance of the 16 factors on population growth has not diminished and there has been little improvement in their status. Thus there is still reason for dire predictions about the future of mankind. There also is reason for hope. There is much greater awareness of the urgency to conserve resources and pro-tect the environment today than at any other time in history. Moreover, there is evidence that the rate of human population growth is slowing. The global popu-lation growth rate fell from 1.30% per annum in 2000 to 1.09% per annum in 2011 (http://www.indexmundi.com/g/g.aspx?v=26&c=xx&l=en). This occurred because the death rate has stabilized at around 8 to 9/1000 people per year in most countries while the birth rate fell from around 22/1000 people in 2000 to 19.15/1000 people in 2011.

Assuming that one or more factors do not impose a drastic restriction on popu-lation growth, the world population is predicted to increase to 9.3 billion by 2050 (Table 2.4). The growth of the population will be largely in Africa (46.2% of global increase) and Asia (42.5% of global increase), but the rate of increase will be much greater in Africa than in Asia—110.6% versus 25.0%. Unfortunately, 98.0% of the global increase in population is expected to occur in less developed countries.

The predicted populations for 2050 in the 20 most populous countries (Table 2.5) indicate that India, not China, will be the most populous country—but China will be the second most populous. The United States will remain in third place; Indonesia will fall from fourth to fifth place, with Nigeria becoming the fifth most populous country. Germany, Iran, and Thailand are predicted to drop from the list of 20 most populous countries by 2050. They will be replaced by Uganda, Kenya, and Tanzania. The 20 most populous countries in 2050 will have an estimated 6242 million people. This is about 68% of the predicted world population—roughly the same percentage as in 2012.

Table 2.5 Estimates of the world's 20 most populous countries in 2050 and their populations.

Country	Population (millions)
India	1692
China	1296
United States	403
Nigeria	390
Indonesia	293
Pakistan	275
Brazil	223
Bangladesh	194
Philippines	155
Congo (Democratic Republic of the Congo)	149
Ethiopia	145
Mexico	144
Tanzania	138
Russia	126
Egypt	123
Japan	109
Vietnam	104
Kenya	97
Uganda	94
Turkey	92

Source: Data taken from United Nations Department of Economic and Social Affairs (2013).

Conclusions

The world population has increased greatly since the beginning of the industrial revolution in about 1750. Efforts to limit population growth popular from the 1950s until the 1980s were not effective and seem to have lost their appeal in most countries. The world population increased from slightly over 3 billion in 1960 to 7 billion in 2012. Population growth rate has begun to decline but the population is still increasing—especially in Africa. World population is projected to reach about 9.3 billion by 2050. Just to maintain the *status quo* globally resource use will need to increase by about 30% by 2050. Of course waste generation and associated pollution also will increase if we continue with the present level of conservation and pollution control. There needs to be improvement in resource use conservation and environmental protection in the production of all goods and services—aquaculture included.

The eNGO perspective

The eNGOs have been concerned with population growth and resource use from the perspective that if global population growth outpaces the earth's renewable resources, greater destruction and degradation of habitat will be the result of providing the products necessary for or desired by humans. There have been pivotal time periods that have determined the direction of the environmental movement. The debates and strategies that were developed and tested to reduce population growth during the period of 1960–1980 are important in understanding why the eNGOs do not appear to be strong advocates for population control.

As stated in this chapter, there has been a great deal of fear generated about overpopulation. This fear sustained through the twentieth century after World War II with a growing mainstream apprehension of the repercussions of unchecked population growth. In 1952, the Planned Parenthood Federation was formed in England to address unmanageable growth in the world's population. Shortly thereafter the UN convened the World Population Conference to bring greater unity to the discussions and quantify and standardize the metrics used to predict and determine population growth.

The United States played a central role in attempting to foster a more globalized ethic on family planning to reduce population growth. In 1960, the US Food and Drug Administration (USFDA) approved oral contraceptives, and in 1961, the US Agency for International Development (USAID) was established. Additionally, the US government passed the Foreign Assistance Act which authorized, among other things, research into family planning around the world. This led to USAID-sponsored programs for purchasing contraceptives for distribution in the developing world.

President Richard Nixon was a staunch believer in the potential harm of population growth and signed a bill to fund The Rockefeller Commission on Population Growth and the American Future. In remarks to the Commission on July 10, 1969, Nixon stated the following:

"One of the most serious challenges to human destiny in the last third of this century will be the growth of the population. Whether man's response to that challenge will be a cause for pride or for despair in the year 2000 will depend very much on what we do today. If we now begin our work in an appropriate manner, and if we continue to devote a considerable amount of attention and energy to this problem, then mankind will be able to surmount this challenge as it has surmounted so many during the long march of civilization" (http://www.population-security.org/rockefeller/001_population_growth_and_the_american_future.htm).

The recommendations of the commission were directed toward increasing public knowledge of the causes and consequences of population change, facilitating and guiding the processes of population movement, maximizing information about human reproduction and its consequences for the family, and enabling individuals to avoid unwanted fertility (Commission on Population Growth and the American Future 1972).

During the same period, other countries were seeking some level of stability in population growth. In India, in the 1970s, Prime Minister Indira Gandhi attempted to decrease the national birthrate by offering men money and transistor radios if they agree to undergo vasectomies. This incentive appeared to be unsuccessful and the government developed other coercive practices, such as the loss of employment for civil servants that have a fourth child (Brown 1984), that resulted in a mass sterilization of men and teenagers. In 1976 alone, over 700 million sterilization procedures occurred (Oldenburg 1977).

It is likely that the events in India sparked a decline in the urgency of action to prevent overpopulation because many could see the potential of how far these actions could be taken. Nevertheless in 1979, China began its "One-Child Policy" which restricted the number of children married urban couples could have to one. Although this policy is relatively simple in principle, culture in China tends to see male offspring as optimal because the family name will be carried on to successive generations. According to Chinese Health Ministry data released in March 2013, 336 million abortions and 222 million sterilizations have been carried out since 1971 (Jian 2013). There are also claims that this practice has led to female child abandonment and infanticide.

As a result of the events occurring in India and China as well as other countries in the 1970s, the USAID issued a policy paper stating that family planning programs will be based on fundamental principles of voluntarism and informed choice.

Another important aspect of population growth prevention measures is the conflict these measures may have with religious beliefs. In the United States, where there is a strong religious lobby, the use of abortion as a family planning measure is staunchly opposed by more conservative groups. In 1984, US President Ronald Reagan announced a policy prohibiting foreign NGOs that receive USAID family planning assistance funding from using their own or other non-USAID funds to provide or promote abortion as a family planning method. This policy was rescinded by US President William J. Clinton in 1993, but reinstated by President George W. Bush in 2001. The policy was rescinded once again by President Barrack Obama in 2009.

In 2004, the UN produced a report titled "World Population 2300," and projected as a medium scenario that world population will peak in 2075 at 9.2 billion and decline over the next 100 years to 8.3 billion. There is a growing recognition by the UN that populations cannot be controlled *per se*, but rather people can be informed about the implications and means to address unwanted parenthood.

A significant backlash often results from proposals for population control. Attenborough (2011) claimed that idea of population control has become "taboo" for all parties including the eNGOs who are most concerned with the sustainability of life on earth. The eNGOs that work on aquaculture issues have not addressed population growth and resource use in realm of population control, but rather consumption. Many of the large eNGOs attribute the most pressing environmental concerns to overconsumption in the developed world. However, the growing middle classes in Brazil, Russia, India, and China (BRIC countries) reveal that consumption is no longer a problem of the United States, Europe, and Japan. The overarching claim is that "sustainability" is primarily dependent on reducing consumption. However, it is difficult to envision how an eNGO with a Wal-Mart partnership is able to make inroads on decreasing consumer consumption as most seafood buyers' business models require expansion and potentially greater consumption.

It is unclear how successful eNGOs will be at reducing the consumption of products generated through aquaculture. It is clear that the rate of aquaculture growth will continue to rise as both consumption and global population increase. Further popular aquaculture products such as salmon and shrimp that are of the most concern to eNGOs will remain in retail stores at the cheapest price possible until generational shifts in consumption are achieved. This shift seems unlikely if the focus of aquaculture product consumption rests on retailer partnerships developed by the eNGOs.

References

Attenborough, D. 2011. This heaving planet. *New Statesman*. April 25, 2011:28–32.

Brown, C. H. 1984. The forced sterilization program under the Indian emergency: results in one settlement. *Human Organization* 43:49–54.

Brown, L. R., G. Gardner, and B. Halweil. 1998. Beyond Malthus: sixteen dimensions of the population problem. Worldwatch Paper 143, Worldwatch Institute, Washington, DC.

Commission on Population Growth and the American Future. 1972. Population and the American Future. Commission Report. John D. Rockefeller, 3rd, Chairman. March 27, 1972.

Ehrlich, P. R. 1971. *The Population Bomb*. New York: Ballantine Books.

Gilbert, G. (editor). 1993. *Thomas Malthus an Essay on the Principle of Population*. New York: Oxford University Press.

Janssen, S., M. L. Liu, S. Ross, and N. Badgett (editors). 2012. *The World Almanac and Book of Facts 2012*. Crawfordsville: R. R. Donnelly.

Jian, M. 2013. China's brutal one-child policy. *The New York Times*. May 22, 2013:A27

Kharas, H. and G. Gertz. 2010. The new global middle class: a cross-over from west to east. In C. Li, editor, *China's Emerging Middle Class: Beyond Economic Transformation*, pp. 32–54. Washington: Brookings Institution Press.

King, T. K. (editor). 2011. Time almanac 2011. Encyclopedia Britannica, Chicago, Illinois.

Mazoyer, M. and L. Roudart. 2006. *A History of World Agriculture: From the Neolithic Age to the Current Crisis*. New York: Monthly Review Press.

Oldenburg, P. 1977. *Briefing Materials on the Indian Parliamentary Elections, 1977*. New York: India Council of the Asia Society.

Reher, D. S. 2004. The demographic transition revisited as a global process. *Population, Space, and Place* 10:19–41.

Tanton, J. H. 1994. End of the migration epoch? *The Social Contract* 4:162–196.

Chapter 3

World food production

A dependable supply of food is essential for the well-being of humans. We must consume a diet that contains enough energy and nutrients for growth when we are children and adolescents and for maintenance when we become adults (Whitney and Rolfes 2011). Those who do not receive an adequate diet become malnourished, susceptible to diseases, or may die of starvation.

Nutritional requirements are complex consisting of minimum average daily intakes of energy, protein and essential amino acids, fats and specific fatty acids, carbohydrates, minerals, and vitamins (Whitney and Rolfes 2011). The quantity of food needed each day varies with age, gender, body size, degree of activity, physiological characteristics, etc., but most individuals should have a food energy intake between 2000 and 3000 cal/day. The calories should be provided by the three major classes of macronutrients in the following proportions: protein, 10–35%; fat, 20–35%; carbohydrates, 45–60%. The daily requirements for macronutrients will vary with caloric intake, but for a 2000 cal/day diet, 50 g protein, 65 g fat, and 300 g total carbohydrate were recommended by the US Department of Agriculture (USDA) and US Department of Health and Human Services (USDHHS) as a general guideline for adults and children of 4 years or older (USDA and USDHHS 2010).

Hunting and gathering wild plants and animals is no longer an important source of nutrients in human diets, and other than for wild-caught fisheries products, nearly all human foods originate from agriculture. Although human nutrient requirements can be satisfied with plant products, vegetarians often consume eggs and dairy products to assure adequate protein. Most humans, however, eat a diet containing plant products, eggs and dairy products, and meat. Animal feeds used in production of eggs, dairy products, and meats contain grain and other plant-based feedstuffs. Thus part of the output of plant agriculture is used as livestock feed. Moreover, in recent years, there has been interest in using plant crops, for example, sugarcane, soybeans, and corn, to make biofuels. This practice competes directly with use of these crops for livestock feed and human food.

Aquaculture, Resource Use, and the Environment, First Edition. Claude E. Boyd and Aaron A. McNevin.
© 2015 John Wiley & Sons, Inc. Published 2015 by John Wiley & Sons, Inc.

In 2010, world agricultural production was estimated to be about 7670 Mt or 1121 kg/cap (http://faostat.fao.org). The food supply is not equally distributed; on average, people in wealthy countries consume considerably more food per capita than do those in poor countries. The current global food situation is not ideal as discussed in Chapter 2, and more than 1 billion people are malnourished. They are malnourished, however, not because of a global deficit in food production, but for reasons of poverty, drought, civil conflicts, and failure of food distribution systems.

The world population is expected to increase by slightly over 30% by 2050. However, because of the growing middle class, the world food demand is expected to increase at a more rapid pace than the population. Supplying enough food will be a greater challenge in the future than it has been in the past.

There have been predictions beginning with Mathus (Gilbert 1993) and continuing to the present of impending starvation because of the growing population. These predictions fortunately have not come to pass because agricultural production has increased at a slightly greater rate than has population growth. Greater food production has been possible because of improved varieties of plants and breeds of animals, increased use of fertilizers and pesticides, greater mechanization, increased use of irrigation, other technological advances, and to a lesser extent, increased agricultural production area (Cassman 1999; Matson and Vitousek 2006).

The gains in production have required large inputs of nonrenewable resources and caused soil, water, and air pollution that negatively affect natural and agricultural ecosystems alike (Tilman 1999; Clay 2004; Foley et al. 2005; Reijnders and Soret 2003). Concerns about world food shortages persist and the sustainability of many agricultural practices often are questioned.

In this chapter, world food production will be discussed with particular emphasis on the role of fisheries and aquaculture.

Agricultural production

The FAO maintains annual estimates of crop production by country (http://faostat.fao.org). There is some variation in the way quantities of different crops are reported by FAO. Production of grains and soybeans is for the amounts of seeds harvested. Sugar sources are for the raw product processed for sugar rather than sugar yield, but cotton production is reported as lint yield. Other plant and animal products are reported as the edible portions, for example, meat yield, volume of milk, and weight of eggs. Fiber crops such as cotton are included in the database even though they are not used directly as human or animal food. Fisheries production (including seaweed) is included in a separate database.

Total global production of all agricultural products and of the top 20 products (in terms of quantity produced) was obtained at 5-year intervals from the FAO database (Table 3.1). The top 20 products include grains, vegetables, cotton lint, and other plant crops. It also includes milk, eggs, and several meats. The top 20 products differed slightly across dates (1961–2010), but only 23 products have appeared in the top 20 and 18 of these products were present in the top 20 throughout.

Total agricultural production and the combined production of the top 20 crops are presented along with per capita production (Fig. 3.1). It is interesting to note

Table 3.1 Agricultural products in the top 20 crop production groups between 1961 and 2010.

Plant products		Animal products
Grains	**Fruit**	**Meat**
Rice[a]	Grapes[a]	Cattle meat[a]
Wheat[a]	Apples[a]	Pig meat[a]
Corn (maize)[a]	Bananas	Chicken meat[a]
Legumes	Mangos, mangosteens and guavas	Sheep meat[a]
Soybeans[a]	**Other**	**Products**
Sugar sources	Potatoes[a]	Cow milk[a]
Sugarcane[a]	Sweet potatoes	Buffalo milk[a]
Sugar beets	Tomatoes[a]	Hen eggs[a]
Fiber	Fresh vegetables[a]	
Cotton lint[a]	Cassava	

Source: FAO. FAOSTAT. Production. 2013. Accessed: 10:15/2013. URI: http://faostat3.fao.org/faostat-gateway/go/ to/download/Q/*/E.
[a]Included on all annual lists.

that the top 20 crops made up 80% or more of total production. Total agricultural production rose by a factor of 2.98 between 1961 and 2010—from 2,567 to 7,670 Mt, while combined production of the top 20 crops increased from 2,127 to 6,147 Mt over the same period—a factor of 2.89. The increases in production were fairly steady over time ($R^2 = 0.989$ for total production and 0.990 for the top 20 crops) and averaged about 104 Mt/year and 82 Mt/year for total production and combined production of the top 20 crops, respectively.

Human population rose from 3080 million in 1961 to 6840 million in 2010—a factor of increase of 2.2. The rate of increase in agricultural production has exceeded the rate of population growth, and the per capita amount of agricultural production has exhibited a slightly upward trend for both total agricultural production ($R^2 = 0.836$) and combined production of the top 20 crops ($R^2 = 0.718$).

Slightly more than 80% of the total production of food crops was contributed by plants. In 2010, the major crop in terms of production quantity was sugarcane—1685 Mt. The three grain crops, wheat, corn, and rice, had combined

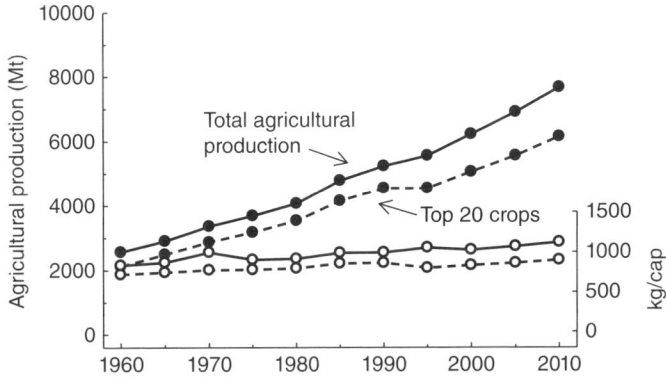

Figure 3.1 Global production of all agricultural crops and of the top 20 crops with per capita supply of each category from 1961 to 2010. *Source:* FAO. FAOSTAT. Production. 2013. Accessed: 10/15/2013. URI: http://faostat3.fao.org/faostat-gateway/go/to/download/Q/*/E.

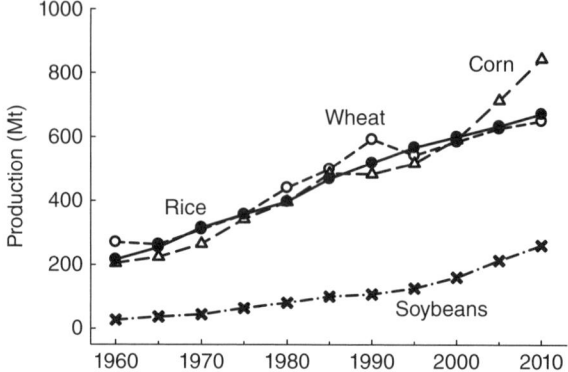

Figure 3.2 Global production of wheat, rice, corn, and soybeans from 1961 to 2010. *Source:* FAO. FAOSTAT. Production. 2013. Accessed: 10/15/2013. URI: http://faostat3.fao.org/faostat-gateway/go/to/download/Q/*/E.

production of 2167 Mt. Animal products accounted for 1016 Mt with the largest amount being from milk and eggs; meat production was 293 Mt.

Plant production

The increase in food production illustrated in Fig. 3.1 has resulted from increase in production of many crops as illustrated in Fig. 3.2 for grain crops and soybeans. Production of these four crops has increased between 1961 and 2010 by the following factors: corn, 4.12; rice, 3.11; wheat, 2.40; soybeans, 9.67.

Yield per unit area also has been increasing (Fig. 3.3). The factors for increase in yields were corn, 2.60; rice, 2.34; wheat, 2.54; soybeans, 2.11. Thus, the increase in production can be partially attributed to an increase in yield.

Grains and soybeans are produced in many countries and the top 10 producer countries are listed along with yields (Table 3.2). China is the largest producer of

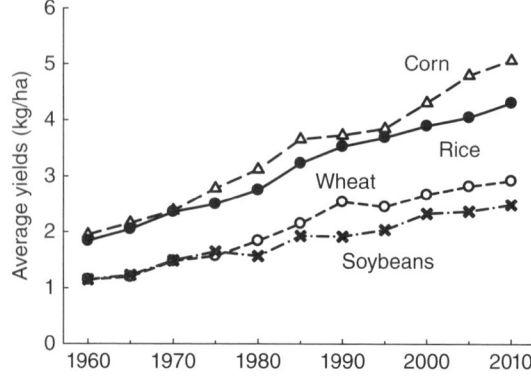

Figure 3.3 Global average yields of wheat, rice, corn, and soybeans from 1960 to 2010. *Source:* www.fas.usda.gov/psdonline.

Table 3.2 Top 10 grain- and soybean-producing countries in 2010 by production and yield.

Country	Production (Mt)	Yield (t/ha)
Wheat		
World	**651.1**	**2.98**
China	115.2	4.75
India	80.8	2.84
United States	60.1	3.12
Russia	41.5	1.91
France	38.2	7.04
Australia	27.9	2.04
Pakistan	23.9	2.65
Germany	23.8	7.37
Canada	23.7	2.80
Turkey	17.0	2.14
Rice		
World	**449.3**	**4.25**
China	137.0	6.55
India	96.0	3.36
Indonesia	35.5	4.67
Bangladesh	31.7	4.06
Vietnam	26.4	5.55
Thailand	20.3	2.88
Philippines	10.5	3.69
Burma	10.5	2.35
Brazil	9.3	4.83
Japan	7.7	6.51
Corn		
World	**829.1**	**5.10**
United States	316.2	9.59
China	177.2	5.45
Brazil	57.4	4.16
Argentina	23.6	6.74
India	21.7	2.53
Mexico	21.0	3.00
France	13.8	9.00
Ukraine	11.9	4.50
Canada	11.7	9.74
South Africa	10.9	3.82
Soybeans		
World	**264.7**	**2.57**
United States	90.6	2.92
Brazil	75.5	3.12
Argentina	49.0	2.68
China	15.1	1.77
India	9.8	1.05
Paraguay	8.4	2.92
Canada	4.4	2.94
Bolivia	2.1	2.10
Ukraine	1.7	1.62
Uruguay	1.6	1.79

Source: FAO. FAOSTAT. Production. 2013. Accessed: 10:15/2013. URI: http://faostat3.fao.org/faostat-gateway/go/to/download/Q/*/E.

wheat and rice, while the United States is the leader in corn and soybean production. Of the leading rice-producing countries, nine of ten are in Asia; Brazil in South America is the ninth largest rice producer. Soybean production is dominated by countries in the Americas.

Several of the leading grain- and soybean-producing countries fall well below the global averages for yield. Some of the differences in yield possibly are related to climate and especially to rainfall; nevertheless the differences suggest that there is still considerable opportunity to increase world grain and soybean production by better production practices to improve yields. Of course the yields of most other crops also probably can be increased by application of improved technology.

Considerable amounts of agricultural products are used as animal feedstuffs. In 2006, an estimated 737 Mt of the total production of 1991 Mt of grain crops (about 37%) were used as animal feeds (http://www.fas.usda.gov/psdonline/psdQuery.aspx). An even greater percentage of soybean production was used in animal feeds.

There also is considerable conversion of agricultural crops to biofuels. For example, in Brazil, ethanol made from sugarcane has long been an important part of the country's energy supply. The production of biofuels such as biodiesel and ethanol tripled between 2000 and 2007 (http://www.ers.usda.gov/AmberWaves/November07/Features/Biofuels.htm). The United States is attempting to become less dependent upon petroleum-based fuels but the climate in most regions of this country is not suitable for sugarcane production. The US government has promoted ethanol production from corn through subsidies to farmers to cultivate this crop. The percentage of US corn production converted to ethanol has climbed from about 5% in 2001 to nearly 40% in 2010 (Fig. 3.4).

There are two huge flaws in the US corn for ethanol policy. The ethanol yield from corn is low in comparison to that of sugarcane and sugar beets (Fig. 3.5), and the use of corn and other grains for ethanol production competes with the use of these crops for human and animal feed resulting in greater food and feed costs. More

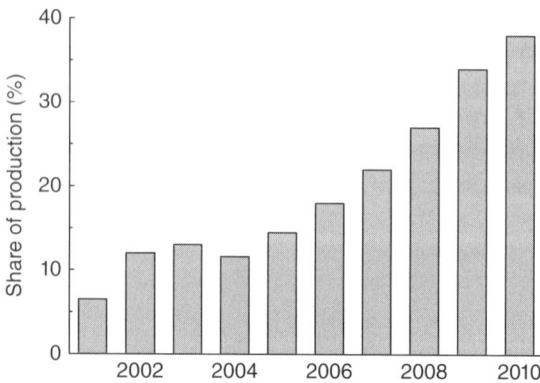

Figure 3.4 Percentage of corn used for ethanol production in the United States. *Source:* FAO. FAOSTAT. Production. 2013. Accessed: 10/15/2013. URI: http://faostat3.fao.org/faostat-gateway/go/to/download/Q/*/E.

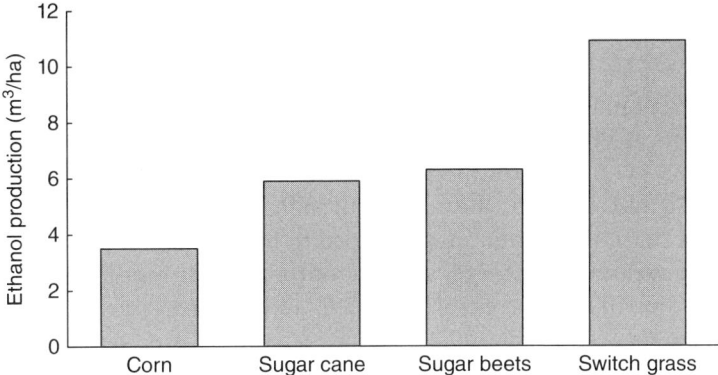

Figure 3.5 Ethanol yield of selected crops. *Source:* FAO (2011).

research should be done on the use of switchgrass and similar crops as a source of biofuel. Switchgrass has promise to be an even better alternative than sugar beets or sugarcane for producing ethanol (Fig. 3.5).

Animal production

There also has been an increase in production of milk, eggs, and meat as illustrated using data for milk and meat (Fig. 3.6). Meat production increased by a factor of 4.11 between 1961 and 2010 (Fig. 3.6)—remember that fisheries production is not combined with meat production by agriculture but reported separately in another database. The steady trend of increase of meat during this period had an R^2 value of 0.985. Production of these foods per capita almost doubled during the period. This suggests that people are consuming more meat in their diet. Despite the increase in global average meat consumption that reached about 120 kg/cap annually in 2007,

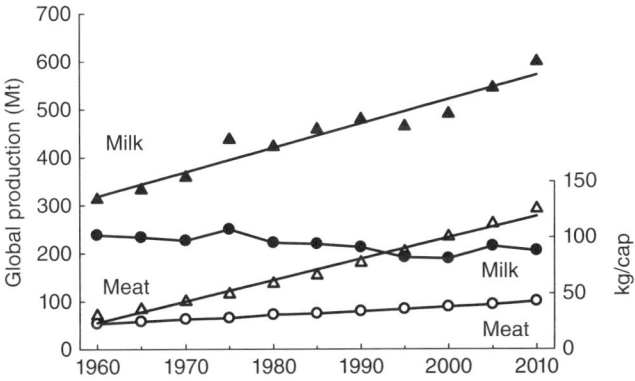

Figure 3.6 Global production of milk and meat with per capita supply of each from 1961 to 2010. *Source:* FAO (2011).

there is a great difference in meat consumption between developing and developed countries. McMicheal et al. (2007) estimated that average daily meat intake was 47 g/cap in developing countries versus 224 g/cap in developed countries.

Milk production also has steadily increased since 1961 ($R^2 = 0.932$) but not as drastically as meat production. There was about 1.91 times more milk produced in 2010 than in 1961. This factor of increase is slightly less than the rate of population increase—2.22-fold. In 1961, the amount of milk available annually per capita was 102 kg but in 2010, the amount had dropped to 88 kg (Fig. 3.6).

The great increase in meat, egg, and dairy products has resulted from intensification of production in highly controlled systems and requires large amounts of feed. A recent estimate suggested that 80% of animal agriculture production was conducted in "so-called" industrial systems (http://www.fao.org/ag/magazine/0612sp1.htm).

Food sufficiency

Agricultural production is not spread across the world in the same geographical distribution as is the human population. Much of the food produced in rural areas is transported to cities in the same or other countries. Most people depend upon others to produce some or all of their food, and the production areas may be far away. Food production also may be adversely affected by droughts, diseases, or other phenomena in a particular region. Thus food shortages usually result from natural, logistical, political, or economic "bottlenecks" in specific areas rather than from lack of food on a global basis.

There is no single criterion for defining undernourishment in humans but daily caloric intake often is used as a way of assessing the status of nutrition in different countries. The average caloric intake ranged from 3770 cal/cap/day in the United States to 1590 cal/cap/day in Eritrea in 2009—the world average was 2800 cal/cap/day. Thus most individuals in some countries take in far too many calories resulting in a high incidence of obesity, while in other countries, many people do not get enough food and suffer from hunger. A recent study by FAO, WFP (World Food Program), and IFAD (International Fund for Agricultural Development 2012) estimated that there are about 870 million chronically undernourished people in the world. In spite of the increase in agricultural production per capita on a global basis during recent decades, many people are denied access to a sufficient quantity of food daily and suffer from hunger and malnutrition.

Future issues

The great increase in agricultural productivity that has occurred in the past 50 years was not without negative consequences (Clay 2004). The increase has resulted in much greater use of resources to include energy, fertilizers, water, and to lesser extent, land. There also has been a great increase in negative environment impacts on the world's ecosystems such as erosion, water pollution, deforestation, loss of biodiversity, and carbon emissions.

The population will increase by around 30% by 2050, and at the current rate of food consumption, the agricultural output will need to increase proportionally to maintain the *status quo*. However, with the projected increase in the middle class the demand for food will be even greater. There is much concern over the ability of the world to provide the amount of food required by 2050 without seriously depleting nonrenewable resources and causing unprecedented ecological disturbances.

Fisheries production

Fisheries products are important to the world food supply and to the protein supply in particular. Fisheries products account for 16.6% of the world population's animal protein intake and 6.5% of all protein consumed in 2009 (FAO 2012). In addition, fisheries products are used to make fish meal and oil used in animal feeds—especially aquaculture feeds (Tacon and Metian 2008).

Fisheries production includes fish, crustaceans, molluscs, seaweed, and other species groups. Production of seaweed and other aquatic plants is excluded from most presentations of world fisheries data or presented separately. This custom will be followed here.

China is by far the leader in aquaculture production and some feel that the production statistics from China are inflated. Because of this, FAO reports sometimes make assessments of fisheries production with and without the Chinese data. The global estimates that include the production of China will be used in this discussion.

Capture fisheries

World capture fisheries production was about 30 Mt in 1960 and it steadily increased to about 90 Mt (Fig. 1.1). Since the late 1980s, annual capture fisheries production has fluctuated slightly but there has been no trend of increase, for example, annual capture fisheries production was 92.4, 92.1, 90.0, 90.3, 89.7, 89.6, 88.6, and 90.4 Mt for the years 2004 to 2011, respectively. There seems to be little prospect for increasing global capture fisheries because many species are being fished beyond sustainable limits.

Capture fisheries are mainly marine. In 2011, inland fisheries production was 11.5 Mt or 12.7% of the total capture fisheries. The capture fishery also is comprised mostly of fish—in 2007, fish comprised 91% and 84% of the inland and marine capture fisheries production, respectively. About 20 Mt/year of the marine capture fisheries is small pelagic fish used for fish meal and oil production rather than for human consumption

The top 10 fishing countries (in terms of quantity) for marine and freshwater species are listed in Table 3.3. China was the leading country in both categories. Most of the capture fishery in Chile and Peru is for small pelagic fish used to manufacture fish meal and oil. Six of the eight countries responsible for the greatest capture of food fish are in Asia. Six of the major inland fishery countries also are in Asia. Uganda and Tanzania in Africa border on Lake Victoria and most of their capture production

Table 3.3 Top 10 countries for inland and marine capture fisheries in 2009.

Country	Production (Mt)
Marine	
China	14.8
Peru	7.4
Indonesia	5.0
United States	4.3
Japan	4.2
India	4.1
Chile	3.6
Russia	3.4
Philippines	2.6
Myanmar	2.5
Inland	
China	2.25
Bangladesh	1.06
India	0.95
Myanmar	0.81
Uganda	0.45
Cambodia	0.36
Indonesia	0.32
Nigeria	0.30
Tanzania	0.28
Brazil	0.24

Source: ftp://ftp.fao.org/FI/STAT/summary/default.htm and FAO (2011).

is from that lake. The major inland fishery in Cambodia is the large lake known as Tonle Sap.

The top 10 marine capture species are given in Table 3.4. However, many marine species are captured and those listed only comprise about 30% of the total catch.

The per capita supply of capture fisheries products for human consumption is declining; it was around 12 kg/cap in 2000 but had fallen below 10 kg/cap by 2011.

Table 3.4 Top 10 marine capture species in 2010.

Species	Production (Mt)
Anchoveta	4.21
Alaska pollock	2.83
Skipjack tuna	2.52
Atlantic herring	2.20
Chub mackerel	1.60
Largehead hairtail	1.34
Sardine	1.22
Japanese anchovy	1.20
Yellowfin tuna	1.17
Atlantic cod	0.95

Source: ftp://ftp.fao.org/FI/STAT/summary/default.htm and FAO (2011).

Assuming no increase in capture fisheries production, the supply for human consumption would be about 7.5 kg/cap by 2050. Because of aquaculture, the per capita food fish supply has increased from 15.9 kg in 2000 to 17.2 kg in 2009.

Aquaculture

Aquaculture production increased slowly from 1950 until the late 1980s (Fig. 1.1). However, the growing gap between capture fisheries production and world demand for fisheries products has allowed aquaculture to expand dramatically since the late 1980s. It is particularly interesting to look at how aquaculture production increased by continent and region from 1950 to 1970 to 2010 (Fig. 3.7). There has been an increase on all continents, but the increase in Africa lags other continents. This is unfortunate because Africa is projected to double in population by 2050 and will need much more animal protein.

Freshwater aquaculture is considerably greater than marine (including brackish-water) aquaculture (Table 3.5), and the two combined produced 63.6 Mt in 2011. This represented about 41% of total fisheries production, but aquaculture accounted for 48.6% of fisheries production for human consumption.

The increase in production of some individual species groups has been quite impressive as shown in Fig. 3.8 for Atlantic salmon, tilapia, and whiteleg shrimp. The production of these species by aquaculture grew slowly for years, but since the mid-1990s tilapia and Atlantic salmon have experienced phenomenal growth because they became a popular food in developed countries. For example, in 2002, annual consumption of tilapia meat in the United States was 0.14 kg/cap but in 2011, the per capita consumption had risen to 0.58 kg. Interestingly, tilapia consumption in 2011 was 0.08 kg/cap less than in 2010.

Shrimp aquaculture increased some between 1980 and 1990, but as a result of worldwide issues with viral diseases, growth stagnated during the 1990s (Fig. 3.8). With the development of better broodstock, methods for producing specific pathogen-free postlarvae, improved biosecurity, more intensive culture methods, and shift of the majority of shrimp culture from black tiger prawn *Penaeus monodon* to whiteleg shrimp *Litopenaeus vannamei*—also known as Pacific white shrimp—the production of marine shrimp by aquaculture exploded during the first decade of the twenty-first century. Shrimp is quite popular with consumers in developed nations; the annual per capita consumption of shrimp meat in the United States increased from 1.68 to 1.90 kg between 2002 and 2011. During this time, shrimp consistently has been the most popular seafood item (Table 3.6).

Most of the world's channel catfish *I. punctatus* production had been conducted in the southern United States and sold almost exclusively in the domestic market until the Chinese began producing this species in the early 2000s. A tremendous research effort has been devoted to the culture of this species because it is popular with US consumers. Production increased rapidly from less than 5000 t in 1970 to 300 000 t in 2003 (Fig. 3.9); but since then production has declined by more than 50% to 136 123 t in 2012. Consumption of channel catfish has declined in the United States (Table 3.6) as consumption of tilapia and the Vietnamese catfish

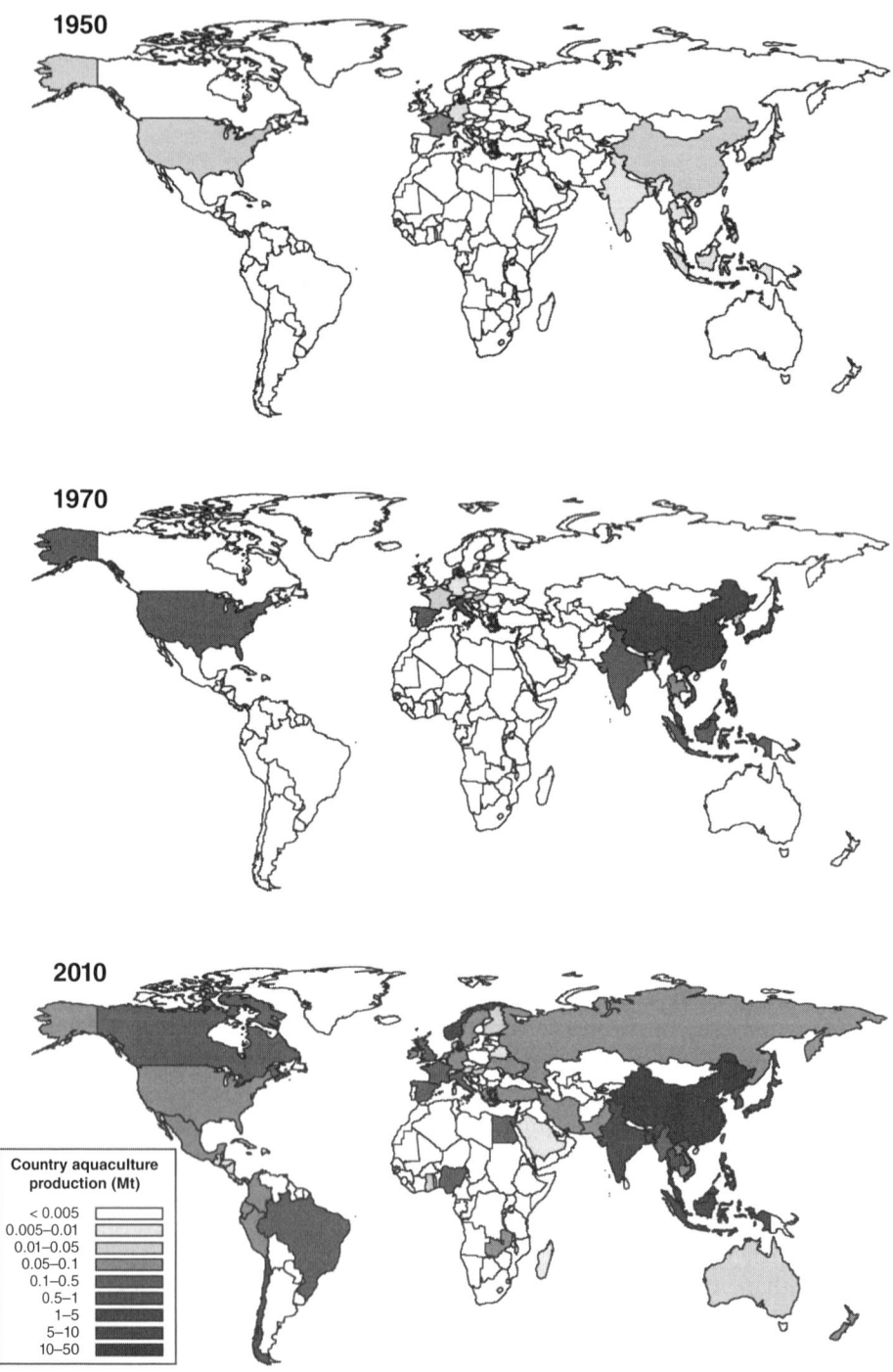

Figure 3.7 Aquaculture growth by continent and region from 1950 to 2010. *Source:* (http://www.agecon.msstate.edu/what/farm/budget/catfish/php) and FAO (2011).

Table 3.5 Aquaculture production summary for 2004 and 2011 showing amount for human consumption.

	Production (Mt)	
	2004	**2011**
Capture		
Inland	8.6	11.5
Marine	83.8	78.9
Subtotal	92.4	90.4
Aquaculture		
Inland	25.2	44.3
Marine	16.7	19.3
Subtotal	41.9	63.6
Total	134.3	154.0
Human consumption	104.4	130.8

Source: Modified from FAO (2012).

(*Pangasius*) has increased. This clearly illustrates how market forces can influence aquaculture production. Channel catfish production has declined mainly because it has not competed well against tilapia and *Pangasius*, but export of channel catfish to the United States from China has provided some additional competition.

Channel catfish production was initiated in several other countries beginning in the late 1970s. There was little production in these countries until the early 2000s, but 2005 world production was more than twice US production (Fig. 3.9). China now produces as much channel catfish as the United States.

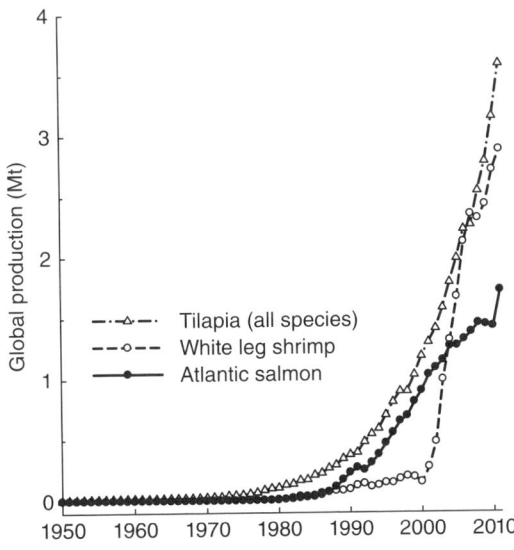

Figure 3.8 Global production by aquaculture of tilapia, whiteleg shrimp, and Atlantic salmon from 1950 to 2010. *Source:* FAO (2011) and FAO. FAOSTAT. Production. 2013. Accessed: 10/15/2013. URI: http://faostat3.fao.org/faostat-gateway/go/to/download/Q/*/E.

Table 3.6 Top 10 fisheries products in the United States in 2002 and 2011.

Species group	Amount (kg/cap)	
	2002	**2010**
Shrimp	1.68	1.90
Canned tuna	1.40	1.18
Salmon	0.92	0.88
Pollock	0.71	0.59
Tilapia	0.14	0.58
Pangasius[a]	–	0.28
Catfish	0.50	0.25
Cod	0.30	0.23
Crab	0.25	0.23
Clams	0.26	0.15

Source: Modified from Loke et al. (2012).

[a]*Pangasius* was not on the list in 2002. The 10th species in 2002 was flatfish (0.18 kg/cap).

The Atlantic salmon *Salmo salar* is the primary coldwater aquaculture species. The two major Atlantic salmon producing countries are Norway and Chile, but Canada, United States, Iceland, Ireland, Australia, and a few other countries have salmon farming. Although salmon aquaculture started in the early 1960s, it did not become a significant activity until the early 1980s. Salmon farming experienced phenomenal expansion between 1990 and 2005. Production in 2010 reached almost 1.5 Mt (Fig. 3.8).

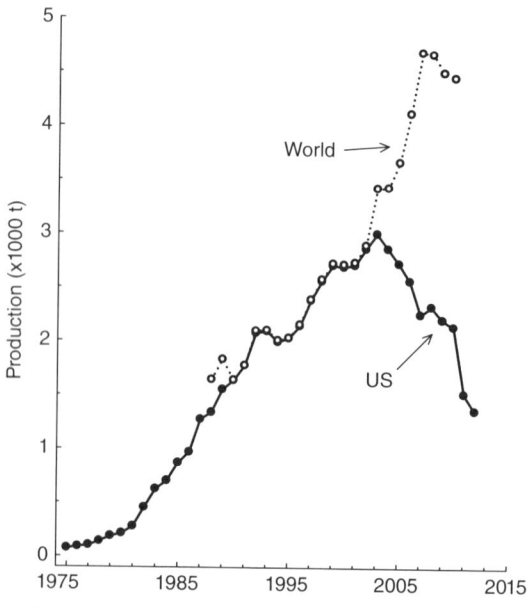

Figure 3.9 Production of channel catfish in the United States from 1975 to 2012 (http://www.agecon.msstate.edu/what/farm/budget/catfish/php) and for world from 1988 to 2010 (www.fao.org/fishery/statistics). *Source:* FAO (2011) and FAO. FAOSTAT. Production. 2013. Accessed: 10/15/2013. URI: http://faostat3.fao.org/faostat-gateway/go/to/download/Q/*/E.

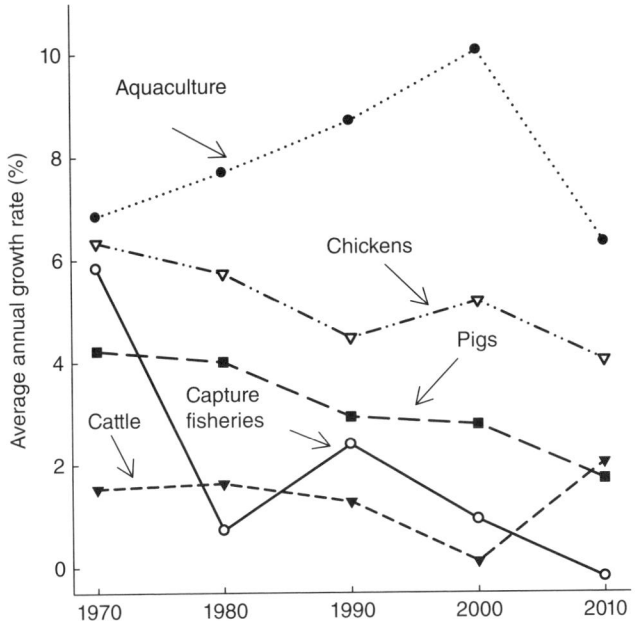

Figure 3.10 Average annual growth rates at 10-year intervals for major meat products. *Source:* FAO. FAOSTAT. Production. 2013. Accessed: 10/15/2013. URI: http://faostat3.fao.org/faostat-gateway/go/to/download/Q/*/E.

Most types of food production have been increasing as the world's population grew—the capture fishery is the only major meat source for which a decline occurred between 1970 and 2010 (Fig. 3.10). Aquaculture has increased faster than other types of meat production despite a decline in growth rate between 2000 and 2010.

The top 10 aquaculture-producing countries in 2011 included only two non-Asia members—Norway and Egypt (Table 3.7). China dwarfs the rest of the world's

Table 3.7 Top 10 aquaculture countries in 2011 compared with 1990 production.

	Production (Mt)	
Country	**1990**	**2011**
China	6.482	36.734
India	1.017	4.649
Vietnam	0.160	2.672
Indonesia	0.500	2.305
Bangladesh	0.193	1.309
Thailand	0.292	1.286
Norway	0.151	1.008
Egypt	–	0.920
Myanmar	–	0.851
Philippines	0.380	0.745

Source: ftp://ftp.fao.org/FI/STAT/summary/default.htm and FAO (2011).

Table 3.8 Total and per capita supply of fisheries products for human consumption by FAO region for 2010.

Region	Production Total (Mt)	Production Aquaculture (Mt)	Per capita supply (kg/year)
World	125.6	59.9	18.4
Africa	9.1	1.288	9.1
Americas			
North	8.2	0.656	24.1
Latin	5.7	1.92	9.9
Asia	85.4	53.30	20.7
Europe	16.2	2.52	22.0
Oceania	0.9	0.184	24.6

Source: Modified from FAO (2012).

production, but even if the data from this country have been inflated two- or three-fold, China still would be the major producer. It also is interesting to note the growth of aquaculture between 1990 and 2011 in the top 10 countries. Production has increased by threefold or more in most countries. Vietnam and Norway have had particularly impressive increases as a result of expansion of *Pangasius* and salmon farming, respectively.

The total production of food fish by continent is provided in Table 3.8. Asia produces 65.8% of total fisheries products and 88.9% of aquaculture products. China alone accounts for 32.7 Mt of aquaculture production and 17.05 Mt of capture fisheries production—44.0% of total production and 62.3% of aquaculture production.

Inland aquaculture is primarily fish production while marine and brackishwater aquaculture is mainly mollusc and crustacean production. Most of the molluscs are produced in marine water while the shrimp production is mainly in brackishwater (Table 3.9). The greatest part of the molluscan production is oysters, and the shrimp production is mainly whiteleg shrimp. As in terrestrial food production, aquaculture production is shifted to a few species that comprise most of the market.

The 10 major aquaculture species are listed in Table 3.10. These 10 species had a combined production of 30.87 Mt or 51.6% of total production of animals by aquaculture in 2010. Six of the top 10 species are carps. Several species such as whiteleg shrimp, Japanese carpet shell, Nile tilapia, and Atlantic salmon are popular with consumers in developed countries.

Aquaculture contributes over 60% of the production of nine of the 10 species groups popular with consumers in developed countries (Table 3.11). Although shrimp aquaculture provides only 30% of total shrimp production, the percentage is probably greater for internationally traded shrimp.

Aquaculture products have a high value—slightly greater than other major meats besides cattle meat and much more than typical plant crops (Fig. 3.11). Despite the comparatively low production of aquaculture products compared to some agricultural products, the total global value of aquaculture products is exceeded only by chicken, cattle, and pig meat and rice (Fig. 3.12).

Table 3.9 World aquaculture production by species group for 2010.

Environment and species group	Quantity (Mt)
Freshwater	
Carps, barbels, and other cyprinids	24.32
Miscellaneous fish	6.01
Tilapias and other cichlids	3.50
Other	3.04
Total	36.87
Brackishwater	
Shrimp and other crustaceans	3.79
Other	0.96
Total	4.75
Marine	
Molluscs	13.99
Marine fishes	2.71
Other	1.56
Total	18.26
Grand total	**59.88**

Source: ftp://ftp.fao.org/FI/STAT/summary/default.htm.

Table 3.10 Top 10 aquaculture species by total production in 2010.

Species and common names	Production (Mt)
Ctenopharyngodon idellus (grass carp)	4.34
Hypophthalmichthys molitrix (silver carp)	4.12
Catla catla (Indian carp)	3.87
Ruditapes philippinarum (Japanese carpetshell clam)	3.60
Cyprinus carpio (common carp)	3.44
Penaeus vannamei (whiteleg shrimp)	2.72
Hypophthalmichthys nobilis (bighead carp)	2.59
Oreochromis niloticus (Nile tilapia)	2.54
Carassius carassius (crucian carp)	2.22
Salmo salar (Atlantic salmon)	1.43

Source: ftp://ftp.fao.org/FI/STAT/summary/default.htm.

Table 3.11 Contribution of aquaculture to production of 10 species groups popular in international trade.

Species group	Production (Mt/year)		Aquaculture contribution (%)
	Total	Aquaculture	
Shrimp	4.27	1.29	30.3
Salmon	2.10	1.31	62.4
Channel catfish	0.29	0.29	100.0
Rainbow trout	0.51	0.49	96.1
Pangasius	1.00	1.00	100.0
Tilapias	2.19	1.51	68.8
Oysters	4.50	4.32	95.9
Clams	4.26	3.43	80.6
Mussels	1.71	1.44	84.5
Scallops	1.97	1.23	63.3

Source: FAO (2011).

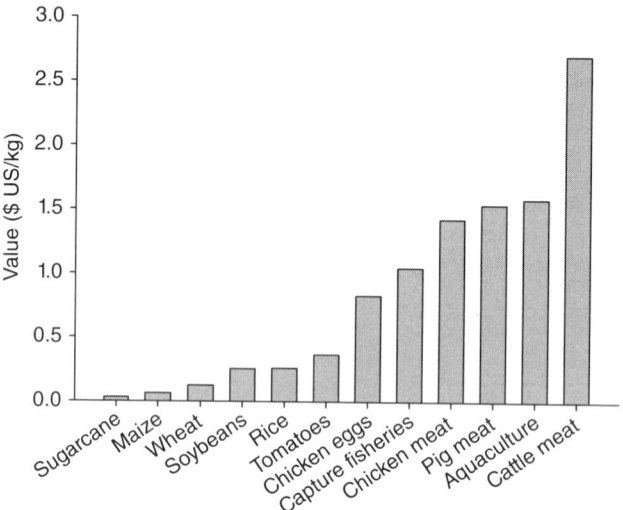

Figure 3.11 Average value of some major food products. *Source:* FAO (2012).

The farm-gate value of aquaculture products—seaweed included—was US$125 billion in 2010 (http://faostat.fao.org). There are various estimates of the multiplier effect of aquaculture products in the economy, but most are around 1.5 to 3.0. Assuming a value of 3.0, the aquaculture sector might have a total value of US$375 billion. The world's gross domestic product was about US$63 trillion in 2010 (http://siteresources.Worldbank.org/DATASTATISTICS/Resources/GDP.pdf), of which aquaculture possibly contributed as much as 0.60%.

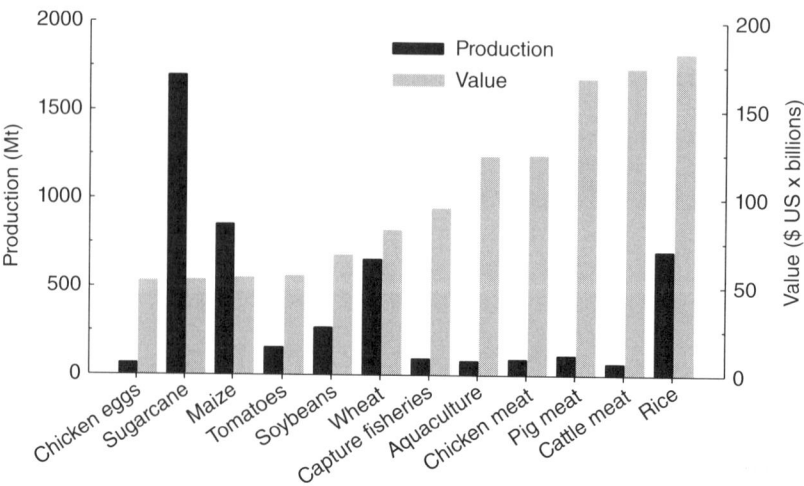

Figure 3.12 Global production and value of some major food products. *Source:* FAO (2011).

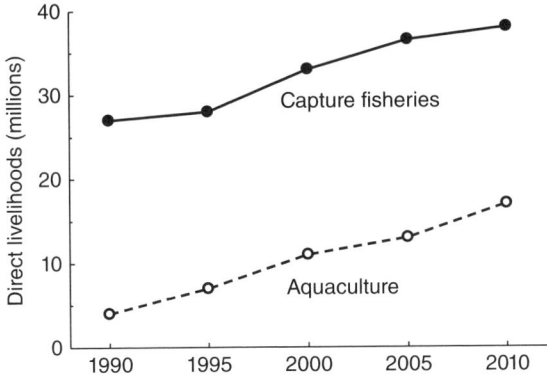

Figure 3.13 Estimated global employment by capture fisheries and aquaculture. *Source:* FAO (2011).

Aquaculture provided employment for about 17 million in 2010 while about 38 million were employed by capture fisheries (Fig. 3.13). However, aquaculture resulted in about 50% of fisheries production for human consumption and about 40% of all fisheries production. Thus, the output of aquaculture per worker is considerably greater than for capture fisheries both in units of production and economic value.

The contribution of aquaculture to the total world employment of about 3 billion in 2010 (Table 3.12) was only about 0.56%. Aquacultural employment was 1.65% of agricultural employment. Nevertheless the number of jobs in aquaculture increased about threefold between 1990 and 2010 (Fig. 3.13).

Future issues

The production of natural waters is not likely to increase so the entire increase in demand for fisheries products in the future must be provided through aquaculture. Aquaculture production in 2050 would have to increase to 93 Mt just to maintain the *status quo* for consumption (Table 3.13). The actual demand is likely to be even more, because as discussed earlier, the increasing middle class will have more money to spend on fisheries products.

Table 3.12 World employment estimates.

Sector	2000	2010
Agriculture	1057	1033
Industry	533	672
Services	1022	1333
Total	**2612**	**3038**

Source: Made from data found in International Labour Organization (2012).

Table 3.13 Estimated increase in demand for edible fisheries products to maintain *status quo* in 2050.

Year	World population (billion)	Fisheries products consumed (Mt/year)		
		Total	Capture	Aquaculture
2009	6.91	117.8	62.7	55.1
2050 projection for *status quo*	9.15	156	63	93

Aquaculture yields will have to increase and there will be expansion of the production area in ponds. Although it would be possible to greatly increase production per unit area and avoid increasing the production area (Verdegem and Bosma 2009), it is unlikely that much of the domestic fish production in poor countries will be greatly intensified. In addition to more land and water use, there will be a greater use of energy, fertilizers, and feed in aquaculture.

Seaweed production

Seaweed is used mainly for extraction of iodine, agar, and carrageenan, but a few species are consumed directly by humans. Total seaweed production in 2010 was about 19 Mt and about 95% was from aquaculture (FAO 2012). The production of seaweed has increased drastically since the 1960s primarily as a result of expansion of seaweed aquaculture (Fig. 3.14). The major seaweed-producing countries are China, Japan, South Korea, North Korea, Philippines, and Indonesia (Roesijadi et al. 2010).

International trade

Agricultural products

In 2009, more than 1000 Mt of agricultural products were traded internationally—about 14% of total agricultural production (http://faostat.fao.org). The major traded

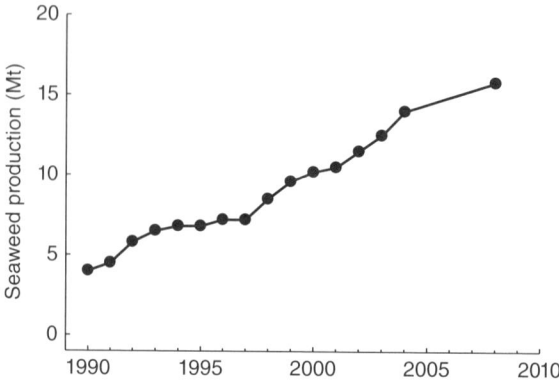

Figure 3.14 Global seaweed production from 1990 to 2010. *Source:* FAO (2011, 2012).

products were grains, soybeans, and processed grain and soybean products. About 7% of milled rice, 11% of corn, 19% of wheat, and 36% of soybeans produced in 2010 were exported from producing countries. Chicken meat and sugar also were major, internationally traded products. The FAO database includes water, nonalcoholic beverages, and wine in the list of traded products; water and nonalcoholic beverages were among the top 20 products.

The share of total agricultural imports and exports to total imports and exports of merchandise vary greatly among countries. The shares for the 20 top and 20 bottom countries are listed (Table 3.14). Several countries appear in both lists because they export a large amount of one or more products and import much of one or more other products. Notice also that the top and bottom agricultural import/export countries tend to be small. Large countries such as China, India, and the United States have large imports and exports of agricultural products, but the share of total merchandise is relatively low, 4–6%. The United States imported a huge amount of agricultural products valued at US$98 billion in 2011. But it exported products valued at US$136 billion and had a net surplus in agricultural trade with the world. Agricultural products made up only 4.5% and 10.5% of the value of imports and exports, respectively. Thus agriculture did not prevent the United States from having a huge foreign trade deficit because total merchandise imports (US$2.2 trillion) far exceeded total merchandise exports (US$1.3 trillion).

In addition to transportation of agricultural products among countries, there is considerable movement of agricultural products within a country. For example, in the United States, grains produced in the Midwest are shipped throughout the country. A good example is the huge amounts of grain shipped from the midwest to the southeast for use in chicken feed.

Fisheries and aquaculture products

There is an increasing amount of aquaculture products in world trade (Fig. 3.15) and export quantities exceeded 30 Mt in 2009. The value of the world fisheries—captured plus aquacultured products—was greater than for any agricultural product exported from less developed countries in 2009 (Fig. 3.16). Although data could not be found on the contribution to export of aquaculture to total fisheries products, it probably is about 50% of the total. Thus aquaculture products alone likely exceed the value of any exported agricultural crop.

The percentage of total fisheries production involved in international trade is greater than for agricultural products in general. The percentage has risen from about 30% in 1976 to nearly 40% in 2010 (FAO 2012). The major exporters and importers of fisheries products in 2008 are given in Table 3.15. The major exporter was China but it does not overshadow the other leading countries to the extent that it does for aquaculture production. Japan and the United States were the major importers. Each of these countries imported about twice the quantity of fish and fisheries products as the third leading country. China, the United States, and Denmark were in the top 10 countries for both imports and exports.

Table 3.14 Agriculture share of total export and import merchandise value for top 20 countries in each category.

Country	Share (%)
Exports	
Malawi	96.07
Burundi	94.14
Paraguay	86.06
Guinea-Bissau	84.95
Ethiopia	81.46
Djibouti	81.32
Afghanistan	74.00
Nicaragua	73.96
Eritrea	67.80
Comoros	64.91
Saint Vincent & Grenadines	64.89
Uruguay	64.32
Belize	57.91
Liberia	55.61
Kenya	55.53
Sao Tome and Principe	55.15
New Zealand	52.57
Vanuatu	48.89
Cote d'Ivoire	48.58
Argentina	47.83
Imports	
Djibouti	97.57
Somalia	52.28
Timor-Leste	46.67
Kiribati	41.63
Mauritania	35.28
Benin	34.78
Haiti	32.27
Zimbabwe	31.92
Sao Tome and Principe	30.87
Samoa	30.68
Comoros	30.67
Tuvalu	30.44
Liberia	30.20
Niue	29.56
Guinea	28.83
Cape Verde	28.47
Sierra Leone	27.43
Yemen	26.75
Guinea-Bissau	25.32
Senegal	25.25

Source: FAO. FAOSTAT. Production. 2013. Accessed: 10:15/2013. URI: http://faostat3.fao.org/faostat-gateway/go/to/download/Q/*/E.

The United States is a major importer of fisheries products. In 2011, this country imported 2.42 Mt of edible fisheries products valued at US$16.6 billion. The single largest commodity was shrimp—575 100 t valued at US$5.2 billion. It exported 1.44 Mt of edible fishery products with a value of US$5.12 billion—most from capture fisheries (http://www.st.nmfs.noaa.gov/st1/index.html).

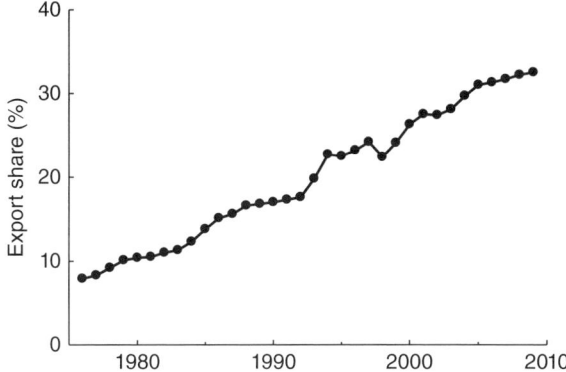

Figure 3.15 Export share of total aquaculture and capture fisheries production (for human food). From www.fao.org/fishery/statistics.

Conclusions

World agricultural production has increased at a rate greater than that of the human population and there is slightly more food per capita today than in 1960. This increase in agricultural production resulted mainly from advances in technology that allowed greater yields, but production area increased by about 10%. Production of agricultural products must increase by at least 30% just to maintain the *status quo* in 2050.

Fisheries products account for 16.6% of animal protein intake by humans and 6.5% of all protein consumed. However, capture fisheries have apparently reached their limit of around 90 Mt/year—less than necessary to meet the global demand. Aquaculture supplements the capture fisheries and currently supplies roughly half of

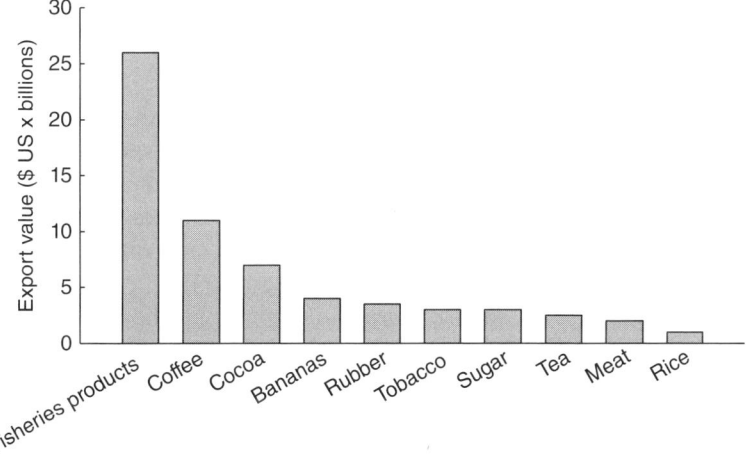

Figure 3.16 Total export value in 2009 of fisheries products and other selected commodities from less developed countries. From www.fao.org/fishery/statistics and http://faostat.fao.org.

Table 3.15 Top 10 exporting and importing countries for fisheries products in 2010.

Country	(US$ billion)
Exports	
China	13.27
Norway	8.82
Thailand	7.13
Vietnam	5.11
United States	4.66
Denmark	4.15
Canada	3.84
Netherlands	3.56
Spain	3.40
Chile	3.39
Imports	
United States	15.50
Japan	14.97
Spain	6.64
China	6.16
France	5.98
Italy	5.45
Germany	5.04
United Kingdom	3.70
Sweden	3.32
Republic of Korea	3.19

Source: Modified from FAO (2012).

the world supply of products for human consumption—about 55.1 Mt in 2009. The future growth of fisheries production must come entirely from aquaculture and by 2050, production of farmed aquatic animals will need to be about 93 Mt to maintain the *status quo* with respect to current seafood consumption patterns.

The eNGO perspective

Environmental impacts of food production at a broad level are somewhat of a challenge for many eNGOs because they have chosen to work in specific priority regions of the world. Areas that are high in biodiversity, have high rates of habitat degradation, or a combination of these traits are the typical attributes of the regions where eNGOs focus. At one time when the primary mode of environmental conservation was the development of protected areas or preserves, the "place-based" strategy was logical and allowed for succinct means of promoting the work of a particular eNGO. Increasingly, eNGOs are broadening their work to address food production concerns through their business engagement efforts. Food production is typically handled in a similar manner among various eNGOs—through some arrangement to work more closely with businesses and industry. Some examples of the common eNGOs working more broadly on the sustainability of food production and their specific departments are World Wildlife Fund—Market Transformation; Conservation

International—Center for Environmental Leadership in Business; Environmental Defense Fund—Corporate Partnerships Program.

The eNGO business engagement efforts in food production targets large buyers, aggregators, or sellers of a variety of products. The traditional and most defensible mechanism of protected area development in priority regions does not lend itself well to business engagement, particularly for food products as buyers and sellers are constantly changing suppliers and product lines in an effort to capitalize on consumer trends and interests. That is businesses may source product from a particular region of the world that is designated as a priority region for an eNGO, but the eNGO likely has not prioritized all of the regions where that business is procuring product. Reconciling this aspect of industry partnerships with internal strategy development of a particular eNGO can be challenging because of the varying magnitude of size of these regions. For example, WWF defines their regional focus by "Eco-regions." A specific Eco-region can include many countries such as the Coral Triangle Eco-region (portions of Indonesia, Malaysia, the Philippines, Papua New Guinea, Timor Leste and Solomon Islands) or fairly specific geography such as the Chihuahuan Desert (specific region of the southern United States and northern Mexico). It is interesting to note that in some cases, eNGOs will work outside of their defined regional focus if it allows a path of entry to a business partnership. For example, WWF sought engagement with Costco Wholesale Corporation on several fronts in various priority areas, but as it turned out, the flagship project dealt with shrimp aquaculture in southern Thailand outside of any geographic focus.

It is not surprising that there exist some internal organizational struggles to transpose business engagement on priority places. However, most eNGOs have placed importance on specific food production systems that have the greatest impact on their priority places. For example, Conservation International has a business partnership with Monsanto to help conserve areas of Brazil's Atlantic Forest and Cerrado. Although Monsanto is a large global agro-business with investments in most parts of the world, Conservation International negotiated the specific regional work.

One of the greatest benefits to working closely with corporate partners is access to current practices and performance of a particular industry. These data are invaluable to eNGOs, primarily because the academic literature is seldom representative of current practices of industry—that is, unless the academic researcher works closely with the industry or is an active consultant for the sector. The data shared with eNGOs by companies can be used to benchmark other companies and gain some perspective on production impact of specific environmental criteria.

Because business engagement on food production at a large scale is relatively new for eNGOs, there is little data about how these groups stand on certain issues. Of course there are straightforward stances on rain forest deforestation, rare species habitat and slash and burn farming, but rarely will one be informed of an eNGOs view on how much fertilizer is too much, how much water used per unit of production is acceptable, or what the intensity of production should be per unit area of cultivation. It was once easy for eNGOs to claim that organic food production was the clear choice for consumers, but with the recognition of the large land resources required for organic production, many eNGOs are leaning toward more intensive forms of production. This slow and often reticent shift by some eNGOs makes it much more controversial to set clear performance targets for businesses and industry in the food production sector.

References

Cassman, K. G. 1999. Ecological intensification of cereal production systems: yield potential, soil quality, and precision agriculture. *Proceedings of the National Academy of Science* 96:5952–5959.

Clay, J. W. 2004. *World Agriculture and the Environment*. Washington: Island Press.

FAO (Food and Agriculture Organization). 2012. *The State of World Fisheries and Aquaculture*. Rome: FAO Fisheries and Aquaculture Department.

FAO (Food and Agriculture Organization), WFP (World Food Program), and IFAD (International Fund for Agricultural Development. 2012. *The State of Food Insecurity in the World 2012*. Rome: FAO.

Foley, J. A., R. DeFries, G. P. Asner, C. Barford, G. Bonan, S. R. Carpenter, F. S. Chapin, M. T. Coe, G. C. Daily, H. K. Gibbs, J. H. Helkowski, T. Holloway, E. A. Howard, C. J. Kucharik, C. Monfreda, J. A. Patz, I. C. Prentice, N. Ramankutty, and P. K. Snyder. 2005. Global consequences of land use. *Science* 309:570–574.

Gilbert, G. (ed.). 1993. *Thomas Malthus an Essay on the Principle of Population*. New York: Oxford University Press.

International Labour Organization. 2012. *Global Employment Trends 2012*. Geneva: International Labour Office.

Matson, P. A. and P. M. Vitousek. 2006. Agricultural intensification: will land spared for farming be land spared for nature? *Conservation Biology* 20:709–710.

McMicheal, A. J., J. W. Powles, C. D. Butler, and R. Uauy. 2007. Food, livestock production, energy, climate change and health. *The Lancet* 370:1253–1263.

Reijnders, L. and S. Soret. 2003. Quantification of the environmental impact of different dietary protein choices. *American Journal of Clinical Nutrition* 78:664S–668S.

Roesijadi, G., S. B. Jones, L. J. Snowden-Swan, and Y. Zhu. 2010. Macroalgae as a biomass feedstock: a preliminary analysis. PNNL-19944, Pacific Northwest National Laboratory, Richland, Washington.

Tacon, A. G. J. and M. Metian. 2008. Global overview on the use of fish meal and fish oil in industrially compounded aquafeeds: trends and future prospects. *Aquaculture* 285:146–158.

Tilman, D. 1999. Global environmental impacts of agricultural expansion: the need for sustainable and efficient practices. *Proceedings National Academy of Science* 96:5995–6000.

USDA (United States Department of Agriculture) and USDHHS (United States Department of Health and Human Services). 2010. *Dietary Guidelines for Americans* 2010, 7th ed. Washington: U. S. Government Printing Office.

Verdegem, M. C. J. and R. H. Bosma. 2009. Water withdrawal for brackish and inland aquaculture, and options to produce more fish in ponds with present water use. *Water Policy* 11:52–68.

Whitney, E. and S. R. Rolfes. 2011. *Understanding Nutrition*, 12th ed. Belmont: Wadsworth Publishing.

Chapter 4

Assessing resource use and environmental impacts

Assessments of resource use and effects of human endeavors on the environment are controversial because certain stakeholders may perceive that they have much to gain or lose from the outcome. Those responsible for conducting these assessments should strive to use information from established databases and documented sources. They also should employ methodology considered to be standard or acceptable by those working in the particular field of study to the extent that it does not stifle innovation.

This chapter provides a discussion of procedures used in evaluating resource use, negative environmental impacts, and accumulation of chemical residues in fish and other species related to aquaculture production.

Sources of information

Much of our knowledge about management inputs for increasing aquacultural production and the negative impacts thereof were acquired through traditional research. These studies usually were not conducted for the purposes of resource conservation or environmental sustainability, instead they were carried out to provide aquaculture producers information for effective management or for complying with governmental regulations and market demands for specific types of products.

Studies to determine amounts of liming materials and fertilizers to apply in ponds typically involve testing several treatment rates, each replicated three or four times in experimental ponds, and applying statistical analyses to ascertain differences among treatments and controls. An example of a fertilization experiment is illustrated in Table 4.1. Water quality and sediment analyses often are conducted to assess effects of management inputs on physical, chemical, and biological variables. Many studies of fertilizers and liming materials have been conducted with a variety of species at different locations, and literature reviews of the research are available (Mortimer 1954; Hickling 1962; Boyd 1990; Boyd and Tucker 1998; Mischke 2012).

Aquaculture, Resource Use, and the Environment, First Edition. Claude E. Boyd and Aaron A. McNevin.
© 2015 John Wiley & Sons, Inc. Published 2015 by John Wiley & Sons, Inc.

Table 4.1 Example of pond fertilization experiment. All ponds were stocked at 2500/ha with fingerling sunfish.

Treatment P_2O_5 (kg/ha/application)[a,b]	Fish production (kg/ha)[c]
0	140[a]
1	260[b]
2	390[c]
4	400[c]
6	410[c]
8	408[c]

[a] Triple superphosphate (46% P_2O_5) applied at 2- to 3-week intervals.
[b] Treatments were replicated three times in 0.1-ha earthen ponds.
[c] Indicated by the same letter did not differ at $\alpha = 0.05$.

Results of pond fertilization and liming studies provide information on typical fertilizer nutrient inputs per unit area, effects of fertilizers on benthic and plankton communities in ponds, increase in fish and shrimp production, and the chemical composition of water in ponds. This information is useful in making optimum fertilizer application recommendations adherence to which will avoid overfertilization to conserve fertilizer nutrients, lessen the likelihood of water quality deterioration in ponds, and reduce the pollution potential of pond discharges.

Pond fertilization research has revealed ways of lessening nutrient inputs to ponds. One example is studies of granular fertilizers applied by broadcasting versus liquid fertilizers (Boyd 1981; Metzger and Boyd 1980). Granular fertilizers quickly settle to the bottom with only 5–10% of phosphorus dissolving, but the fertilizer granules dissolve on the bottom where phosphate is strongly adsorbed by soil. Phosphorus in liquid fertilizers will dissolve almost completely in the water column where it is available to phytoplankton. Fertilizer phosphorus input can be reduced by 50–75% through use of liquid fertilizer. Another example is investigations demonstrating that nitrogen fertilization usually can be suspended or reduced in ponds with a history of several years of nitrogen fertilization because of recycling of nitrogen from organic matter (Swingle et al. 1963; Murad and Boyd 1987; Boyd et al. 2008).

Feeding experiments compare effects of different diets on growth of culture animals. These studies also provide information on the composition of diets and the efficiency with which feed nutrients are converted to biomass of culture animals. Many fish nutrition studies contain information that can be used to estimate proportions of major nutrients that enter waters of the culture unit as wastes (uneaten feed, feces, and metabolic excretions such as carbon dioxide and ammonia).

Feeding experiments also are conducted on the most efficient way of feeding fish and other animals because feed is usually the greatest production cost. These studies have considered various issues as follows:

- Method of applying the feed, for example, broadcast over surface or feeding trays in shrimp culture;
- Type of feed pellet, for example, sinking versus floating;

- Feeding frequency, for example, days per week; multiple versus single daily applications;
- Amount of feed per day, for example, a percentage of estimated body weight daily versus satiation feeding.

Results of experiments are critical to efforts to lessen resource use and reduce negative impacts because overfeeding results in wasted feed, less efficient conversion of feed to biomass of culture animals, and a greater pollution load.

Water quality deteriorates in ponds with feeding, and many studies have focused on determining the relationship between feed input and important water quality variables such as dissolved oxygen, ammonia, phosphorus, and carbon dioxide. The amount of feed that can be safely applied depends upon the ability of a pond or a water body containing cages to assimilate wastes through microbial decomposition, sediment adsorption, nitrification and denitrification, and volatilization (Boyd and Tucker 1998). Many investigations have been conducted on ways of increasing the assimilation capacity of ponds through liming those that are acidic, applying mechanical aeration, using water exchange, and removing sediment. Mechanical aeration is much more efficient both in the amount of dissolved oxygen provided and energy use than is water exchange, and aeration internalizes oxidation of wastes avoiding their discharge in effluents. Methods for estimating the aeration requirements based on feed input have been developed.

Toxicity of aquacultural chemicals to fish and other aquatic organisms has been assessed by determining mortality at different concentrations of the chemicals and estimating the LC50—the concentration that will kill 50% of the organisms during the selected exposure time (Fig. 4.1). Although this procedure gives an indication of relative toxicities of chemicals, the concentration used in aquaculture systems must be much lower (usually about 1–5% of the 96-hour LC50) to avoid harmful effects on culture animals. As a general rule, other organisms in ponds or in waters receiving discharges from culture systems will not differ greatly from aquaculture species with respect to susceptibility to chemicals.

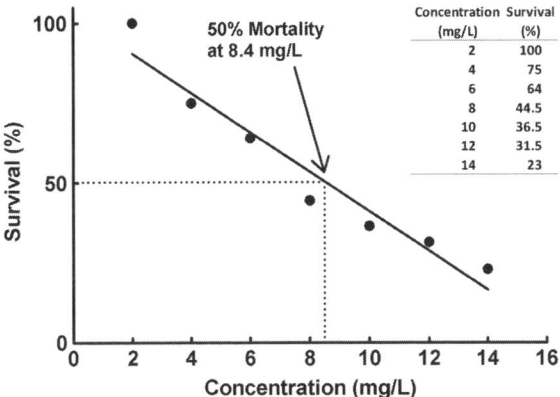

Figure 4.1 Example of using the results of a toxicity test to determine the LC50 of a chemical compound to fish. *Source:* Boyd, C. E., J. Queiroz, J. Lee, M. Rowan, G. N. Whitis, and A. Gross. Environmental assessment of channel catfish, Ictalurus punctatus, farming in Alabama. *Journal of the World Aquaculture Society* 31:511–544. Copyright © 2000, John Wiley & Sons, Inc.

Many studies have considered concentrations of residues in aquaculture species exposed to various chemical substances. Such investigations have been necessary to avoid penalties that could result from inspections of products by government agencies or refusal of products by buyers. Of course only a small number of samples of all products that enter an importing country can be tested for residues, and the probability of finding residues increases with greater sample number.

In some countries, considerable efforts have been made to assess the quality of aquaculture effluents, determine their impacts on receiving waters, and provide suggestions on how to improve effluent quality and lessen impacts. These studies often have been made in collaboration with governmental agencies responsible for developing and enforcing water quality regulations. In the late 1990s, the USEPA initiated a rule-making process for aquaculture effluents. This agency asked the USDA to organize an aquaculture effluent task force consisting mainly of aquaculture scientists to serve as a source of information about aquaculture knowledge to EPA scientists and their consultants.

Environmental advocacy groups often consider information generated by the aquaculture sector to be bias and intended to minimize criticism of the industry. This may be true for some general articles on aquaculture and the environment, but most aquaculture research was conducted for other purposes and often does not mention environmental or resource use issues. This literature is an excellent source of information to use in the effort to improve the environmental performance and sustainability of aquaculture.

In addition to the aquaculture literature, information pertinent to the environmental impacts of fertilizers, feeds, and agrochemicals is available from the agricultural literature about plant and livestock production. Some techniques that have been used to minimize environmental effects of traditional agriculture can be modified for use in aquaculture. Studies of sanitary chemicals and therapeutants may be found in the literature on sanitary engineering and the literature of veterinary and human medicine, respectively, and can also be applied to aquaculture.

Information of the properties, uses, safety hazards, and disposal of chemical compounds are provided by manufacturers. This information can be found on product labels and material safety data (MSD) sheets. There are regions where MSD sheets are not available for some products, and products that are not intended for aquaculture can be purchased and used in culture systems. Both authors have observed instances where inappropriate chemicals were used in aquaculture. Some farmers—particularly rural small-scale ones—may not be informed of the repercussions of certain chemical applications but believe that the products are beneficial.

Several manuals also have particularly useful compilations of information about chemicals. Some examples are The Merck Index; US Pharmacopeia; The Pesticide Manual; Crop Protection Reference; Chemical Rubber Company Handbook of Chemistry and Physics. These manuals are updated at frequent intervals and can be found in many libraries.

Global information on most aspects of agricultural, capture fisheries, and aquacultural production are maintained and updated annually on the FAO websites. These databases can provide information on each major species down to country level. Accumulation of data for many items in the database began in the early 1950s; thus, the large increases in agricultural production resulting from the "green" revolution and the dramatic increase in aquacultural production are included in the databases. Other databases on agricultural, fisheries, and aquacultural production are available but the FAO databases are by far the most comprehensive and widely used.

It is no longer necessary to spend long hours in a university library seeking sources of information. The majority of the important sources can be found online using a standard search engine; Google Scholar is particularly useful for finding information published in refereed journals.

Environmental impact assessment

An Environmental Impact Assessment or EIA is a study to estimate the magnitude of positive or negative environmental impacts that will likely result from a proposed project. In addition, the EIA considers alternatives for a project and assesses the benefits of each. Options are proposed for mitigating or preventing the negative impacts, and it is also common to consider the advantages and disadvantages of each option.

The use of EIAs began in the 1960s, the US National Environmental Policy Act of 1969 created the EIA process for federal projects as a means to "integrate the generation and dissemination of environmental information, and foster collaboration among the diverse set of public and private actors and stakeholders which characterize major, environmentally controversial decisions" (http://www.eoearth.org/article/Environmental_Impact_Assessment). The use of EIAs has become common worldwide for both public and private projects. In the United States, European Union (EU) countries, and in many other countries, an EIA must be conducted and approved before a major infrastructure development or renovation project is initiated. International funding organizations such as the World Bank also require an EIA or some other type of environmental impact study before granting funds for a project (http://www4.worldbank.org/afr/ssatp/Resources/HTML/rural_transport/knowledge_base/English/Module%205%5C5_4a%20Environmental%20Impact%20Assessment.pdf). Thus larger aquaculture projects often require an EIA. Civil engineering firms usually conduct EIAs but some developers may hire individual consultants and form an EIA team for their proposed project.

Formal EIAs must be conducted according to a specific, approved protocol that may differ slightly from country to country. Typically the public must be notified that an EIA is being prepared for a specific project, and all stakeholders including the public are allowed to examine the draft EIA and provide comments. The final EIA must be approved by one or more governmental agencies before the project is allowed to begin. The format outlined in Table 4.2 for conducting the environmental assessment part of an EIA is fairly typical.

In some instances, a sector of the government may waive the necessity for an EIA, but normally the decision to not require an EIA would be limited to small projects and types of projects for which negative environmental impacts have not been associated in the past.

The complete EIA is a process and consists of more than the document that describes the project, alternatives, impacts, and mitigations (Wood 1995). The formal

Table 4.2 Typical format for preparation of an Environmental Impact Assessment (EIA) document.

Step	Purpose
Screening	This task is conducted to determine if an EIA is required for a project.
Description of project	Complete description of site and project (including construction, operations, and decommissioning) with identification of possible environmental impacts.
Alternatives	Examination of alternatives available for all aspects of project that will have negative impacts.
Description of environment	Complete description of the aspects of the environment that could be affected by the project.
Environmental effects	Description of the negative environmental effects that may occur.
Mitigation	Set forth a plan for preventing or mitigating the possible negative environmental impacts.
Nontechnical summary	This summary is a nontechnical description of the environmental assessment and mitigation plans that is understandable by the average lay person.
Special issues	Weaknesses in knowledge related to the assessment are identified.

document often is called an environmental impact statement but the entire process involves key stakeholder and public comment, response to the comments to finalize the document, and approval or disapproval by the appropriate governmental authorities.

The environmental evaluation aspect of an EIA can be used in a less formal manner to determine the possible negative impacts of an existing operation or industry and recommend procedures for lessening these impacts. For example, Boyd et al. (2000) made an environmental assessment of channel catfish farming in Alabama. This study provided a description of the farms and their surrounding environment. Production procedures were outlined in detail. Particular emphasis was given to the volume and quality of effluent, and water conditions in streams above and below farm outfalls. A list of possible negative environmental impacts was developed and best management practices for avoiding or lessening the negative impacts were suggested. A similar study of the possible negative impacts of Vietnamese catfish (*Pangasius*) farming in Vietnam also has been published (Bosma et al. 2009).

Although the EIA can be a powerful tool for identifying negative environmental impacts and proposing alternatives and methods for lessening or preventing the impacts, there is no assurance in many cases that the suggestions of an EIA are actually used by the project developer. Moreover, implementation of suggestions proposed for avoiding negative environmental impacts may not always be effective.

Life cycle assessment

A product can be thought of as having a life cycle because it is produced (birth), used (life), and disposed of or recycled (death). Life cycle assessment (LCA) assesses environmental impacts at all phases of a product's life to include raw material acquisition, processing, manufacturing, packaging, distribution, use, repair and maintenance, and disposal, decommissioning, or recycling. This analysis often is said to consider all environmental impacts of a product from "cradle to grave."

The LCA process is more complex than it appears at first glance because the production of a given product usually requires the use of several other products each requiring additional products for their production. For example, production of electricity often relies on coal as an energy source, but coal must be mined using machinery and fuel and transported by train to the electric generating facility. Electricity production can be analyzed by LCA, but coal, the machinery for mining and transporting coal, and fuel used to power this machinery and the electric power plant itself all have life cycles. The same is true for all other inputs used at the coal-fired electrical generation plant. The LCA is extremely complex because of the overlapping LCAs of inputs, and a complete LCA of a product is seldom made.

The identification of environmental impacts in LCA is not particularly difficult, but it is more problematic to assess the magnitude of the impacts. Usually the LCA only provides estimates of the amount of resources used and the quantities of pollutants released for each component. For example, the contribution of a product to carbon emissions, habitat alteration, sulfur emissions, nutrient pollution, discharge of toxins, etc. can be assessed. Likewise the amounts of resources such as land, water, metals, and fuels can be estimated.

The LCA can provide information useful for comparison among different products, for example, chicken versus beef, traditional automobile versus a hybrid car, bottled water versus potable water from a municipal supply, aquacultured shrimp versus farm-reared tilapia, etc. It also can be used to identify the best environmental alternatives for methods of producing products, for example, imported versus a locally grown food item, feedlot beef versus pasture-grown beef, intensive shrimp production versus semi-intensive shrimp production.

Because of the complexity of LCA, the studies usually do not investigate all possible aspects of the life cycle of products. This can lead to large differences in the identification and quantification of impacts that could drastically influence conclusions about the environmental effects of different products. Authors of LCA reports should be careful to describe the limits of an LCA, and comparisons among different studies be adjusted for differences in scope.

The procedures for conducting an LCA are too complex to be presented here. Documents from the International Standards Organization (ISO) 14000 series are widely used as the basic framework of LCA. There are several books to include Guinée (2002); Horne et al. (2009); and Baumann and Tillman (2004), and at least one journal (*International Journal of Life Cycle Assessment*) devoted entirely to LCA. Software programs also can be purchased for use in LCA and industry and trade associations have databases available to LCA researchers (Rebitzer et al. 2004; Pennington et al. 2004). However, the LCA investigator often will have to rely on his or her knowledge of an industry and upon information provided by their contacts in this industry to obtain all of the data necessary in LCA.

The example of a simplified life cycle flow chart for the production of channel catfish meat (Fig. 4.2) shows the major components of the LCA. Data for components directly associated with the industry would likely be more easily obtained than data

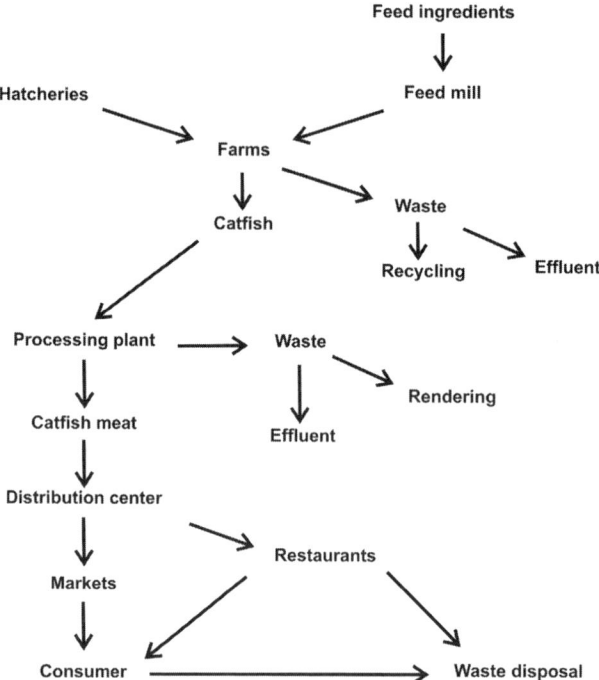

Figure 4.2 A simplified life cycle flow chart for the production of channel catfish meat.

on the supporting processes and the final use and disposal of the product. This is because transportation, storage, and final use and disposal of a product often become intermingled. For example, catfish fillets may be transported along with other meat products from distribution centers to supermarkets and restaurants in the same truck, stored in coolers or freezers with other products, purchased and carried home with other groceries, and prepared along with other items for meals. Likewise, wastes from catfish become intermingled with other household wastes.

Several LCA investigations have been made in aquaculture (Aubin et al. 2006; Mungkung et al. 2006; Ayer and Tyedmers 2009; Papatryphon et al. 2004a, 2004b; Pelletier et al. 2007, 2009; Thrane 2004). These have not been complete LCAs, but the limits were explained. Nevertheless findings of LCAs in aquaculture are sufficient to allow a meaningful comparison with LCAs for other kinds of food including wild fisheries products.

Although LCA seems to have taken on a life of its own—possibly because great and often unnecessary complexity tends to appeal to many scientists—it really does not represent anything novel. The basic process of LCA involves accounting for all of the inputs and outputs for the production of an item. This approach has been used for many years for preparing material, nutrient, and water balance budgets or balances in geochemistry, agriculture, engineering, and many other fields. The effort of assigning impacts also amounts to nothing more than determining the contribution of the production of the particular item to waste loads. The LCA process does not provide any help in determining the actual impacts of these waste loads.

Footprints

Various types of footprints can be calculated for the production of goods and services or for the use of a resource at individual, national, or global levels. Some common examples are the ecological footprint, the carbon footprint, and the water footprint. The ecological footprint in its broadest sense is an indicator of the human demand for goods and services on the world's ecosystems expressed in terms of area (Rees 1992).

In order to calculate the ecological footprint of humans, the land and water space necessary to provide resources, produce food, provide area for all the infrastructure required to provide living space and daily activities of humans, and to treat the resulting waste must be estimated. Wackernagel et al. (1999) presented a simple format for conducting national and global natural capital accounting and converting these data to footprints.

According to Ewing et al. (2010), the ecological footprint represents appropriated biocapacity. Biocapacity is an estimate of the available bioproductive area of the earth. These bioproductive areas have been adjusted to the world average for different land uses. The equivalence factors for different land use are given in Table 4.3. On average, a hectare of primary cropland is 2.21 times more productive than a typical hectare of the earth's surface, while aquatic habitats are only 0.36 times as productive as the average hectare. The equivalence factors convert actual areas of a land use to their global hectare equivalents.

The world biocapacity is approximately 11.2 billion hectares—this is the same as global hectares (Table 4.3). Dividing global hectares of biocapacity by the human population gives the available biocapacity per capita. The actual amount of biocapacity used per capita (the ecological footprint) has been estimated by various means. In 1961, the ecological footprint of humans was about 50% of the earth's biocapacity; by mid-1980s, the two were about equal, but since then the footprint has exceeded the biocapacity and is continuing to rise. Kitzes et al. (2007) gave the global ecological footprint as 14.1 billion global hectares or about 1.25 times the earth's biocapacity. Some estimates by other authors have been as much as 1.5 times greater than the available biocapacity.

The ecological footprint may be estimated for countries, but the world average equivalence factors (Table 4.3) must be adjusted for differences in productivity

Table 4.3 Conversion of land use in billions of surface hectares to billions of global hectares based on relative bioproductivity (equivalence factors).

Land use	Area	Equivalence factors	Area (global ha)
Fishing grounds	2.8	0.36	1.01
Cropland	1.6	2.21	3.54
Grazing land	3.3	0.49	1.62
Forest land	3.8	1.34	5.09
Built-up land	0.2	2.21	0.44
Total	**11.7**		**11.7**

Source: Modified from Ewing, B., A. Reed, S. M. Rizk, A. Galli, J. Kitzes, and M. Wackernagel. 2010. *Calculation Methodology for the National Footprint Accounts, 2010 Edition*. Oakland: Global Footprint Network.

Table 4.4 Ecological footprint and available biocapacity for selected countries.

Country	Ecological footprint (global ha/cap)	Available biocapacity (global ha/cap)
Australia	6.68	14.57
Brazil	2.93	9.63
Chile	3.24	3.74
China	2.13	0.87
Germany	4.57	1.95
Indonesia	1.13	1.32
India	0.87	0.48
Japan	4.17	0.59
Korea	4.62	0.72
Pakistan	0.75	0.40
Russia	4.40	6.62
South Africa	2.59	1.21
Thailand	2.41	1.17
United States	7.19	3.86
Zimbabwe	1.17	0.72
Congo	1.08	12.20
Chad	1.89	3.17
Egypt	2.06	0.65
United Kingdom	4.71	1.34
France	4.91	2.99
World	1.50	1.00

Source: Selected from list of countries given at
http://www.footprintnetwork.org/en/index.php/GFN/page/glossary

among countries. The ecological footprints for selected countries are provided in Table 4.4; values range from 0.75 ha/cap in Pakistan to 7.19 ha/cap in the United States.

Some environmentalists have estimated the ecological footprint for specific activities. Kautsky et al. (1997) reported that the ecological footprint of tilapia cage culture was 10 000 times the area of the cages while the ecological footprint of semi-intensive shrimp ponds was 35–190 times the pond water surface areas. However, such calculations are of limited benefit because they do not take into account the greater productivity of the aquaculture area compared to natural biocapacity. Nevertheless such footprints do reveal that aquaculture requires much space in addition to the actual production area. Of course the same can be said about any type of intensive animal production.

Food miles

Initially food was produced locally, but the variety and quantity of products that could be cultivated or gathered from natural sources in a particular area was limited. Since the early days of recorded history, there are reports of trade in food products between regions and countries. This trade was originally restricted to high-value products but improvements in cargo transportation have removed most limitations on types and amounts of food products that can be traded among regions and countries.

Environmental concern has been expressed about the use of energy and associated carbon emissions necessary for transporting food products (Pretty et al. 2005; Weber and Matthews 2008). Some have suggested reporting the distance that a product has traveled on its label as an aid to consumers seeking "environmentally responsible" food choices. The main problem with food miles is that different types of transportation use different amounts of fuel to transport a unit of food a given distance. For example, the fuel used to ship 1 t of product 1 km by container ship is about four times less than for rail cargo, 11 times less than for heavy truck cargo, and 125 times less than for air cargo. It would be more appropriate to develop some type of index that shows the efficiency with which a product has been transported rather than the distance it has been transported.

Effluent and environmental monitoring

Eutrophication (see Endnotes[1]), sedimentation, and other negative impacts of aquaculture effluents in receiving water bodies are concerns. Government agencies, international lending agencies, aquaculture certification programs, and some seafood buyers are requiring effluent monitoring as discussed in other chapters.

Measuring concentrations and loads of potential pollutants in aquaculture effluents provides data on the pollutional strength of discharges from facilities. This information can be useful in monitoring the volume and pollutional strength of effluent from a facility over time, and limits on concentrations of water quality variables in effluent can avoid undesirably high concentrations in the mixing zone around the outfall. Knowledge of effluent quality is of much less value in determining the overall effect of aquaculture effluents on the receiving water body.

In some situations, water bodies may be able to assimilate waste introduced in aquaculture effluents, and changes in water quality may not occur—except possibly in the mixing zone. In other situations, aquaculture facilities could maintain small concentrations and loads of pollutants in effluent, avoid elevation of water quality variables in the mixing zone, and still cause water quality deterioration in the receiving water body. A good example is effluent discharge into a lake with long hydraulic retention time. The nutrients entering the lake are not flushed out rapidly, and over time, nutrients may gradually increase resulting in greater eutrophication.

Many water bodies receive effluents from one or more other sources in addition to aquaculture. Deteriorating water quality in a water body receiving aquaculture effluent may not be the result of pollutants from aquaculture but from other industries. Of course aquaculture effluents can contribute to pollution even though they are not the main source. A good example can be provided by the authors' experiences with cage aquaculture in Lake Yoja in Honduras. This natural lake has many activities on its catchment—hotels around the lake, town and several villages, cattle farming, mining, and possibly others. The lake had a Secchi disk visibility of about 7 m in 1980, and by 1990, the value had fallen to around 4 m. Thus the lake had become more eutrophic before cage culture was initiated in the late 1990s. Cage culture is releasing nutrients into the lake and contributing to eutrophication, but removal of cage culture from the lake would not cause an end to nutrient enrichment.

The effects of aquaculture effluents on receiving water bodies usually cannot be assessed by monitoring alone. Studies of the receiving water body to establish the present status of chemical, physical, and biological variables for comparison with past conditions and setting a reference against which future conditions may be compared if necessary. Moreover, the kinds and amounts of pollutants discharged into the water body by each effluent source (including aquaculture) must be assessed. Such studies are beyond the capabilities of aquaculture producers and would require a high degree of collaboration among the different stakeholders and considerable financial input and scientific expertise.

Indicators

A fairly substantial number of indicators have been developed for assessing efficiency of resource use and for expressing and comparing pollution loads (Boyd et al. 2007; Volpe et al. 2010). Some of these indicators will be used in other chapters, and the purpose here is to present and explain some of the more important indicators.

The feed conversion ratio or FCR is likely the most commonly used indicator in aquaculture. It is an estimate of the amount of feed needed to obtain 1 kg live, net weight of the culture species:

$$FCR = \frac{Feed(kg)}{Net\ production(kg)}. \tag{4.1}$$

For example, if a pond was stocked with tilapia weighing 200 kg, a total of 7830 kg feed was applied, and 4550 kg of fish were harvested the FCR would be:

$$FCR = \frac{7830}{(4550 - 200)} = 1.80. \tag{4.2}$$

An alternative to the FCR is the feed conversion efficiency (FCE)—an estimate of the live weight production obtained from 1 kg of feed. The FCE is simply the inverse of the FCR. In the example of FCR = 1.8, the FCE would be 0.56 (1 ÷ 1.8).

Protein is expensive, and efficient use of this ingredient should be an important objective in feed manufacturing and feeding practice. The ratio of protein applied in feed to aquatic animal production is the protein conversion ratio (PCR). It can be calculated as follows:

$$PCR = FCR\frac{\%\ Feed\ protein}{100}. \tag{4.3}$$

The PCR is the amount of feed protein necessary to produce 1 kg of the culture species. In the tilapia example above, assume that the feed contains 32% protein. The PCR would be [1.8 × (32 ÷ 100)] or 0.58 kg protein per kilogram live fish.

The feed protein is converted to aquatic animal protein for human consumption. The efficiency of this conversion can be estimated by the protein efficiency ratio (PER):

$$PER = FCR \times \frac{\% \text{ Feed protein}}{\% \text{ Protein in culture species}}. \tag{4.4}$$

This indicator reveals the kilograms of feed protein necessary to give 1 kg aquatic animal protein. Of course the PER does not take into account protein that is contained in processing offal.

In the tilapia example, it will be assumed that the tilapia are 14% protein (live weight basis). The PER will be 4.11 revealing that 4.11 kg of feed protein were necessary to produce 1 kg tilapia protein.

Aquaculture feed often contains marine fish meal. The fish meal ratio (FMR) indicates the quantity of fish meal necessary to produce 1 kg of aquatic animal:

$$FMR = FCR \times \frac{\% \text{ Fish meal in feed}}{100}. \tag{4.5}$$

Tilapia feed often contains about 6% fish meal. Thus, in the example, the FMR is 0.108 [1.8 × (6% ÷ 100)] or 1 kg tilapia required 0.108 kg fish meal.

The production of 1 kg of fish meal requires about 4.5 kg live pelagic fish. Thus it is popular to estimate the fish in–fish out ratio (FIFO), or the quantity of live fish needed to produce 1 kg of the aquaculture species:

$$FIFO = FMR \times 4.5. \tag{4.6}$$

The FIFO for the example is 0.49 (0.108 × 4.5). Thus the production of 1 kg tilapia requires only 0.49 kg live fish. The FIFO ratio also should include fish oil in feed and is discussed thoroughly in Chapter 8.

The indicators discussed above are intimately related to FCR. Reducing the FCR improves all of the ratios. This principle is illustrated in Table 4.5 with data for tilapia

Table 4.5 Influence of FCR on other feed use efficiency ratios for *Oreochromis aureus* provided a 32% protein feed containing 90% dry matter and 6% fish meal.

FCR	FCR$_d$	WPR	PCR	PER	FMR	FIFO
2.5	9.0	2.00	0.80	5.72	0.15	0.68
2.25	8.1	1.78	0.72	5.14	0.14	0.63
2.00	7.2	1.55	0.64	4.57	0.12	0.54
1.75	6.3	1.32	0.56	4.00	0.10	0.45
1.50	5.4	1.10	0.48	3.42	0.09	0.40
1.25	4.5	0.88	0.40	2.86	0.075	0.34

Source: Boyd, C. E., C. S. Tucker, A. McNevin, K. Bostick, and J. Clay. 2007. Indicators of resource use efficiency and environmental performance in fish and crustacean aquaculture. *Reviews in Fisheries Science* 15:327–360. Copyright © 2007, Taylor & Francis.
FCR, feed conversion ratio; FCR$_d$, dry weight feed conversion ratio; WPR, waste production ratio; PCR, protein conversion ratio; PER, protein efficiency ratio; FMR, fish meal ratio; FIFO, fish in-fish out ratio.

Table 4.6 Feed use efficiency values for several common aquaculture species.

Variable	Species				
	Salmo salar	*Ictalurus punctatus*	*Oreochromis aureus*	*Oncorhynchus mykiss*	*Penaeus monodon*
FCR	1.0	2.2	1.8	1.2	2.0
Crude protein in feed (%)	43	28	32	45	35
Fish meal in feed (%)	30	2	6	25	19
DMR	3.21	7.62	6.48	4.15	6.64
WPR	0.62	1.72	1.37	0.82	1.53
PCR	0.43	0.62	0.58	0.54	0.70
PER	2.32	4.13	4.11	3.46	3.70
FMR	0.3	0.044	0.108	0.30	0.38

Source: Boyd, C. E., C. S. Tucker, A. McNevin, K. Bostick, and J. Clay. 2007. Indicators of resource use efficiency and environmental performance in fish and crustacean aquaculture. *Reviews in Fisheries Science* 15:327–360. Copyright © 2007, Taylor & Francis.
FCR, feed conversion ratio; DMR, dry matter ratio; WPR, waste production ratio; PCR, protein conversion ratio; PER, protein efficiency ratio; FMR, fish meal ratio.

Oreochromis aureus production. The indicators also vary among different selected aquaculture species (Table 4.6).

Feed is mostly dry matter (usually about 90%) while culture species are much lower in dry matter content (typically around 25%). The aquaculturist purchases air dry feed and sells live weight animals, so the differences in moisture content between feed and culture animals is not an issue. The difference is significant from an ecological perspective because the dry matter in feed is not converted to animal biomass nearly as efficiently as the FCR or FCE seem to indicate.

Boyd et al. (2007) illustrated the importance of the difference in use of wet and dry weight calculation of the feed conversion by calculating a dry matter-based feed conversion ratio (FCR_d) as follows:

$$FCR_d = FCR \times \frac{\% \text{ DM in feed}}{\% \text{ DM in culture animals}}. \tag{4.7}$$

In the case of the FCR of 1.8 for tilapia in our example, the resulting FCR_d is 6.48 [$1.8 \times (92 \div 25)$]. It actually requires 6.48 kg dry matter in feed to produce 1 kg dry matter in tilapia. The difference, 5.48 kg dry matter, enters the culture system in wastes.

Aquaculture production almost always is reported on a live weight basis. Thus the ratio of waste (dry matter basis) to live weight production or waste production ratio (WPR) also may be computed. The appropriate formula is

$$WPR = (DMR - 1) \times \frac{\% \text{ dry matter in culture species}}{100}. \tag{4.8}$$

In our tilapia example, the WPR would be 1.37, or 1.37 kg of wastes (dry matter basis) would result from each kilogram of live weight production.

Conclusions

There are many sources of information on aquaculture production in the literature that can be used in assessing resource use and environmental impacts of aquaculture. However, each situation is different, and the specific environmental conditions and nature of farm operations must be assessed in order to ascertain the most likely impacts that will occur. An EIA designed for a particular situation is the most reliable way of identifying possible negative impacts and recommending measures for avoiding these impacts or reducing them to a more acceptable level. The LCA procedure is popular because of its complexity and all inclusive nature. However, a complete LCA is difficult to conduct and there are many arbitrary decisions that must be made in determining what to include, what to omit, where to begin, and where to end. Moreover, an LCA does not provide any more guidance on identifying environment impacts and the methods for reducing these impacts than does an EIA.

The use of various types of footprints is a convenient way of comparing resource use and certain broad environmental impacts among different production activities. The use of food miles is not so useful in evaluating transportation impacts because of varying efficiencies in different modes of cargo transportation.

Feed is the most expensive input in many types of aquaculture, fish meal and oil are included in feed, and feed is a source of wastes. Indicators of feed use efficiency, demand on live marine fish for feeds, and waste generation per unit of feed can be useful in comparing among species.

The eNGO perspective

There was a time when it was wistful to think of cooperation between the aquaculture industry and the eNGOs. Commercial aquaculture developed rapidly and like any growing industry mistakes were made and negative impacts realized. The large, US-based eNGOs began to mobilize with the alarms raised by grassroots groups against aquaculture development in their regions. However, between 2004 and 2012, the industry and eNGOs began collaborating to develop eco-label standards through a series of discussion forums called Aquaculture Dialogues (see Chapter 14). The engagement was welcomed by some and challenged by others on both the industry and eNGO sides. The bulk of the collaboration was to generate standards that would minimize or eliminate negative impacts of aquaculture. It became apparent that although there were many unbiased research articles generated to enhance knowledge for the benefit of the aquaculture industry and the advancement of science in the field, there was little information on how one compares farms in a side-by-side manner or what constituted a bad farm, a good farm or an excellent farm. In many respects, Boyd et al. (2007) was the desired material that eNGOs assumed was published in the scientific literature.

One of the major hurtles that was necessary to overcome was to develop civilized working relationships among the various stakeholders in the Aquaculture Dialogues. It is of interest to note that of all the relationships fostered by the junior author during his time managing several of the Aquaculture Dialogues, there was no more antagonistic stakeholder group than the academic researchers and professors. The eNGOs were viewed by many academics as too unskilled to understand the information that was being presented or not capable of engaging in a meaningful debate because of a lack of experience. It is no wonder that many of the eNGOs developed better relationships with industry than with academia because academia also tends to have a similar view of those within the commercial aquaculture industry. Indicators such as those mentioned in this chapter allowed a more objective, comparative evaluation of species, culture methods, and farms than could be obtained from stakeholder opinions.

One of the guiding principles in WWF's Aquaculture Dialogues was to develop a measure (indicator) to evaluate farms among themselves, but also to develop maximum tolerances on these measures (standards). Unlike most of the eNGOs engaged in the Aquaculture Dialogues, WWF was committed to the notion that the best means of causing a large-scale environmental improvement was to concentrate on the main impacts of aquaculture production. This was challenging for other eNGOs because there was a potential for missing some aspect and the organization being viewed as endorsing something that was not "bulletproof".

There is a general trend among eNGOs in acceptance of the various types of aquaculture (Table 4.7). In general, eNGOs do not consider the aquaculture production in Asia acceptable because of reports of chemical use and contaminated products. Cage culture in general is considered unacceptable. Culture of a species in the United States is typically given high level of acceptance primarily because of the lobbying of US producers and the magnitude of research and familiarity with those conducting the research on US production. The dependency on wild fish or the higher the trophic status of a species the more unacceptable product becomes. This simplification is not a guidepost that is formally used by the eNGOs and there is much more detail typically examined, but in general Table 4.7 gives the reader some understanding of how

Table 4.7 Simplified guide to understanding the perception and relative acceptance of the most important aquaculture species by mainstream eNGOs.

Species	Produced in Asia	Produced in cages	Wild fish dependency	Produced in United States	Total
Clams	0	0	0	1	1
Channel catfish	0	0	0	1	1
Mussels	0	0	0	1	1
Oysters	0	0	0	0	0
Scallops	0	0	0	0	0
Trout	0	0	−1	1	0
Abalone	0	−2	0	1	−1
Pangasius	−1	0	0	0	−1
Tilapia[a]	0 or −1	−1 or 0	0	0 or 1	**−2 to 1**
Salmon	0	−1	−1	0	**−2**
Shrimp	−1	0	−1	0	**−2**

Species with a total score of 0 or 1 are considered most acceptable. A score of −1 is considered acceptable, and a score of below −1 is considered unacceptable.
[a] Note that tilapia aquaculture is variable because of the different types of culture systems and the many regions of the world where it is cultured

any aquaculture species would be relatively perceived by the broader mainstream eNGOs. It is important to point out that the only two eNGOs that work on aquaculture and do not use or produce generalized seafood recommendations (stop light cards) are Sustainable Fisheries Partnership and WWF-US (many other WWF offices in other countries make these recommendations and guides). This is partly in recognition that aquaculture operations will have specific impacts in the specific site they are located, but also recognition that simplistic generalizations that may be logical for capture fisheries do not work well in characterizing aquaculture species or countries where these species are raised.

More detailed information on eNGO concerns related to specific resource uses will be provided in their respective chapters that follow.

References

Aubin, J., E. Papatryphon, H. M. G. Van der Werf, J. Petit, and Y. M. Morvan. 2006. Characterization of the environmental impact of a turbot (Scophthalmus maximus) re-circulating production system using Life Cycle Assessment. *Aquaculture* 261: 1259–1268.

Ayer, N. W. and P. H. Tyedmers. 2009. Assessing alternative aquaculture technologies: life cycle assessment of salmonid culture systems in Canada. *Journal of Cleaner Production* 17:362–373.

Baumann, H. and A.-M. Tillman. 2004. *The Hitch Hiker's Guide to LCA. An Orientation in Life Cycle Assessment Methodology and Application.* Oxford: SwePub.

Bosma, R. H., C. T. T. Hanh, and J. Potting. 2009. Environmental impact assessment of the pangasius sector in the Mekong Delta. Wageningen University, Wageningen, The Netherlands.

Boyd, C. A., P. Pengseng, and C. E. Boyd. 2008. New nitrogen fertilization recommendations for bluegill ponds in the southeastern United States. *North American Journal of Aquaculture* 70:308–313.

Boyd, C. E. 1981. Solubility of granular inorganic fertilizers for fish ponds. *Transactions of the American Fisheries Society* 110:451–454.

Boyd, C. E. 1990. *Water Quality in Ponds for Aquaculture.* Auburn: Auburn University, Alabama Agricultural Experiment Station.

Boyd, C. E. and C. S. Tucker. 1998. *Pond Aquaculture Water Quality Management.* Boston: Kluwer Academic Publishers.

Boyd, C. E., C. S. Tucker, A. McNevin, K. Bostick, and J. Clay. 2007. Indicators of resource use efficiency and environmental performance in fish and crustacean aquaculture. *Reviews in Fisheries Science* 15:327–360.

Boyd, C. E., J. Queiroz, J. Lee, M. Rowan, G. N. Whitis, and A. Gross. 2000. Environmental assessment of channel catfish, *Ictalurus punctatus*, farming in Alabama. *Journal of the World Aquaculture Society* 31:511–544.

Ewing, B., A. Reed, S. M. Rizk, A. Galli, J. Kitzes, and M. Wackernagel. 2010. *Calculation Methodology for the National Footprint Accounts, 2010 Edition.* Oakland: Global Footprint Network.

Guinée, J. B. (editor). 2002. *Handbook on Life Cycle Assessment, Operational Guide to the ISO Standards.* Dordrecht: Kluwer Academic Publishers.

Hickling, C. F. 1962. *Fish Cultures.* London: Faber and Faber.

Horne, R., T. Grant, and K. Verghese. 2009. *Life Cycle Assessment: Principles, Practice, and Prospects*. Collingwood: CSRIO Publishing.

Kautsky, N., H. Berg, C. Folke, J. Larsson, and M. Troell. 1997. Ecological footprint for assessment of resource use and development limitations in shrimp and tilapia aquaculture. *Aquaculture Research* 28:753–766.

Kitzes, J., A. Peller, S. Goldfinger, and M. Wackernagel. 2007. Current methods for calculating national ecological footprint accounts. *Science for Environment and Sustainable Society* 4:1–9.

Metzger, R. J. and C. E. Boyd. 1980. Liquid ammonium polyphosphate as a fish pond fertilizer. *Transactions of the American Fisheries Society* 109:563–570.

Mischke, C. C. 2012. *Aquaculture Pond Fertilization: Impacts of Nutrient Input on Production*. Ames: Wiley-Blackwell.

Mortimer, C. H. 1954. *Fertilizers in Fish Ponds*. London: Her Majesty's Stationery Office.

Mungkung, R. T., H. A. Udo de Haes, and R. Clift. 2006. Potentials and limitations of life cycle assessment in setting ecolabelling criteria: a case study of Thai shrimp aquaculture product. *International Journal of Life Cycle Assessment* 11:55–59.

Murad, A. and C. E. Boyd. 1987. Experiments on fertilization of sportfish ponds. *Progressive Fish-Culturist* 49:100–107.

Papatryphon, E., J. Petit, and H. M. G. Van der Werf 2004a. The development of Life Cycle Assessment for the evaluation of rainbow trout farming in France. In Proceedings 4th International Conference on Life Cycle Assessment in the Agri-feed Sector, pp. 73–80, October 6–8, 2003, Horsens, Denmark.

Papatryphon, E., J. Petit, S. J. Kaushik, and H. M. G. Van der Werf. 2004b. Environmental impact assessment of salmonids feeds using Life Cycle Assessment. *Ambio* 33:316–323.

Pelletier, N. L., N. W. Ayer, P. H. Tyedmers, S. A. Kruse, A. Flysjo, G. Robillard, F. Ziegler, A. J. Scholtz, and U. Sonesson. 2007. Impact categories for life cycle assessment research of seafood production systems: Review and prospectus. *International Journal of Life Cycle Assessment* 12:414–421.

Pelletier, N., P. Tyedmers, U. Sonesson, A. Scholz, F. Ziegler, A. Flysjo, S. Kruse, B. Cancino, and H. Silverman. 2009. Not all salmon are created equal: Life cycle assessment (LCA) of global salmon farming systems. *Environmental Science and Technology* 43:8730–8736.

Pennington, D. W., J. Potting, G. Finnveden, E. Lindeijer, O. Jolliet, T. Rydberg, and G. Rebitzer. 2004. Life cycle assessment Part 2: current impact assessment practice. *Environment International* 30:721–739.

Pretty, J. N., A. S. Ball, T. Lang, and J. I. L. Morison. 2005. Farm costs and food miles: an assessment of the full cost of the UK weekly food basket. *Food Policy* 30:1–19.

Rebitzer, G., T. Ekvall, R. Frischknecht, D. Hunkeler, G. Norris, T. Rydberg, W.-P. Schmidt, S. Suh, B. P. Weidema, and D. W. Pennington. 2004. Life cycle assessment Part 1: framework, goal and scope definition, inventory analysis, and applications. *Environment International* 30:701–720.

Rees, W. 1992. Ecological footprints and appropriated carrying capacity: what urban economics leaves out. *Environment and Urbanization* 4:121–130.

Swingle, H. S., B. C. Gooch, and H. R. Rabanal. 1963. Phosphate fertilization of ponds. *Proceedings Southeastern Association of Game and Fish Commissioners* 17:213–218.

Thrane, M. 2004. Life cycle assessment in the agri-food sector. In N. Halberg, editor, Proceedings for the 4th International Conference, October 6–8, 2003, Bygholm, Denmark, pp. 78–88. Danish Institute of Agricultural Sciences Report, Tjele, Denmark.

Volpe, J. A., M. Beck, V. Ethier, J. Gee, and A. Wilson. 2010. Global Aquaculture Performance Index. Victoria: University of Victoria.

Wackernagel, M., L. Onisto, P. Bello, A. Callejas Linares, I. S. López Falfán, J. Méndez Garcia, A. I. Suárez Guerrero, and M. G. Suárez Guerrero. 1999. National natural capital accounting with the ecological footprint concept. *Ecological Economics* 29:375–390.

Weber, C. L. and H. S. Matthews. 2008. Food-miles and the relative climate impacts of food choices in the United States. *Environmental Science and Technology* 42:3508–3513.

Wood, C. 1995. *Environmental Impact Assessment, A Comparative Review*. Essex: Longman Scientific & Technical.

Endnotes

1. Eutrophication is mentioned many places in this book, and the phenomenon should be explained for some readers. Eutrophication can be defined as: the accumulation of nutrients in a water body—especially a lake—causing dense growth of planktonic algae and other organisms the respiration and decay of which results in dissolved oxygen depletion.

2. The word eutrophication is from eutrophic meaning nutrient rich as opposed to oligotrophic meaning nutrient impoverished. Any essential plant nutrient can limit plant growth if present in short supply. But, in most freshwaters, coastal waters, and oceanic waters, nitrogen or phosphorus or both are the main limiting nutrients. Introduction of these two nutrients in pollution often leads to excessive phytoplankton blooms and eutrophication.

3. Total phosphorus concentrations in oligotrophic, mesotrophic (the intermediate situation), and eutrophic water bodies typically are <15 µg/L, 15–50 µg/L, and >50 µg/L, respectively. Respective concentrations of total nitrogen are <500 µg/L, 500 to 1000 µg/L, and >1000 µg/L.

4. Chlorophyll *a* concentration usually is <500 µg/L, 500–1000 µg/L, and >1000 µg/L in oligotrophic, mesotrophic, and eutrophic waters, respectively.

5. Corresponding Secchi disk visibilities are: mesotrophic, >5 m; mesotrophic, 2.5 to 5 m, eutrophic, <2.5 m.

Chapter 5

Land use

Most freshwater aquaculture and brackishwater culture of shrimp and fish is conducted in ponds and other land-based facilities. Ponds require considerable land area because they are operated at relatively low production levels, while raceways, other flow-through units, and water-recirculating systems with more intensive production use comparatively little land. The most publicized land use issue in aquaculture is the practice of siting shrimp ponds in mangrove habitats or other coastal wetlands of high ecological value. Other major land use concerns are conversion of agricultural land to ponds, disruption of traditional hunting and gathering areas by installation of ponds, salinization of land and water resulting from use of saline groundwater to supply ponds for inland culture of marine or brackishwater species, and failure to restore the land where aquaculture ponds are abandoned. Disposal of sediments from ponds on nearby land also can result in ecological nuisances.

Land use in aquaculture is not limited to space for culture facilities. In feed-based aquaculture, considerable land is devoted to corn, rice, wheat, soybeans and other crops to provide feed ingredients. Some land also is used to service culture facilities by providing space for offices, storage buildings, shops, staging areas, parking lots, etc.—this also is true for many of those farming systems for which grow-out units are installed entirely within water bodies.

This chapter provides a brief overview of world land use and the associated environmental issues. The amount of land used and the negative impacts of aquaculture on terrestrial ecosystems are discussed. In addition, the relative productivity of land use for agriculture and aquaculture is compared.

Major land uses

The global surface area is 510.1 million km^2 of which 361.1 million km^2 or 70.8% is covered by water (http://www.infoplease.com/ipa/A0001763.html). Areas of the

Aquaculture, Resource Use, and the Environment, First Edition. Claude E. Boyd and Aaron A. McNevin.
© 2015 John Wiley & Sons, Inc. Published 2015 by John Wiley & Sons, Inc.

Table 5.1 Areas of the continents.

Continent	Area (km^2)
Africa	30 065 000
Antarctica	13 209 000
Asia (including Middle East)	44 579 000
Australia (includes Oceania)	8 112 000
Europe	9 938 000
North America (includes Central America and Caribbean)	24 474 000
South America	17 819 000

Source: Obtained from World Bank database (http://search.worldbank.org/data?qterm=arable+land+language=EN+format=).

continents are provided in Table 5.1. Antarctica is not permanently inhabitable and it has no aquaculture or forestry. The other continents have a combined surface area of 135.4 million km^2. However, some of this area is covered by inland water, is mostly frozen, or is otherwise unsuitable for most human uses.

According to FAO (http://faostat.fao.org), in 2011, the world land area was 130.03 million km^2 and there were 4.578 million km^2 of inland water—a slightly lower total continental area than given in Table 5.1. According to FAO statistics, the agricultural area in 2011 was about 49.12 million km^2—13.96 million km^2 of arable land, 1.57 million km^2 of permanent crops, and 33.59 million km^2 of permanent meadows and pastures—consisting of 37.8% of the land surface. Forests and woodlands covered 40.28 million km^2—31.0% of the land surface. The FAO data also reveal that the total, global agricultural area was about 44.60 million km^2 in 1961—an increase of only about 10% in the past 30 years. Food production has, however, increased almost threefold (see Chapter 3).

Although there are other land use categories, the ones mentioned above are the major ones in terms of extent. There is of course the urban area that presently occupies about 3% of the land surface area or about 4.47 million km^2 (http://www.earth.columbia.edu/news/2005/story03-07-05.html). Much of the rest of the land surface consists of deserts and frozen regions unsuitable for agriculture, forests, or permanent human habitation.

Total areas, water areas, forested land, total agricultural land, and arable land for several countries are provided in Table 5.2. There is considerable variation in the sizes of the countries and the amounts of inland water, forests and woodlands, and agricultural areas, but when averaged there is little difference in the percentages of forested and agricultural areas between developing countries and developed ones selected for inclusion in Table 5.2.

Data for total land, cropland, pasture and rangeland, total forest, urban development, and other uses in the United States are provided for the period 1954 to 2007 (Table 5.3). It is interesting that the amount of agricultural land has declined by about 313 946 km^2 or 7.06%, and forested land has increased by 226 654 km^2 or 9.10% during the last five decades. The amazing statistic, however, is that urban area has increased by 69 976 km^2 and was larger in 2007 by a factor of 3.26 than it was in 1954. However, the population of the United States increased from 163 to

Table 5.2 Major land uses for selected countries.

Country	Area (×1000 km²)				
	Total	Inland water	Forest	Agricultural	Other
Developing countries					
Bangladesh	144	13.8	14.5	91.5	24.2
Brazil	8515	56	5309	2645	505
China	9600	273	1930	3243	4154
Egypt	1001	5.6	0.67	36.9	957.83
India	3287	314	678	1800	495
Mexico	1964	20	656	1028	260
Pakistan	796	25.2	18.2	263	489.6
Thailand	513	2.1	189	198	123.9
Developed countries					
Australia	7741	59	1539	4090	2053
Canada	9985	891	3101	676	5317
France	549	1.4	157	293	97.6
Germany	357	8.6	111	169	68.4
Italy	301	7.2	87.6	139	67.2
Japan	378	13.5	249	46	69.5
United States	9629	482	3021	4035	2088

Source: Adapted from http://search.worldbank.org/data?qterm=arable+land+language=EN+format=.

301 million during the same time period—a 1.84-fold increase. This illustrates the great shift in population from rural to urban areas that is occurring worldwide.

Lubowski et al. (2006) assessed major land use categories in the United States for 2002. They reported that the country had a total land area of 9.31 million km². The major uses were as follows: forestland, 2.63 million km² (28.8%); grassland pasture and open rangeland, 2.38 million km² (25.9%); cropland, 1.79 million km² (19.5%); special uses (mainly parks and wildlife conservation areas), 1.20 million km² (13.1%); miscellaneous other uses, 0.923 million km² (10.1%); urban land, 0.243 million km² (2.6%).

It is interesting to note that an estimated 1.5–2.0% of the world's total land surface—19 500 to 26 000 km² is devoted to roads and parking lots. In urban areas,

Table 5.3 Major land uses in the conterminous United States between 1954 and 2007.

Year	Area (×1 million km²)					
	Total land	Cropland	Pasture and range land	Forest and woods	Other	Urban
1954	9.200	1.884	2.565	2.490	2.186	0.075
1964	9.171	1.798	2.592	2.961	1.702	0.118
1974	9.161	1.882	2.419	2.906	1.813	0.141
1982	9.167	1.899	2.415	2.652	1.998	0.203
1992	9.159	1.861	2.392	2.622	2.046	0.238
2002	9.162	1.787	2.374	2.635	2.125	0.241
2007	9.162	1.652	2.484	2.717	2.064	0.245

Source: Boyd, C. E. and M. Polioudakis. 2006. Land use for aquaculture production. *Global Aquaculture Advocate* 9:64–65.

30–60% of land use is for transportation infrastructure (http://people.hofstra.edu/geotrans/eng/ch8en/conc8en/ch8c3en.html).

Deforestation is a major reason for land use changes and ecological perturbations but it is not a recent phenomenon. Humans have been clearing forests for millennia to provide agricultural land and to obtain firewood and construction materials (Williams 2006). It is impossible to know exactly how much forested area existed in the pre-agricultural era but Matthews (1983) placed the estimate at 46.28 million km^2 of closed forest and 15.23 million km^2 of open woodland—about 47.3% of the earth's land area. Williams (2000) reports that the original forested area has been reduced by 7.01 million km^2 of closed forest and 2.13 million km^2 of open woodland. This is a reduction in wooded area of 14.8% and the coverage of the earth by forest is—by this calculation—presently around 52.37 million km^2—considerably more than the FAO estimate of 40.275 million km^2. Despite which estimate is true, much deforestation has occurred and greatly changed the world's landscape.

Deforestation continues but the efforts of eNGOs and governments have resulted in more sustainable forest management in many countries. The United States is a good example; before European settlers arrived, forest apparently covered about 4.05 million km^2 or about 43.5% of the land area. Since the mid-1600s about 1.21 million km^2 were cleared—mainly during the nineteenth century—for agriculture. A major effort to reforest marginal farmland and to manage forests sustainably was implemented and the total forest area has remained fairly constant for the last 100 years; actually Table 5.3 suggests it has increased since 1950 (http://www.nationalatlas.gov/articles/biology/a forest.html).

Most wetland habitats are forests, and the story of wetland loss in the United States is similar to that of forests. In the early 1600s, the area to become the conterminous United States had an estimated wetland area of 0.894 million km^2, but by the mid-1980s, only about 0.417 million km^2 remained—they were drained to create fertile agricultural land and provided construction timber (Dahl and Allord 1996). Beginning in the 1950s, the US government subsidized draining of wetlands and about 2226 km^2 were lost per year until the mid-1970s. Beginning in the mid-1970s, the awareness of the importance of wetlands began to increase and the government started to change its policies on wetlands. From the mid-1970s to the mid-1980s, wetland loss declined to about 1175 km^2/year (Dahl and Johnson 1991). Once the government began efforts to restore wetlands in the late 1980s, the loss of wetlands began to decline and restoration efforts provided additional wetland area; at the end of 2009, there were 0.446 million km^2 of wetlands in the United States (Dahl 2011).

The most extensive examples of deforestation and wetland destruction currently are found in developing countries and especially in tropical and subtropical areas. Many of these countries are beginning to realize the importance of protecting forests and wetlands, and hopefully better conservation policies will be forthcoming. The success with reforestation and wetland restoration in the United States and other developed countries reveal that governmental policies can have a huge effect on conservation of these natural resources.

Land use in aquaculture

Land-based aquaculture facilities are comprised mainly of ponds. The land area needed for ponds includes the land to be covered by water as well as the area for embankments, water supply and distribution systems, and effluent discharge canals or pipelines. In addition, space is needed to support service functions—the two major ones being feed production and waste treatment (Boyd et al. 2007; Boyd and Polioudakis 2006).

Facilities

Pond production typically is reported in weight of production per unit water surface area. But for raceways, tanks, and indoor water-recirculating systems that have highly intensive production within a small area, production normally is reported on a culture volume or water flow-through basis rather than surface area. Cages, net pens, and plots for shellfish and seaweed do not use land for culture area because they are installed in lakes, the sea, or other permanent water bodies—these types of culture have support facilities located on land.

The land or water area necessary per unit of production varies tremendously. Highly intensive water-recirculating systems may produce 1–2 million kg of fish per hectare of culture unit (Timmons et al. 2001). Rainbow trout production in raceways may produce 100 000 to 200 000 kg/ha, while production in ponds usually will be between 100 and 10 000 kg/ha.

As mentioned above, ponds have various earthwork structures to facilitate water management. Farms also have access roads, parking lots, storage areas, staging areas, space for administrative and service buildings, buffer areas, etc. These structures and facilities require dedicated land. Approximate calculations from maps of Alabama catfish farms with watershed ponds suggest that land dedicated solely to aquaculture is typically about 25% more than pond water surface area. Large watersheds—usually about 10 times pond water surface area—supply runoff to ponds, but they were not included in the land use estimates for aquaculture because these watersheds typically have other uses—forestry or pasture land primarily.

Embankment ponds for channel catfish culture in Mississippi typically have only 10–15% more land than used for water surface area, and the land area devoted to supporting pond culture decreases slightly as farm size increases (Keenum and Waldrop 1988). For a farm with a total land area of 65 ha, 2% of the area was devoted to buildings, parking, feed storage, etc., 13% was for embankments to form ponds, and 85% was for water surface area. The percentages were estimated as 1%, 11%, and 88%, respectively, for a 260-ha farm.

In shrimp culture, canals are used to supply and discharge water at farms, and pre-settling and post-settling basins may be included in the water management infrastructure. Farms of 25 ha or more in size appear to have support areas that are about 15–20% of water surface area, while the support areas may increase up to 50% at small farms (Boyd et al. 2007). A study of *Pangasius* farms in Bangladesh (Ali and

Table 5.4 Major ingredients and typical feed conversion ratios in feeds for common aquaculture species.

| | Ingredient content (%) | | | | |
	Atlantic salmon	Trout	Shrimp	Tilapia	Channel catfish
Soybean meal	14.0	15.0	24.5	38.3	34.5
Cottonseed meal	–	–	–	–	12.0
Corn meal	10.0	–	–	48.8	20.4
Wheat middlings	18.0	27.0	27.5	4.0	20.0
Fish meal	30.0	25.0	19.0	6.0	2.0
Shrimp head meal	–	–	13.5	–	–
Squid meal	–	–	5.0	–	–
Rendered products	–	15.0	–	–	4.0
Oil	24.0	16.0	4.5	1.5	2.0
Feed conversion ratio	1.0	1.2	2.0	1.8	2.2

Source: Boyd, C. E. and M. Polioudakis. 2006. Land use for aquaculture production. *Global Aquaculture Advocate* 9:64–65.

Haque 2011) reported means and standard deviations for farm size as 1.36 ± 1.25 ha, for water surface area as 1.06 ± 1.31 ha, and for dike area as 0.30 ± 0.27 ha.

Feeds

Land also is used in agriculture to produce plant meals and oils for aquaculture feeds. Corn, soybean, peanut, and cottonseed meals, wheat middlings, wheat and rice flour, and vegetable oils are common components of feeds. Land must be specifically dedicated to the production of corn, soybeans, peanuts and certain other seeds used in making meals for feeds. Cottonseed meal and wheat middlings are by-products of cotton fiber and wheat flour production, and vegetable oils are extracted from corn, soybean, peanuts, and other seeds during plant meal manufacturing. No additional land space is necessitated for use of by-products in feeds for aquaculture.

The typical feed ingredients and feed conversion ratios (FCRs) for some common aquaculture species are shown in Table 5.4. Average yields of some common plant meals are provided (Table 5.5).

Table 5.5 Average yields in 2004 in the United States for common plant meals used in aquaculture feeds.

Plant	Seed yield (kg/ha)	Meal yield (kg/ha)
Corn	9413	9413
Soybean	2824	2231
Peanut	3440	1927

Source: Adapted from Boyd and Polioudakis (2006).

Table 5.6 Land required to produce plant meals and live fish needed for fish meal in feed to produce 1 t of some common aquaculture species.

Species	Land area for plant meals (ha)	Live fish for fish meal (kg)
Atlantic salmon	0.074	1350
Trout	0.081	1350
Shrimp	0.220	1710
Tilapia	0.402	486
Channel catfish	0.388	198

Boyd and Polioudakis (2006) present the following equation for calculating the land area necessary to provide plant meals for production of 1 t of live aquatic animals:

$$\text{Land requirement (ha/t)} = \sum \frac{(\% \text{ Meal}/100)(\text{FCR})(1000)}{\text{Meal yield(kg/ha)}} \qquad (5.1)$$

Land areas for producing feed for 1 t of selected species are given in Table 5.6. Areas range from 0.074 ha/t for salmon to 0.402 kg/t for tilapia. The average for the five species is 0.233 ha/t. Notice that for the more carnivorous species, salmon and trout, the land requirement for plant meals is quite low but a lot of fish meal must be included in the feed. The other three species require more plant meal, but less fish meal in their feed.

It is interesting to compare land requirements for plant meals used in aquatic animal feeds with those for livestock feed. A typical feed for swine contains 74.4% corn and 23.4% soybean meal, while broiler chicken feed usually is about 67% corn and 23.7% soybean meal. Typical FCRs for broilers and swine are 1.88 and 2.80, respectively. The land requirements for plant ingredients in feed for 1 t net production are 0.515 and 0.333 ha for swine and broilers, respectively. These land area estimates for producing plant meal ingredients are similar to the ones for channel catfish and tilapia feeds.

Cattle fed grain usually have an FCR around 6 (Schnepf 2011) and the amount of corn needed to produce 1 t of cattle would require 0.637 ha. Corn is a high yielding grain and more land would be needed for many other types of cattle feed.

Total land requirements

More investigation should be directed at determining the usual total land area to water surface area ratio for aquaculture facilities. However, even with the meager information available, it is possible to make approximate calculations for some species. In pond culture of channel catfish in Alabama, production often reaches about 8 t/ha per year (Boyd et al. 2000). Each hectare of water surface for watershed ponds requires about 0.25 ha of surrounding land for operational purposes. Another 3.1 ha of land must be dedicated to producing plant ingredients for feed to produce the fish. Thus the land requirement to support production of 8 t of fish in a 1-ha pond is 3.35 ha, and the total land dedicated to production is 4.35 ha (0.54 ha/t).

Data related to a large pond culture facility for tilapia in Ecuador that produces about 6 t/ha per crop suggested that the total farm area including ponds is about 1.25 times the water surface area—similar to catfish farms in Alabama. The land area of the farm equates to 0.21 ha/t per crop but the land area for feed ingredients is 0.402 ha/t (Table 5.6), bringing total land use to 0.61 ha/t. Pond culture of channel catfish and tilapia has similar values for total land use per crop. However, tilapia are produced in warmer climates than those for channel catfish and it is possible to grow more than one crop per year increasing the yield possible per unit area.

Tilapia production in ponds fertilized with chemical fertilizers or manures typically is about 2 t/ha. Ponds usually are small and the ratio of land area to water area might be as much as 1.5:1. Total land use of 1.33 ha/t in such fertilized ponds is greater than for more intensive, feed-based systems.

High-density cages for tilapia culture can yield up to 100 kg fish/m^3 or 500 kg fish/m^2 in 5-m deep cages (Beveridge 1996; Schmittou 1993). The water surface area requirement for tilapia cage culture is only 2 or 3 m^2/t; however, the land required for plant ingredients necessary for the feed is 0.402 ha/t. Total land requirement for feed-based tilapia culture in ponds is only 1.5 times greater than for cage culture of this species. This results because the land requirement for plant ingredients in feed per unit or production in cage culture is greater than the land (water) area occupied by the cages.

Shrimp producers in Central and South America typically operate large ponds at relatively low stocking densities with yields of only 1000 to 1500 kg/ha per crop. The FCR often is about 1.5 because natural food organisms eaten by shrimp in semi-intensive ponds contribute considerably to growth. Assuming a 1.25:1 land:water surface ratio for production facilities and 0.164 ha/t land for feed plant ingredients, total land use is 1.0 to 1.41 ha/t per crop, respectively. In Thailand, production of shrimp may be 5–8 t/ha per crop in small, heavily aerated ponds where FCR is about 1.5. Assuming a 1.5:1 land:water surface ratio because farms in Thailand have small ponds and 0.165 ha/t for feed plant ingredients, total land use is 0.35 to 0.46 ha/t per crop, respectively. The more intensive systems in Thailand use considerably less land per unit of production than do the semi-intensive farms in Ecuador.

Waste assimilation

Internalizing waste treatment imposes a relatively high direct land cost on pond aquaculture that is evident in the large size of ponds. The large land requirement is the result of ponds functioning both as a crop confinement area and a waste treatment facility. More than 95% of the total area of a channel catfish pond functionally acts as a high-rate, photosynthetic waste treatment lagoon, and less than 5% of its total area is necessary as a "fish-holding" area (Brune et al. 2003). In effect more than 95% of the land and construction costs for ponds can be assigned to waste treatment, and the relatively large land area occupied by a pond aquaculture facility is a price the farmer pays for treating wastes on-site rather than discharging the waste into public water. Engle and Valdarrama (2002) explored other costs (aeration, labor, etc.) associated with internalizing waste treatment in channel catfish ponds and calculated that almost 30% of the total cost of producing channel catfish can be ascribed to internal waste treatment processes. Land area requirements and other costs that result from internalizing waste treatment costs in pond aquaculture limit profitable culture to areas where land is relatively inexpensive.

Greater intensification of production in ponds is possible if wastes are treated external to the culture unit, usually by discharging wastes into public waters. When this happens ecosystem area outside the culture facility is used to assimilate wastes, and in most instances, the aquaculturist does not bear the cost of that treatment.

Relatively few studies have been conducted to determine the external ecosystem area needed to treat aquaculture wastes produced by raceway or net pen culture. Furthermore, requirements will vary depending on the hydrology and biology of the water body into which wastes are discharged. Based on the few studies conducted, it appears that ecosystem support areas between 100 and 300 times the facility area are needed to treat wastes produced in cage and net pen fish culture (Berg et al. 1996; Kautsky et al. 1997; Folke et al. 1998; Brummett 1999). These values for ecosystem waste assimilation area are somewhat larger—but still within the same approximate order of magnitude—as the ratio of "waste treatment area" to "fish-holding" area in channel catfish ponds. This is not coincidence because the same biological and physicochemical processes are responsible for waste treatment whether inside a pond or in the lake, stream, or brackishwater bay into which effluents are discharged or cages or net pens located. Area requirements for waste treatment should therefore be of similar magnitude.

Mechanical aeration is applied in channel catfish culture and many other types of pond aquaculture. This practice increases the amount of waste oxidation per unit of pond area as compared to that of natural water bodies receiving aquaculture effluents. The ecosystem area needed for waste treatment in ponds is a part of the culture system, and the amount of land affected to provide waste treatment is obvious. Waste treatment for cage and net pen culture is external to the system and affects areas not dedicated to aquaculture. The receiving water body area necessary to treat aquaculture wastes is not considered a land use here because of lack of reliable data on treatment area required. Nevertheless assimilation of aquaculture wastes, like that of other kinds of water pollution, must be assimilated in receiving water bodies.

Comparison of land use in aquaculture and agriculture

Feed is widely used in fish and shrimp aquaculture and the proportion of production in systems receiving feed has increased from about 50% in 1980 to around 67% in 2012 (FAO 2012). However, many ponds receive fairly small feed inputs that only supplement natural productivity, and less than 67% of fish and shrimp production is based solely on feeds. According to Alltech (2013), world animal feed production totaled 954 Mt in 2012: poultry, 418 Mt; ruminants, 253 Mt; swine, 218 Mt, aquaculture, 34 Mt; horses, 11 Mt; pets, 20 Mt. Aquaculture is a rather minor feed user in comparison with the major sources of meat, milk, and eggs from agriculture. If we consider an average FCR of 1.64 (from Table 5.4) and an average land use of

0.233 ha/t of aquaculture production for plant meal ingredients in feed (Table 5.6), plant meals for this amount of feed might be around 4.83 million ha (48 300 km^2).

Using a possible overestimation of the support area and earthwork infrastructure for ponds of 50% of water surface area and the estimate of 116 830 km^2 of aquaculture pond surface area globally (Verdegem and Bosma 2009), the total area devoted to pond aquaculture farms is about 175 245 km^2. Total global production of aquatic animals in land-based, flow-through culture systems is likely no more than 10 Mt, and yields might average around 100 t/ha per year. The required land area would be 1000 km^2. Land area needed to service cage culture and other types of culture based in lakes and rivers is relatively small, but there is no way at present to calculate it. Adding to this land necessary for plant meals for use in aquaculture feeds gives 224 545 km^2.

The amount of land used by aquaculture is miniscule compared to agriculture—224 545 km^2 versus 49 200 000 km^2. Agricultural land exceeds land for pond aquaculture by 219 times and aquaculture requires only 0.17% of the earth's land surface area.

In 2010, world agricultural production was 7670 Mt from a land area of 4912 million ha or 1.56 t/ha. Aquaculture production in 2010 was 59.87 Mt, but 13.99 Mt were molluscan shellfish that required no land for culture of feed ingredient. The production of the remaining 45.88 Mt of aquaculture species required 22.45 million ha—2.04 t/ha. It is more reasonable to compare aquaculture with meat, milk, and eggs by agriculture that require 33.586 million km^2 of grazing land and 33% of the cropland area to make feed ingredients (http://www.grida.no/publications/rr/food-crisis/page/ 3565.aspx). About 15.54 million km^2 of land are used for crops; thus approximately 5.13 million km^2 are devoted to production of feed grain and plant meals for livestock feed ingredients. Subtracting the area necessary for aquaculture feeds leaves 5.08 million km^2. Total meat, eggs, and milk production was 891 Mt in 2009, and the area devoted to this production totaled an estimated 38.67 million km^2—a yield of 0.23 t/ha.

Either way the comparison is made—total agricultural production or production of meat, milk, and eggs—the use of land for aquaculture has a greater return in quantity of food than would be realized by using the same amount of land for traditional agriculture.

Land use conversions by aquaculture

Ponds obviously are the main contribution of aquaculture to land use. Ponds must be built on flat or gently sloping land where soil has sufficient clay content to avoid excessive seepage (Yoo and Boyd 1994; Hajek and Boyd 1994). There also must be an abundant supply of water to maintain adequate water levels throughout the growing season. Such land usually is suitable for agriculture or already in agriculture, so in most cases, construction of ponds does not result in deforestation but rather in change of land use from agriculture to aquaculture. In some instances, wetland habitat has been converted to ponds. The most common example has been conversion of mangrove habitat to ponds for milkfish and shrimp culture in Southeast Asia (Boyd

Figure 5.1 Mangrove wetland.

and Clay 1998). Although aquaculture does not have an appreciable influence on the landscape and land use on a global scale, it may cause changes in specific habitat types or other land uses on a local or possibly regional level.

Mangroves

Mangrove forests (Fig. 5.1) are an association of halophytic trees, shrubs, palms, ferns, and other plants growing in brackish and saline tidal waters on mudflats, riverbanks, and coastlines in tropical and subtropical regions (Mitsch and Gosselink 1993). This vegetation lives in the zone inundated by the highest tides and exposed by the lowest tides. There are about 70 species of mangrove plants and the greatest diversity of these species is in Southeast Asia and Australia (Spalding et al. 1997). Mangrove forests have high biodiversity and are nursery areas for many aquatic animals; they also stabilize the coastline and protect against high winds and storm surges (Massaut 1999). They serve as filtering areas to remove pollutants from runoff before it enters the sea, to allow suspended solids to settle, and stabilize the sediment to create new land along the coastline and protect it from erosion by waves, floods, and storms (Boyd 2002).

Mangrove areas are used by local communities as hunting, fishing, and gathering areas (Jothy 1984; Paw and Chua 1991). It has been reported that 80% of the marine fishery in Florida depends in some way on mangroves (Mangrove Working Group 1998).

Estimates of the existing global mangrove area vary (Table 5.7), and the average of the three estimates is 182 920 km^2. Two-thirds of the world's mangrove forest area is in South and Southeast Asia and the Americas. It is estimated that the global mangrove area is about 50% of its extent of 100 years ago (Massaut 1999).

Mangrove forests have many uses: construction material, livestock forage, and for making paper, fiber, dye, cloth, and tannic acid. Major losses of mangrove have resulted because of these uses (Ong 1982; Spalding et al. 1997; Macintosh and

Table 5.7 Estimates of global mangrove areas (km^2).

Region	Mangrove area[a]	Mangrove area[b]	Mangrove area[c]
South and SE Asia	51 766	76 226	75 173
Australasia	16 980	15 145	18 789
The Americas	67 446	51 286	49 096
West Africa	27 100	49 500	27 995
East Africa and the Middle East	5508	6661	10 024
Total area	168 810	198 818	181 077

[a]From IUCN (1983).
[b]From Fisher and Spalding (1993).
[c]From Spalding et al. (1997).

Phillips 1992). Mangrove areas also have been consumed by urban sprawl of many coastal cities, for making salt ponds (Paw and Chua 1991), and for aquaculture ponds (Primavera 1993, 1995; Lahmann et al. 1987).

The use of mangrove areas for aquaculture ponds—especially shrimp ponds—has created a great controversy between environmentalists and aquaculturists. There is no question that aquaculture has been responsible for conversion of some mangrove area to ponds (Boyd 2002), but as Hambrey (1996) pointed out, the greatest loss of mangrove area occurred before the shrimp farming boom in the 1980s and 1990s. It is informative to consider data on historic mangrove loss in Africa and South and Southeast Asia compiled by the World Resources Institute (WRI 1996). Individual nations in the two regions have lost from 0% (Congo) to 85% (India) of their original mangrove areas (Table 5.8). The unweighted average was 61% loss in South and Southeast Asia and 55% in Africa. There is little coastal aquaculture in Africa while South and Southeast Asia host the majority of the world's coastal fish and shrimp

Table 5.8 Estimated loss of original mangrove area in different regions (Based on country).

Region	Country	% Loss	Region	Country	% Loss
South and SE Asia	Bangladesh	73	Africa	Angola	50
	Brunei	17		Congo	00
	India	85		Djibouti	70
	Malaysia	32		Equatorial Guinea	60
	Myanmar	58		Gabon	50
	Pakistan	78		Guinea	60
	Singapore	76		Guinea-Bissau	70
	Thailand	87		Kenya	70
	Vietnam	62		Madagascar	40
	Unweighted average	61		Mozambique	60
				Somalia	70
				Tanzania	60
				Zaire	50
				Unweighted average	55

Source: Adopted from Boyd (2002).

farming. This observation certainly does not support the notion that aquaculture is responsible for the majority of mangrove area decline.

A survey of former land use at shrimp farms in Thailand revealed that 13.7% of farms were located in former mangrove areas (Mcintosh and Phillips 1992). Another survey in Thailand (Menasveta 1997) reported that 17.5% of shrimp farms were on former mangrove land. This study indicated that other reasons for mangrove decline in Thailand were clearing for production of charcoal for cooking, agriculture, salt farms, mining, and infrastructure development. Conversion of mangrove to aquaculture ponds was a major factor in reducing Philippine mangrove reserves from 448 000 ha in 1968 to 110 000 ha in 1988, but most of the conversion was to milkfish ponds rather than shrimp ponds (Chua 1992; Boyd 2002). A study by the Asian Development Bank and Network of Aquaculture Centres in Asia-Pacific (ADB/NACA 1997) included 5000 aquaculture farms in 12 countries. This effort showed that 41.9% of extensive farms were built in former mangrove areas, and only 18.6% of semi-intensive and 19.0% of intensive farms were in such areas.

The development of Global Information Systems (GIS) technology has allowed more accurate assessments of changes in land use patterns than previously possible. Béland et al. (2006) used GIS to study the conversion of mangrove to shrimp farms in the Giao Thuy district of Vietnam. Results suggest that 63% of mangrove area apparent in 1986 had been replaced by shrimp farms in 2001. Although 440 ha of mangrove trees had disappeared, the mangrove area increased by 441 ha between 1992 and 2001 because of mangrove reforestation efforts in the vicinity of the shrimp farms.

Another GIS study was conducted in Tra Vinh province of Vietnam (Thu and Populus 2007). The total area of mangrove was 21 221 ha in 1965 and 12 797 ha in 2001 in spite of 5784 ha of mangrove reforestation. Most of the loss in mangrove occurred between 1995 and 2001 and was attributed mainly to shrimp farming.

In Ecuador, the mangrove area was estimated at 204 000 ha in 1969 before shrimp farming began. By 1992, there were 120 000 ha of shrimp farms in this country, but 162 000 ha of mangrove area still were present (CLIRSEN 1997). If the loss of 42 000 ha is attributed solely to shrimp farming, 35% of farms would have resulted from mangrove clearing. The study, however, estimated that only 15–20% of the mangrove loss—6300 to 8400 ha—was caused by shrimp farm construction.

An investigation in four Mexican states—Nayarit, Sinaloa, Campeeche, and Chiapas—containing large mangrove areas revealed that mangrove had been replaced by shrimp farms, but concluded that the extent of the damage has not been determined (Páez-Osuna et al. 2003). A study of land cover changes between 1984 and 1999 over a 442 779 ha area in the Ceuta coastal lagoon system in Sinaloa State, Mexico (Alonso-Pérez et al. 2003) revealed a decline in mangrove area from 7583 to 7217 ha (a loss of 366 ha) and an increase in shrimp ponds from 0 to 3192 ha. However, none of the shrimp ponds were built in mangrove areas. It is interesting that the dry forest area declined by 0.32%/year (from 169 051 to 159 268 ha)—not much less than the rate of mangrove loss of 0.39%/year.

Shrimp farming and other forms of coastal pond aquaculture are not the major cause of mangrove destruction worldwide or even in major shrimp-producing countries. Rossenberry (1991) reported that about 10 000 km^2 of coastal lowland had been converted to shrimp ponds. Assuming these lowlands were all covered by mangrove forest, it would only equal about 5.9% of the 168 810 km^2 of mangrove estimated to be present in 1983 (IUCN 1983). Boyd and Clay (1998) estimated that less than 10% of the global mangrove resources were converted to shrimp farms. Of course there are, no doubt, many instances where aquaculture has been a major contributor to local or country-level mangrove loss. The most notable conversions of mangrove to aquaculture possibly occurred in the Philippines and Indonesia (Chua 1992).

The rate of mangrove loss to aquaculture has been greatly reduced since the 1990s. Governments have passed regulations to protect mangrove, mangrove reforestation programs have been initiated, and aquaculture certification programs have standards for mangrove protection. However, the main reason that aquaculture no longer significantly infringes on mangrove forests is that aquaculturists learned that these areas were not good sites for aquaculture ponds. The soils are usually acidic and high in organic matter content; construction is difficult because the areas are low and poorly drained. Because of poor drainage, pond bottoms cannot be dried adequately between crops making it difficult to implement biosecurity procedures or oxidize bottom soil (Boyd 2002).

Reclamation of abandoned ponds

Another land use concern about pond aquaculture is whether or not land of abandoned ponds in mangrove areas or in agricultural areas can be returned to its former use. If the dams of abandoned ponds in mangrove areas are broken to restore the former connection to coastal water, mangroves will regrow naturally or the process can be accelerated by transplanting mangrove plants from a nursery.

A study near a single village in Southwestern Bangladesh (Ali 2006) revealed that 274 ha (79%) of the village's prime quality rice fields were converted to shrimp farms between 1985 and 2003. This conversion resulted in degradation of the soils for rice farming. In former coastal riceland areas, the salinity of the soils in the bottoms of abandoned ponds will decline in response to leaching by rainfall, and application of calcium sulfate will accelerate sodium removal and improve soil texture (Boyd 1995).

In inland, nonsaline agricultural areas ponds can be drained and immediately used for agriculture. When the price of corn increased in the United States in response to government subsidies to encourage corn production as a raw material for ethanol production for fuel, some channel catfish farmers in Mississippi drained their ponds and planted the bottoms with corn as shown in Fig. 5.2.

Farmers are very clever worldwide in taking advantages of government subsidies. Governments would be wise to carefully analyze the potential benefits of subsidies—something that the US government did not do with respect to corn subsidies to encourage ethanol production as an alternative fuel. Of course politics often play a bigger role than scientific reasoning or economic practicality when it comes to subsidies that can influence votes.

Figure 5.2 Left: Bottom of former Mississippi channel catfish farm prepared for planting corn. Right: corn crop in former channel catfish pond. Courtesy of Travis Brown and David Cline.

Waste disposal on land

Deep areas of ponds tend to fill in as a result of erosion of pond earthwork by waves, rain, and aerator-generated water currents (Boyd 1995). Sedimentation is particularly heavy in ponds filled by turbid water from estuaries or rivers. Farmers may remove sediment from ponds, water supply reservoirs, and canals (Boyd 1995, 2000). Disposal of this sediment may cause ecological nuisances on the land, and erosion of sediment disposed on land may cause turbidity and sedimentation in nearby surface waters. The disposal of sediment from ponds supplied with brackishwater or seawater can be particularly troublesome; this sediment has a salt burden that will leach out and potentially cause groundwater or surface water salinization (Boyd et al. 1994).

Sediment from ponds can be used as landfill or earthwork but it usually is mainly silt and sand and not ideal for repairing the embankments of ponds. In China, sediment from freshwater ponds often is applied to land as fertilizer.

During construction of ponds earthfill may be sourced from nearby areas and transported to construction sites. The creation of barrow pits or denuded areas with a high erosion potential is not uncommon. Of course these unsightly, highly erodible, degraded areas can be avoided by reclamation of the earthfill borrow area as part of the construction effort.

Aquaculture farms also generate wastes, and ecological nuisances may be created nearby by improper trash disposal. Recycling of wastes, incineration, and landfills can be used to properly dispose of wastes at aquaculture facilities.

Conclusions

The world's continental area is covered with inland water (3.5%) or forests and woodland (31%), devoted to agriculture (37.8%), or consumed by urban sprawl (2–3%). The remaining portion of the earth's surface (about 25%) is devoted to other uses or not suitable for forests, woodland, and agriculture. The main changes in land use involve deforestation for agriculture, firewood, and construction materials. The increasing urban area usually is on former agricultural land.

Land for aquaculture—for facilities and feed ingredients—is a truly small fraction of the global land surface (about 0.17%). The land for aquaculture has come primarily from former agricultural land but some has resulted from clearing mangroves and other coastal or freshwater wetlands. In the future, land for the expansion of aquaculture likely will come from agricultural land but the intensity of agricultural production is increasing, and this will lessen the amount of land required for expansion. Of course the use of feeds for aquaculture will continue to increase and the extra land for feed ingredient production may partially offset the land savings resulting from intensification. At present, feed use by aquaculture is about 3.6% of total animal feed use. Thus even if aquaculture production doubles or triples in the future, it still will have a comparatively small requirement in comparison to terrestrial animal agriculture for land to produce feed ingredients.

Although aquaculture is not a major land use and does not appreciably influence the world's landscape, it can have significant effects on local land use patterns and cause ecological perturbations. Thus efforts should be made to integrate aquaculture into local settings in a manner that minimizes conflicts over land use and reduces negative environmental impacts and changes in the landscape.

The eNGO perspective

Although a portion of the eNGO community is concerned with amount of land used to produce agricultural feed ingredients, the primary focus is on the reduction of wildlife habitat through land conversion specifically for aquaculture production. As stated in this chapter, the conversion of coastal mangrove wetlands to shrimp aquaculture ponds has been a widely criticized practice. The eNGO community has concentrated much of their efforts on publicizing this practice both during the time that the clearing of mangroves was commonplace in shrimp aquaculture and to a lesser extent, presently. The conversion of mangrove forests has fueled a precautionary approach toward acceptability of many types of aquaculture. The fundamental principle that is of concern to eNGOs is the threat of biodiversity loss through habitat change. As will be discussed in Chapter 11, gradual trends in biodiversity at the macro-environmental scale are difficult to identify and quantify. Consequently attribution of biodiversity impacts to aquaculture operations is challenging. It should be noted that in protected areas around the world where no land conversion is permitted, the cause of biodiversity change can be similarly difficult to identify. Thus, eNGOs attempt to bring a greater awareness to the environmental perturbations that can occur through aquaculture development and habitat loss. The effects of habitat loss most commonly identified by conservationists are the reduction of mating and breeding areas, reduction in nursery grounds for juvenile organisms, elimination of migration routes, water diversion and removal of vegetation that sequesters carbon dioxide from the atmosphere. These effects are not limited to aquatic ecosystems but also to terrestrial ones.

The eNGO community has found itself in a dilemma with regard to habitat loss caused by aquaculture in that more land conversion is likely to result from the expansion of extensive aquaculture than from a heightened intensity of production. Although more efficient in production per unit area of land, industrialized aquaculture requires greater resource use for inputs such as feed, energy, and potentially water use. Much of the eNGO community, particularly in Europe, has embraced the organic agriculture movement. There is a challenge with transposing organic agriculture methodologies onto aquaculture in general because organic practices tend toward lower inputs and mimicry of the natural environment.

The most natural of environments in aquaculture are cage operations in public water bodies which are perceived as unacceptable to most eNGOs as will be discussed in later chapters. Thus pond aquaculture systems that are closed from discharge or draining and recirculating aquaculture systems are the preferred or more tolerable types of aquaculture. The latter requires a higher level of sophistication and is a smaller proportion of the share of aquaculture production, particularly in less developed regions of the world. Pond aquaculture at an extensive scale will require more land conversion (including conversion of agricultural land) than any other type of aquaculture. With this perspective, the widespread adoption of organic or low intensity pond aquaculture could have dramatic changes on habitat availability. For example, the United States imported about 500 million kg of shrimp in 2010. To produce this much shrimp by certified organic methods would likely require about 3 125 000 ha as compared to about 52 000 ha of intensive aquaculture ponds producing two crops per year (Boyd and McNevin 2012).

With the understanding that aquaculture is a major livelihood generator in many parts of the world and a very large international business, it becomes apparent that the eNGO community likely will need to determine what the priorities are to maximize conservation outcomes in the form of habitat protection. This prioritization will need to be in the form of some level of tolerance for high input and low land conversion for industrialized aquaculture, or low input and high land conversion for extensive and organic production. A common argument on the adherence to extensive and organic production is that it is a sustainable endeavor if consumption habits are lowered. Although this is an understandable perspective, one must consider the increasing trends in per capita consumption of seafood around the world and determine if these rates can be reversed. Moreover, the time necessary to initiate and facilitate this consumption shift likely will play a large role in the rate of habitat loss attributed to aquaculture for the foreseeable future.

References

ADB/NACA (Asian Development Bank and Network of Aquaculture Centers in Asia-Pacific). 1997. *Final Report on the Regional Study and Workshop on Aquaculture Sustainability and the Environment (RETA 5534)*. Bangkok, Thailand: NACA.

Ali, A. M. S. 2006. Rice to shrimp: land use/land cover changes and soil degradation in Southwestern Bangladesh. *Land Use Policy* 23:421–435.

Ali, H. and M. M. Haque. 2011. Impacts of *Pangasius* aquaculture on land use patterns in Mymensingh district of Bangladesh. *Journal of Bangladesh Agricultural University* 9:169–178.

Alltech. 2013. *Global Feed Summary 2012*. Nicholasville: Alltech.

Alonso-Pérez, F., A. Ruiz-Luna, J. Turner, C. A. Berlanga-Robles, and G. Mitchelson-Jacob. 2003. Land cover changes and impact of shrimp aquaculture on the landscape in the Ceuta coastal lagoon system, Sinaloa, Mexico. *Ocean and Coastal Management* 46:583–600.

Béland, M., K. Goïta, F. Bonn, and T. T. H. Pham. 2006. Assessment of land-cover changes related to shrimp aquaculture using remote sensing data: a case study in the Giao Thuy District, Vietnam. *International Journal of Remote Sensing* 27:1491–1510.

Berg, H., P. Michelson, C. Folke, N. Kautsky, and M. Troell. 1996. Managing aquaculture for sustainability in tropical Lake Kariba, Zimbabwe. *Ecological Economics* 18:141–159.

Beveridge, M. C. M. 1996. *Cage Culture*, 2nd ed. Oxford: Fishing News Books.

Boyd, C. E. 1995. *Bottom Soils, Sediment, and Pond Aquaculture*. New York: Chapman and Hall.

Boyd, C. E. 2002. Mangroves and coastal aquaculture. In R. R. Stickney and J. P. McVey, editors, *Responsible Marine Aquaculture*, pp. 145–157. Oxon: CABI Publishing.

Boyd, C. E. and J. Clay. 1998. Shrimp aquaculture and the environment. *Scientific American* 278:42–49.

Boyd, C. E. and A. A. McNevin. 2012. An early assessment of the effectiveness of aquaculture certification and standards. Report to RESOLVE, Washington, DC.

Boyd, C. E. and M. Polioudakis. 2006. Land use for aquaculture production. *Global Aquaculture Advocate* 9:64–65.

Boyd, C. E., P. Munsiri, and B. F. Hajek. 1994. Composition of sediment from intensive shrimp ponds in Thailand. *World Aquaculture* 25:53–55.

Boyd, C. E., C. S. Tucker, A. McNevin, K. Bostick, and J. Clay. 2007. Indicators of resource use efficiency and environmental performance in fish and crustacean aquaculture. *Reviews in Fisheries Science* 15:327–360.

Boyd, C. E., J. Queiroz, J. Lee, M. Rowan, G. N. Whitis, and A. Gross. 2000. Environmental assessment of channel catfish, *Ictalurus punctatus*, farming in Alabama. *Journal of the World Aquaculture Society* 31:511–544.

Brummett, R. E. 1999. Integrated aquaculture in sub-Saharan Africa. *Environment, Development, and Sustainability* 1:315–321.

Brune, D. E., G. Schwartz, A. G. Eversole, J. A. Collier, and T. E. Schwedler. 2003. Intensification of pond aquaculture and high rate photosynthetic systems. *Aquacultural Engineering* 28:65–86.

Chua, T. E. 1992. Coastal aquaculture development and the environment, the role of coastal area management. *Marine Pollution Bulletin* 25:98–103.

CLIRSEN (Centro de Levantamiento Integrado por Recursos Sensores). 1997. Estudio multitemporal de camaroneras, manglares y salinas, 1995. *Aquacultura del Ecuador* 18:27–35.

Dahl, T. E. 2011. *Status and Trends of Wetlands in the Conterminous United States 2004 to 2009*. Washington, DC: US Department of the Interior, US Fish and Wildlife Service.

Dahl, T. E. and G. J. Allord. 1996. History of wetlands in the conterminous United States. In J. D. Fretwell, J. S. Williams, and P. J. Redman, compilers, *National Water Summary on Wetland Resources*, pp. 19–26. Washington, DC: US Geological Survey, Paper 2425.

Dahl, T. E. and C. E. Johnson. 1991. *Status and Trends of Wetlands in the Conterminous United States, mid-1970s to mid-1980s*. Washington, DC: US Department of the Interior, US Fish and Wildlife Service.

Engle, C. R. and D. Valdarrama. 2002. The economics of environmental impacts in the United States. In J. R. Tomasso, editor, *Aquaculture and the Environment in the United States*, pp. 240–270. Baton Rouge: US Aquaculture Association/World Aquaculture Society.

FAO (Food and Agriculture Organization). 2012. *The State of World Fisheries and Aquaculture*. Rome: FAO Fisheries and Aquaculture Department.

Fisher, P. and M. D. Spalding. 1993. Protected areas with mangrove habitat. Draft Report, World Conservation Monitoring Centre, Cambridge, UK.

Folke, C., N. Kautsky, H. Berg, A. Jansson, and M. Troell. 1998. The ecological footprint concept for sustainable seafood production: a review. *Ecological Applications* 8:63–71.

Hajek, B. F. and C. E. Boyd. 1994. Rating soil and water information for aquaculture. *Aquacultural Engineering* 13:115–128.

Hambrey, J. 1996. Comparative economics of land use options in mangrove. *Aquaculture Asia* 1:10–14.

IUCN (International Union for Conservation of Nature). 1983. In P. Saenger, E. J. Hegerl, and J. D. S. Davie, editors, *Global Status of Mangrove Ecosystems*, Gland: International Union for Conservation of Nature and Natural Resources.

Jothy, A. A. 1984. Capture fisheries and the mangrove ecosystem. In J. E. Ong and W. K. Gong, editors, *Productivity of the Mangrove Ecosystem: Management Implications*, pp. 129–141. Penang: Universiti Sains Malaysia.

Kautsky, N., H. Berg, C. Folke, J. Larsson, and M. Troell. 1997. Ecological footprint for assessment of resource use and development limitations in shrimp and tilapia aquaculture. *Aquaculture Research* 28:753–766.

Keenum, M. E. and J. A. Waldrop. 1988. Economic analysis of farm-raised catfish production in Mississippi. Mississippi Agricultural and Forestry Experiment Station Technical Bulletin 155, Mississippi State University, Mississippi.

Lahmann, E. J., S. M. Snedaker, and M. S. Brown. 1987. Structural comparisons of mangrove forest near shrimp ponds in southern Ecuador. *Interciencia* 12:240–243.

Lubowski, R. N., M. Vesterby, S. Bucholtz, A. Baez, and M. J. Roberts. 2006. Major land used in the United States, 2002. Economic Information Bulletin No. (EIB-14), United States Department of Agriculture, Washington, DC.

Macintosh, D. J. and M. J. Phillips. 1992. Environmental issues in shrimp farming. In H. de Saram and T. Singh, editors, *Shrimp '92, Proceedings of the 3rd Global Conference on the Shrimp Industry*. Kuala Lumpur: INFOFISH.

Mangrove Working Group. 1998. Coastal shrimp aquaculture and mangrove forests. Part 1: a background report. Global Aquaculture Alliance, St. Louis, Missouri.

Massaut, L. 1999. Mangrove management and shrimp aquaculture. Department of Fisheries and Allied Aquacultures, and International Center for Aquaculture and Aquatic Environments, Auburn University, Alabama.

Matthews, E. 1983. Global vegetation and land use: new high-resolution data bases for climate studies. *Journal of Climate and Applied Meteorology* 22:474–487.

Menasveta, P. 1997. Intensive and efficient shrimp culture system—the Thai way—and save mangroves. *Aquaculture Asia* 2:38–44.

Mitsch, W. J. and J. G. Gosselink. 1993. *Wetlands*, 2nd edition. New York: Van Nostrand Reinhold.

Ong, J. E. 1982. Mangroves and aquaculture in Malaysia. *Ambio* 11:252–257.

Páez-Osuna, F., A. Gracia, F. Flores-Verdugo, L. P. Lyle-Fritch, R. Alonso-Rodríguez, A. Roque, and A. C. Ruiz-Fernández. 2003. Shrimp aquaculture development and the environment in the Gulf of California ecoregion. *Marine Pollution Bulletin* 46: 806–815.

Paw, J. N. and T. E. Chua. 1991. An assessment of the ecological and economic impact of mangrove conversion in Southeast Asia. In L. M. Chou, T. E. Chua, H. W. Khoo, P. E. Lim, J. N. Paw, G. T. Silvestre, M. J. Valencia, A. T. White, and P. K. Wong, editors, *Towards an Integrated Management of Tropical Coastal Resources*, pp. 201–212. ICLARM Conference Proceedings 22, National University of Singapore, Science Council of Singapore, and International Center for Living Aquatic Resources Management, Philippines.

Primavera, J. H. 1993. A critical review of shrimp pond culture in the Philippines. *Reviews in Fisheries Science* 1:151–201.

Primavera, J. H. 1995. Mangroves and brackishwater pond culture in the Philippines. *Hydrobiologia* 295:303–309.

Rossenberry, B. 1991. *World Shrimp Farming 1990*. San Diego: Aquaculture Digest.

Schmittou, H. R. 1993. *High density fish culture in low volume cages*. Singapore: American Soybean Association.

Schnepf, R. 2011. U.S. Livestock and poultry feed use and availability: Background and emerging issues. CRS Report for Congress, Congressional Research Service, Washington, DC.

Spalding, M., F. Blasco, and C. Field. 1997. *World Mangrove Atlas*. Okinawa: International Society for Mangrove Ecosystems.

Thu, P. M. and J. Populus. 2007. Status and changes of mangrove forest in Mekong Delta: case study in Tra Vinh, Vietnam. *Estuarine, Coastal and Shelf Science* 71:98–109.

Timmons, M. B., J. M. Ebeling, F. W. Wheaton, S. T. Summerfelt, and B. J. Vinci. 2001. *Recirculating Aquaculture Systems*. Ithaca: Cayuga Aquaculture Ventures.

Verdegem, M. C. J. and R. H. Bosma. 2009. Water withdrawal for brackish and inland aquaculture, and options to produce more fish in ponds with present water use. *Water Policy* 11:52–68.

Williams, M. 2000. Dark ages and dark areas: global deforestation in the deep past. *Journal of Historical Geology* 26:28–46.

Williams, M. 2006. The role of deforestation in earth and world-system integration. In A. Homborg, J. R. McNeill, and J. M. Alier, editors, *Rethinking Environmental History: World-system History and Global Environmental Change*, pp. 101–122. Lanham: Alta Mira Press.

WRI (World Resources Institute). 1996. A guide to the global environment. Database diskette and users guide. Washington, World Resources Institute.

Yoo, K. H. and C. E. Boyd. 1994. *Hydrology and Water Supply for Aquaculture*. New York: Chapman and Hall.

Chapter 6

Water use by aquaculture systems

Water is a major component and physiological necessity of all living things; in addition, aquatic species live in water. An adequate volume of high quality water is necessary to initially fill holding and culture units at aquaculture facilities, and to replace water that evaporates, seeps, or is intentionally discharged from them. Aquaculture is more water intensive than most kinds of terrestrial agriculture. Nevertheless total water use in aquaculture is less than consumptive use because much of the water entering culture units is ultimately discharged downstream.

This chapter provides a brief overview of global water availability and use, discusses the use of water in aquaculture, and contrasts aquaculture water use with other uses of water.

Availability of freshwater

Although humans use brackishwater and ocean water for many purposes, saline water is not considered in most assessments of water use because it is not potable, not useful for crop irrigation, and many other applications, and it is inexhaustible. Humans primarily use water from streams, natural lakes, man-made reservoirs, and underground aquifers, but in a few countries, desalination of seawater is a major source of freshwater.

The water cycle or hydrologic cycle (Fig. 6.1) controls the amounts of water in the earth's crust, on its surface, and in the atmosphere. Some of the precipitation on land masses falls directly onto the surface of inland water bodies and some strikes bare soil and rocks. The rest is intercepted by above-ground objects such as vegetation or buildings, but when the interception capacity of these objects is filled, water drips off or flows to the ground. Water infiltrates the soil and when the rate of delivery of water into the ground exceeds the infiltration rate, water accumulates on the land surface. Once depressions on the surface are full, water flows downslope and

Aquaculture, Resource Use, and the Environment, First Edition. Claude E. Boyd and Aaron A. McNevin.
© 2015 John Wiley & Sons, Inc. Published 2015 by John Wiley & Sons, Inc.

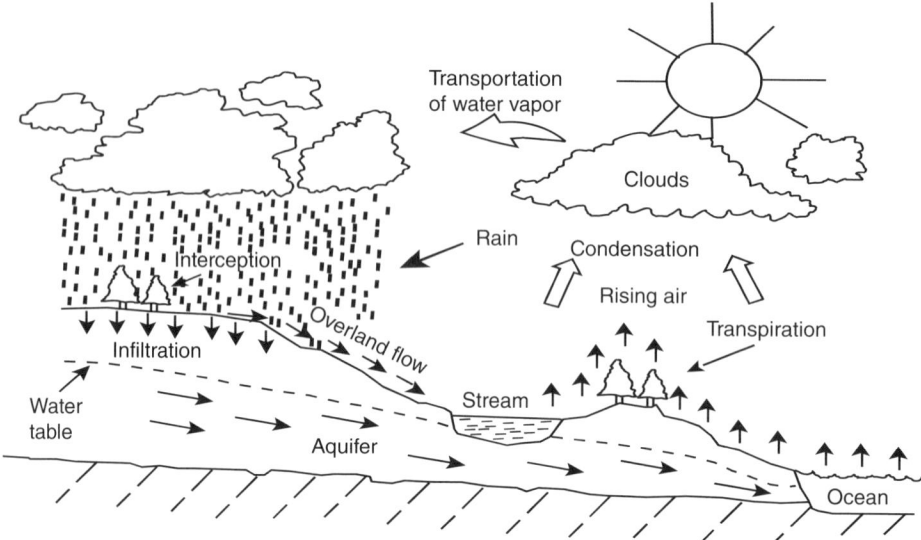

Figure 6.1 The hydrologic cycle or water cycle.

enters temporary or permanent streams. Some of the water infiltrating the soil is retained as soil moisture in the pore spaces but the rest percolates downward until it reaches an impermeable layer. Water that accumulates in voids within geological formations above impermeable layers is called groundwater, and these saturated zones are referred to as aquifers. The top of a saturated layer is known as the water table.

Water evaporates from moist surfaces, and soil moisture is transpired into the air by plants—the two processes usually are referred to collectively as evapotranspiration. Stream flow consists of water that runs over the land surface (overland flow) and groundwater that seeps into streams where their bottoms cut below the top of the water table—groundwater intrusion into streams is called base flow. Stream flow represents the total runoff from the land and most stream flow eventually enters the ocean.

The distribution of the earth's water is provided in Table 6.1; most of the water is ocean water, and most of the freshwater is frozen. The earth always has the same total amount of water, and because of hydrologic cycling, the volume of water contained in glaciers, icebergs, polar ice, streams, lakes, reservoirs, soil, and groundwater aquifers does not change much from one year to the next. This signifies that water storage on and in the land masses is more or less at equilibrium with precipitation and evapotranspiration. Thus, the amount of stream flow should be equal to the amount of precipitation falling on the land minus the amount of evapotranspiration from the land (Baumgartner and Reichel 1975).

Removing water from aquifers with wells will lower the water table and lessen base flow of groundwater into streams. Likewise removing water from lakes will decrease lake overflow and diminish downstream flow. The water withdrawn for human use is not lost from the hydrologic cycle; it is either discharged back into water bodies or evaporates. Of course water may be polluted during use and degrade receiving water bodies. Evaporation, however, is a water purification process.

Table 6.1 Volumes of ocean water and different compartments of freshwater. Renewal times are provided for selected compartments.

Compartment	Volume (km³)	Proportion (%)	Renewal time
Oceans	1 348 000 000	97.40	37 000 years
Freshwater			
Polar ice, icebergs, and glaciers	27 818 000	2.01	16 000 years
Groundwater (800–4000 m depth)	4 447 000	0.32	–
Groundwater (to 800 m depth)	3 551 000	0.26	300 years
Lakes	126 000	0.009	1–100 years
Soil moisture	61 100	0.004	280 days
Atmosphere (water vapor)	14 400	0.001	9 days
Rivers	1070	0.00008	12–20 days
Plants, animals, humans	1070	0.00008	–
Hydrated minerals	360	0.00002	–
Total freshwater	36 020 000	2.60	–

Source: Modified from Baumgartner and Reichel (1975) and Wetzel (2001).

Moreover, groundwater becoming base flow or removed by wells is replaced by infiltration of rainfall into aquifers and the water table remains relatively constant from year to year unless water removal by wells exceeds natural recharge by infiltration. Therefore, an estimation of global stream flow by taking the difference in precipitation and evapotranspiration on the world's land mass is probably the most reliable way of estimating the amount of annually available and renewable freshwater for human use (Fig. 6.2).

Baumgartner and Reichel (1975) estimated precipitation on the earth's land mass as 111 100 km³/year, evapotranspiration as 71 400 km³/year, and stream flow (total runoff) as 39 700 km³/year. The distribution of runoff by continent is provided in Table 6.2 along with the total runoff available per capita. Notice that the sum of the

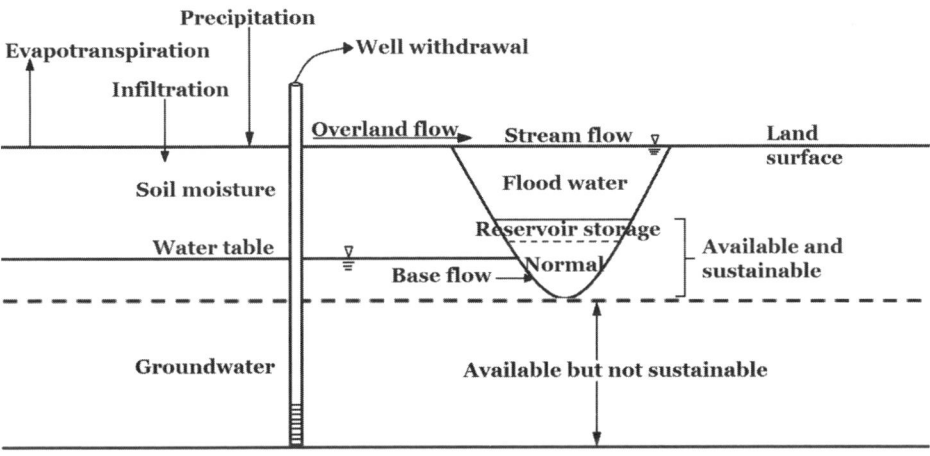

Stream flow = Precipitation - Evapotranspiration = Overland flow + Base flow
Sustainable freshwater = Base flow + Normal flow + Reservoir storage

Figure 6.2 Illustration of annual, renewable, and available freshwater.

Table 6.2 Total runoff by continent.

Continent	Total runoff	
	(km³/year)	(m³/cap/year)
Oceania	911	35 869
Africa	3950	4980
Europe	6874	9444
North America	7496	15 541
South America	12 380	35 808
Asia	12 500	3408

Source: Adapted from http://faostat.fao.org.

runoff for the continents is 44 111 km³/year—slightly more runoff than indicated by Baumgartner and Reichel (1975). The data in Table 6.2 for annually renewable freshwater by continent were derived by summing estimates of annual renewable freshwater for each country in a continent and included groundwater. This procedure probably resulted in the overestimation where water flowed from one country into another or possibly too much groundwater was considered annually renewable. Of course, the other possibility is that Baumgartner and Reichel underestimated precipitation or overestimated evapotranspiration. The point is that estimates of annually renewable freshwater are only approximate.

Flood stage flow—about 75% of total runoff—is not available for human use unless it is impounded behind dams on rivers. Moreover, only about 70% of the available water is geographically accessible (Postel 1996). Dams constructed on major rivers throughout the world have a volume of 6700 km³ (White 2010), all of which is considered accessible. The global volume of accessible, renewable freshwater is estimated to be 13 512 km³/year (Table 6.3); this estimate agrees well with an earlier estimate of 12 500 km³/year (Postel 1996). Of course, in some regions, people capture and store rainwater from roofs and other structures in cisterns for domestic use, and this water is not included in the estimate of accessible, renewable freshwater.

Water footprint

The water footprint is a measure of the volume of water used by an individual, a family, a business or industry, a nation, or the world. The water in water footprints

Table 6.3 Global supply of accessible, renewable, freshwater.

Component of freshwater	Volume (km³/year)
Total runoff	39 700
Flood stage flow (75% of total)	29 250
Non-flood stage flow (assume 100% available)	10 450
Accessible (70% of available)	7315
Reservoir storage (assume 100% available)	6197
Accessible, renewable supply	13 512

may be separated into three types (http://www.waterfootprint.org/?pages=files/NationalWaterAccountingFramework). The volume of rainwater evaporated during the production process is referred to as green water in water footprints, and this term applies almost exclusively to agriculture. Surface water and groundwater that evaporated as a result of its use by humans is called blue water, while the term gray water is reserved for the volume of water used to dilute anthropogenic pollution. Gray water is of lower quality than normal freshwater and it is impaired for some purposes. In addition, the water footprint for countries may be separated into water used within the country (internal water footprint) and water used in other countries to produce imported goods and services (external water footprint).

The concept of virtual water or embedded water also is applied in water footprints. Virtual water is an economic concept referring to the amount of water used in making a product or providing a service but not included in the product or service (Hoekstra 2008). Importation of products involves importing virtual water, but real water was used in the exporting country to make the product. Exportation of products has the opposite effect. There is a huge international trade in virtual water—particularly with respect to food and other products involving agricultural raw materials. The virtual water contents of some common products are provided in Table 6.4. A 150-g hamburger has an embedded water content of 2400 L—most of this water was used in growing the grain or grass needed to produce the beef and is green water.

Some examples of country-level water footprints are provided (Table 6.5). The total water footprint for the world averages 1243 m^3/cap/year, but for the selected countries in Table 6.4—seven developed and eight developing—the total water footprint varied from 702 m^3/cap/year in China to 2483 m^3/cap/year in the United States. Developed countries in Table 6.5 had an average water footprint of 1833 m^3/cap/year as compared to 1242 m^3/cap/year for developing countries. The domestic water footprint of developed countries was more than threefold larger than for developing countries, and the industrial water footprint was almost eightfold greater for the developed countries. The agricultural water footprint was about the same for developed and developing countries; however, in developing countries, only about 10% of

Table 6.4 Virtual water content of selected products.

Animal products	Virtual water (L)	Plant products	Virtual water (L)
Chicken, 1 kg	3900	Chocolate, 1 kg	24 000
Pork, 1 kg	4800	Rice, 1 kg	3400
Beef, 1 kg	15 500	Sugar, 1 kg	1500
Cheese, 1 kg	5000	Wheat bread, 1 kg	1300
Pair of shoes of cow leather	800	Corn, 1 kg	900
Milk, 250 mL	250	Banana, 1 kg	860
		Potato, 1 kg	250
		Cotton tee shirt (one)	2000
		Coffee, 125 mL	140
		Wine, 125 mL	120
		Tea, 125 mL	15

Source: Adapted from Hoekstra (2008).

Table 6.5 Water footprints and their components for selected countries and the global average water footprint for the period 1997–2001. External footprints in parentheses.

Country	Total	Domestic	Agricultural		Industrial	
Developing countries						
China	702	26	605	(40)	71	(6)
Bangladesh	897	16	875	(29)	6	(3)
India	980	38	921	(14)	21	(2)
Egypt	1096	66	919	(197)	111	(10)
Pakistan	1217	21	1182	(63)	14	(2)
Brazil	1381	70	1242	(87)	69	(18)
Mexico	1440	139	1198	(361)	103	(72)
Thailand	2222	30	2131	(144)	61	(41)
Average	**1242**	**51**	**1134**	**(117)**	**57**	**(19)**
Developed countries						
Japan	1152	136	779	(614)	237	(129)
Australia	1393	341	777	(41)	275	(211)
Germany	1545	66	1038	(604)	441	(213)
France	1875	105	1331	(517)	439	(182)
Canada	2049	279	1238	(252)	532	(166)
Italy	2333	138	1868	(1,039)	327	(151)
United States	2482	217	1459	(267)	806	(197)
Average	**1833**	**183**	**1213**	**(476)**	**437**	**(178)**
Global average	**1243**	**57**	**1067**	**(160)**	**119**	**(40)**

Table header spanning: **Water footprints (m³/cap/year)**

Source: Adapted from Hoekstra and Chapagain (2007).

the agricultural water footprint was external, while in developed countries, almost 40% was external.

The average water footprint for developing countries (Table 6.5) was 4.1% domestic, 91.3% agricultural, and 4.6% industrial, while in developed countries, it was 10.0% domestic, 66.2% agricultural, and 23.8% industrial. The world averages for major components of the water footprint resemble those of developing countries more than those of developed countries.

In 2005, the total global water footprint was estimated at 9086 km³/year, of which 8362 km³/year (92%) was attributed to agriculture (Table 6.6). The green water

Table 6.6 The global water footprint for 2005.

Type of water	Agricultural	Industrial	Domestic	Total
Green	6684	–	–	**6684**
Blue	945	38	42	**1025**
Gray	733	362	282	**1377**
Total	8362	400	324	**9086**

Table header spanning: **Water footprint (km³/year)**

Source: Modified from Hoekstra, A. Y. and M. M. Mekonnen. 2012. The water footprint of humanity. *Proceedings of the National Academy of Sciences* 109:3, 232–3, 237.

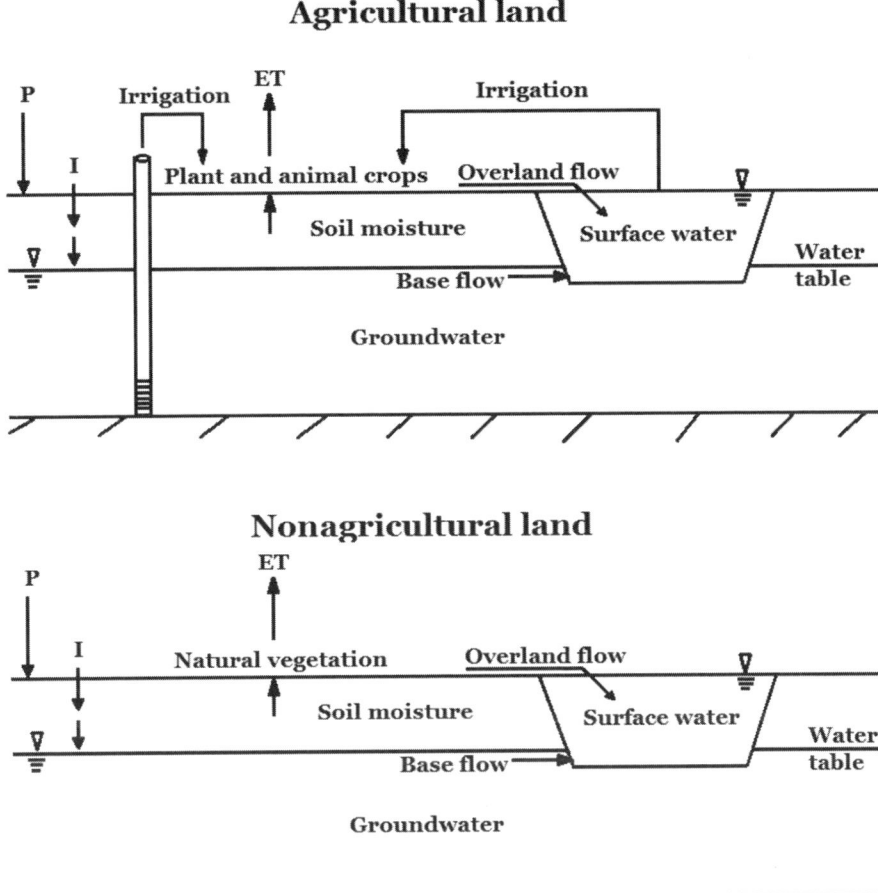

Figure 6.3 The water cycle for agricultural and nonagricultural land.

component of the agricultural water footprint was 6684 km³/year or about 80% of the total (Hoekstra and Mekonnen 2012). In other words, green water inflates agricultural water use by fivefold. There is a reason to argue for exclusion of green water from agricultural water use calculations. In Fig. 6.3, the water cycle for agricultural land and nonagricultural land is depicted to illustrate that the green water would have evaporated from the land anyway had crops not been present. Moreover, precipitation lost from agricultural land by evapotranspiration is not included in the amount of available, renewable freshwater.

If one excludes green water in the global water footprint and projects water use until 2050—assuming no change in water use or availability—there appears to be plenty of water for the future (Table 6.7). Including the rainwater used in crop evapotranspiration as available, renewable freshwater would lead to the same

Table 6.7 Comparison of projected world water footprints and estimates of available, renewable freshwater with green water used in agriculture excluded.[a]

Year	World population (billions)	Global water footprint (km³/year)
2005	6.45	2402
2015	7.7	2868
2025	8.0	2979
2050	9.2	3426

[a] Assuming no change in per capita consumption between 2005 and 2050.

conclusion as does omitting green water. However, there is not plenty of water for the future for various reasons that will be discussed later in this chapter. It should be emphasized that agriculture is the world's major water user whether green water is included or not. Agriculture is responsible for about 77% of global blue and gray water use.

The total amount of global agricultural production was 7670 Mt in 2010 and water use was 8362 km³ (including green water) or 1678 km³ (excluding green water). This equates to 988 m³/t of agricultural production if green water is counted and 215 m³/t if green water is not counted.

Groundwater is a major source of water in many regions—especially where there is not a dependable supply of surface water. However, if groundwater is removed faster than it can be recharged by infiltration and percolation of rainwater, water table levels fall and the supply of groundwater shrinks. One highly publicized instance of the decline of a water table is for the Ogallala Aquifer that occurs over parts of eight states in the High Plain region of the United States. The saturated thickness of this aquifer has been reduced by as much as 50% in some areas mainly as a result of crop irrigation (Buchanan et al. 2009). There are many reports from many other countries about the decline in local or regional groundwater supplies (Changming et al. 2001; Shah et al. 2003; Konikow and Kendy 2005).

Water usage from inland, freshwater lakes also causes their levels to decline when water is removed faster than the natural rate of water renewal. A highly published example is the Aral Sea which has drastically declined in water level as a result of water removal for irrigation (Cai et al. 2003; Aladin et al. 2005).

The world has as much water as it ever did because water is continuously cycled. Moreover, people are not using nearly all of the water that is available in the world. Water scarcity that is common in many countries results from geographical, seasonal, and year-to-year patterns in rainfall relative to population. Arid regions naturally have scant water supplies, and even in well-watered places, there are wet and dry seasons. Droughts may result in much less water than normally available in both arid and humid climates. Add high population densities, regions with inadequate water infrastructure and management, and political instability, and water shortages can be quite common and severe in many parts of the world. For example, in the southern United States, rainfall is typically between 100 and 150 cm/year, but many municipalities that never experienced water shortages during drought years in the past now do so because of greater population and water use (Boyd et al. 2009).

Table 6.8 Water use in countries with different per capita levels of gross domestic production (GDP).

Country	GDP/year/cap in 2010 (US$)	Water footprint (m^3/year/cap)
India	4148	980
South Africa	8929	931
China	10 027	702
Mexico	13 500	1441
France	35 250	1875
United States	46 569	2483

Source: Adapted from https://pwt.sas.upenn.edu/php_site/pwt71/pwt71_form.php.

Water use per capita will likely increase in the future because of the rapid increase in the middle class worldwide. People tend to use more water when they become wealthier (Table 6.8). However, it is difficult to make projections on how much the increase in water use will be. There are many initiatives worldwide to encourage water conservation, and hopefully, as people become wealthier and better educated, they will become more thoughtful and realize the need to conserve water. However, at least in the short term, water use per capita likely will increase in developing countries as the middle class becomes more numerous.

Water conservation is essential in the future because there is little opportunity to increase the supply of water. The construction of large river impoundments is the most practical engineering approach to providing more water. Because of environmental concerns, especially effects of impoundments on downstream flow, it is unlikely that many more large river dams will be built in the future. It should be remembered, however, that conserving water in a water-plentiful region will not offset a water shortage in a water-scarce region.

Water use in aquaculture

Ponds are the major production system for freshwater aquaculture (Verdegem and Bosma 2009). Water use in ponds as in other water bodies may be illustrated with the hydrologic equation (Leopold 1974):

$$\text{Inflows} = \text{Outflows} \pm \text{Change in volume stored (V)}. \tag{6.1}$$

There are several inflows and outflows to a pond (Yoo and Boyd 1994):

Inflows	Outflows
Precipitation on water surface (P)	Evaporation from surface (E)
	Seepage out (S_{out})
Runoff from land surface (R)	Harvest biomass (B)
Seepage in (S_{in})	Overflow (O)
Additions from wells, streams,	Intentional discharge for water exchange
reservoirs, etc. (A)	and draining for harvest (D)

The amount of water removed in biomass is small. Most aquaculture species are about 75% water (Boyd 1990); thus, 1 t of biomass contains about 0.75 m^3 water. This small amount can be ignored in most assessments.

The separate evaluation of seepage in and seepage out usually is problematic, but values for net seepage (S_n) normally can be obtained without great difficulty (Boyd 1982; Yoo and Boyd 1994). In ponds, net seepage usually is a water loss—the main exception is excavated ponds whose bottoms may be below the top of the water table.

A hydrologic equation specifically for ponds can be written as:

$$P + R + A = (E + S_n + O + D) \pm V. \tag{6.2}$$

In pond culture, overflow typically occurs following heavy rainfall events and sometimes water may be drained from ponds and replaced with new water in an effort to improve water quality. Seepage under pond embankments, like overflow, passes downstream, and seepage through pond bottoms enters the groundwater. A pond presents a constant free water surface for evaporation, but without a pond on the site, evapotranspiration from the land in most climates often would have been limited by a lack of moisture. Thus ponds usually have evaporation rates greater than evapotranspiration rates of nearby terrestrial habitats. At harvest time, ponds normally are drained to facilitate capture of the culture animals, and the effluent discharged downstream. Of course, if ponds are not drained for harvest, they remain full or nearly full resulting in overflow in response to natural water inputs. In essence, it does not matter whether or not ponds are drained for harvest with respect to annual discharge volume. The difference is in the timing of the discharge; undrained ponds overflow in response to rainfall patterns, but most of the discharge from annually drained ponds is at harvest time. Ponds on a watershed also influence the hydrograph of downstream flow, because during a storm event, water often enters ponds faster than it flows out. This temporary detention in ponds flattens the peaks of downstream hydrographs (Schoof and Gander 1982). A large number of ponds on a stream catchment also may lessen flood stage height in the stream (Boyd and Chaney 2013; Chaney et al. 2012).

Total water use in flow-through production systems is huge compared to that of ponds. The typical water requirement for trout culture is about 80 000 m^3/t trout (Yoo and Boyd 1994). Water passes through such systems rapidly, and the difference in amounts of inflow and outflow is harvest biomass and evaporation. However, the culture area is exceeding small compared to ponds. A 200-m^2 raceway may produce as many fish as can be produced in a 10 000-m^2 pond—a concentration factor of 50 for production, but with only one-fiftieth as much water surface exposed for evaporation.

Cages for aquaculture usually are installed in natural lakes, reservoirs, or streams. The only water consumed by cage culture is the water removed in harvested biomass.

Total water use in an aquaculture pond or other production system was defined by Boyd (2005) as follows:

$$\text{Total water use} = P + R + S_i + A. \tag{6.3}$$

Table 6.9 Assumptions used in calculating unit water budgets for channel catfish ponds in the United States.

Hydrologic data

P = 145 cm/year onto pond
R = 50 cm/year from watershed

E = 117 cm/year from pond
S_n = 60 cm/year from pond

Pond data

Pond area = 1 ha
Pond depth when filled to intake of overflow structure = 150 cm.
For embankment ponds, watershed area was negligible.
For watershed ponds, watershed:pond area ratio = 10:1.
Ponds did not overflow after rains and there was no inward seepage.
No rainfall or inward seepage occurred while filling.
Water level in embankment ponds was maintained 10 cm below intake of overflow
   (except possibly after rain events); thus, pond depth = 140 cm.
No replacement water applied to watershed ponds.
Ponds were drained when water levels were 10 cm below drains.
Production averaged 6000 kg/ha/year.

Source: Modified from Boyd (2005).

Boyd (2005) also defined consumptive water use as the decrease in surface runoff that the production system imposes upon its watershed, because less runoff equates to less surface water for downstream purposes. In addition, all freshwater withdrawn from aquifers by wells should be included as a consumptive use, because it lowers the water table lessening base flow into nearby streams. It also represents possible competition with other groundwater users in the area. These are areas where water use by aquaculture is suspected of contributing to long-term aquifer drawdown or excessive drawdown during drought (Boyd et al. 2007).

Water that seeps from ponds must be replaced by additions of surface water or groundwater from wells. Water seeping under dams often is absorbed by soil to infiltrate as does water seeping through the bottoms of ponds. Although seepage from ponds may become groundwater, it often enters a shallow aquifer of little direct use to humans. Because of this Boyd (2005) considered it prudent to consider seepage loss as a consumptive use.

Water use calculations will be made for ponds in the Blackland Prairie region of Alabama (USA) where channel catfish farming is a major crop. The assumptions for the calculations are provided in Table 6.9. The calculations are for a 1-ha water surface area unit and the quantities of water are given in centimeters of depth. A depth of 100 cm (1 m) over 1 ha is 1 ha-m or 10 000 m^3.

Embankment ponds

Embankment ponds (Fig. 1.6) typically do not have significant watershed areas. The amount of water to initially fill an embankment pond (A_f), therefore, can be estimated from average depth and area. The quantity of water added to maintain water levels (A_l) is

$$A_l = (E + S_n) - P. \tag{6.4}$$

Using assumptions presented in Table 6.9, total water use can be estimated as:

$$
\begin{aligned}
\text{Total water use} &= A_f + A_i \\
&= A_f + [(E + S_n) - P] \\
&= 140 + [(117 + 60) - 145] \\
&= 172 \text{ cm.}
\end{aligned}
\tag{6.5}
$$

If annual pond discharge is less than the amount of surface runoff that would have occurred from the original area now occupied by the pond water surface, a reduction in downstream flow will occur. In our example, the pond did not overflow because it was an embankment pond with a very small watershed and the pond was operated with the water level below the intake of the overflow structure. At draining for harvest, 140 cm of water were discharged—more than the normal discharge of 50 cm/ha of watershed (Table 6.9). However, there was consumptive water use, because pond discharge comprised groundwater which was added to fill the pond and maintain the water level but did not seep out or evaporate. The consumptive water use was 172 cm (140 cm groundwater used to fill the pond plus 32 cm used to maintain the water level) or 17 200 m^3.

Watershed ponds

Suppose that a 1-ha watershed pond such as the one shown in the left panel of Fig. 1.2 is supplied by overland flow from a 10-ha watershed with no addition of water from other sources. Again we will assume no net seepage into the pond. In this case, the watershed is 10 times the pond surface area and the annual amount of runoff entering the pond will be the runoff depth multiplied by the watershed area (50 cm × 10 ha = 500 cm). Total water use in the pond will be

$$
\begin{aligned}
\text{Total water use} &= P + R \\
&= 145 + 500 = 645 \text{ cm.}
\end{aligned}
\tag{6.6}
$$

Had the pond not been present, the 11 ha area (watershed + pond) would have discharged 550 cm/ha (50 cm runoff × 11 ha). The pond discharged less:

$$
\begin{aligned}
\text{Pond discharge} &= P + R - (E + S_n) \\
&= 145 + 500 - (117 + 60) = 468 \text{ cm/ha.}
\end{aligned}
\tag{6.7}
$$

The reduction in downstream flow or consumptive water use was 82 cm ((550 − 468) cm) or 8200 m^3.

Excavated ponds

These ponds are dug in the ground (Fig. 1.5) and they usually are filled with a combination of rainfall and groundwater seepage. They typically are not drained or supplied by surface or well water. Thus we do not believe that they contribute much to consumptive water use.

Flow-through systems, cages, and net pens

Raceways and other types of flow-through systems were estimated to have consumptive water use values of 30–35 m^3/t (Boyd 2005). Cage and net pen culture consumes even less water—only that contained in the harvested biomass. In a few instances, cages and net pens are placed in ponds rather than lakes or streams, and when this is done, water use would be the same as for normal pond culture.

Water use comparisons

Embankment ponds filled with groundwater and used in multiple-batch culture (harvested without draining ponds) of channel catfish in Alabama would have an average consumptive water use of 2867 m^3/t. Channel catfish production in watershed ponds using rainfall and overland flow as a water source would consume an average of 1367 m^3/t—whether drained annually or not. In arid climates, consumptive water use in ponds may be 50% or more than that of ponds in humid areas such as the southeastern United States (Boyd 1986, 1987). The consumptive water use values for channel catfish production in ponds are considerably greater than the global average for agricultural production of 988 m^3/t (green water included) and 215 m^3/t (green water excluded).

The average consumptive use values for channel catfish would be equivalent to a depth of 172 cm for ponds using groundwater from wells and 82 cm for watershed ponds. Water use for embankment ponds was considerably greater while that of watershed ponds was similar to the amount of water applied for irrigation of crops commonly produced in the southeastern United States (Table 6.10). But the economic value of the crop, however, was considerably greater for channel catfish than for the plant crop species.

According to Verdegem and Bosma (2009), there are 8 750 000 ha of freshwater aquaculture ponds globally that produced 20 965 941 t of finfish and about 877 994 t of crabs, freshwater shrimp, and crayfish in 2004—an average yield of 2496 kg/ha. There are no data on the proportion that each of the three hydrologic types of ponds contributes to the total pond area, but most certainly the great majority consists of embankment and watershed ponds. There should be little difference in consumptive water use in ponds as a result of the species cultured or the intensity of culture. Also water use in Alabama ponds is probably intermediate between that of very high rainfall areas and semiarid and arid regions (Boyd 1986, 1987). Therefore, the average annual water use of embankment and watershed ponds in Alabama of

Table 6.10 Comparison of consumptive water use and water value for irrigation of selected crops and production of channel catfish in the southeastern United States.

Crop	Consumptive water use (cm)	Water value (US$/m³)
Rice	123	0.367
Alfalfa hay	107	0.185
Corn	88	0.254
Cotton	83	0.250
Soybeans	80	0.167
Catfish[a]		
Ponds supplied with groundwater from wells	172	0.808
Watershed ponds supplied by overland flow	82	1.693

[a]Pond bank value of $2.315/kg and production of 6000 kg/ha.

127 cm or 12 700 m³/ha will be taken as the global average. Each tonne of production on average would require 5088 m³ of water (12 700 m³/ha ÷ 2.496 t/ha). Thus, water use in pond aquaculture worldwide would be about 111.1 km³/year [21 843 935 t production/year × (5088 m³ ÷ 10^9 m³/km³)].

Flow-through systems accounted for about 7 214 087 t of production in 2004. Using a consumptive water use estimate of 35 m³/t for trout culture (Boyd 2005) and applying it to both types of production systems, water use by flow-through and cage culture would be about 0.25 km³/year. Water use by other types of systems was considered insignificant for purposes of this discussion.

The total consumptive water use by aquaculture would be 111.4 km³/year. This represents only 0.82% of the world's accessible, renewable freshwater. Verdegem and Bosma (2009) made a global evaluation of water use in aquaculture that included green water evaporated from ponds, seepage losses, and water used to make feed ingredients. The resulting estimate of water use was 16 900 m³/t or 429 km³/year— about 3.2% of global, accessible, renewable freshwater. It should be remembered, however, that the water evaporated from ponds and most of the water for making feed ingredients is green water that is not included as a portion of the world's accessible, renewable freshwater supply. Both approaches to estimating aquaculture water use mentioned above lead to the conclusion that aquaculture makes a relatively small demand on the world's supply of accessible, renewable freshwater—especially in comparison with agriculture that uses about 77% of it (green water excluded).

Water availability for aquaculture

By dividing country-level data on annually available and renewable freshwater (Gleick 2009) and freshwater aquaculture production (http://faostat.fao.org), Boyd et al. (2012) calculated the tonnes of aquacultural production realized per cubic kilometer of freshwater. This quantity was called the aquaculture to freshwater ratio (AFR), and it was proposed for use as an indicator of the intensity of aquaculture production relative to the availability of freshwater in a country. The AFR varied from

Table 6.11 Renewable freshwater, freshwater aquaculture production, and freshwater aquaculture production:renewable freshwater ratio (AFR) by FAO region.

Continent	Renewable freshwater (km³/year)	Freshwater aquaculture (t/year)	AFR (t/km³)
Asia	12 461	30 015 550	2409
Asia without China	9649	9 234 485	957
Africa	3950	332 113	84.1
North America	6662	333 219	50.0
Europe	6619	454 501	68.7
Latin America and Caribbean	13 161	408 692	31.1
Oceania	911	1424	1.56
World	43 764	31 545 499	721
World without China	40 952	10 764 434	263

Source: Boyd, C. E., L. Li, and R. Brummett. 2012. Relationship of freshwater aquaculture production to renewable freshwater resources. *Journal of Applied Aquaculture* 24:99–106.

0 to 15 000 t/km³. Country-level AFRs were assigned to classes: no freshwater aquaculture ($n = 35$); low, <100 t/km³ ($n = 80$); medium, 100–1000 t/km³ ($n = 45$); high, >1000 t/km³ ($n = 12$). The AFRs are summarized by continent in Table 6.11. The world average AFR of 721 t/km³ is influenced greatly by the high AFR (2409 t/km³) for Asia—particularly the AFR of 7344 t/km³ for China. Removing China from the calculation reduces AFR for Asia to 957 t/km³ and for the world to 263 t/km³. Africa, North America, and Europe have fairly similar AFRs—these three continents have a combined AFR of 65.0 t/km³. Oceania and Latin America and the Caribbean have a combined AFR of 29.1 t/km³. The averages and standard deviations by AFR class were low, 22.4 ± 26.6 t/km³; medium, 291.8 ± 203.8 t/km³; and high, 4154.8 ± 4619.7 t/km³. Individual country AFRs were skewed heavily to the left (toward lower AFR), indicating considerable scope for expansion within the existing water resource base.

The projected freshwater aquacultural production needed to maintain current world consumption of fisheries products is about 55.1 Mt/year by 2050. This level of production equates to a world AFR of 1259 t/km³. Tripling production in the low-income and food-insecure countries of Latin America and Africa would cause AFR in these regions to increase from 31.1 to 93 t/km³ and from 84.1 to 252 t/km³, respectively. These AFRs are still well below those measured based on current production in high AFR countries such as China, India, Thailand, and Vietnam.

Overall availability of freshwater at country and regional levels does not appear to be a major constraint to meeting future global fish and shellfish demand by increasing aquaculture. These are negative environmental issues related to high levels of freshwater aquaculture production at the country level. These include competition with other water uses, energy use and carbon emissions associated with pumping water, and water pollution from aquaculture effluents (Boyd et al. 2007). Nevertheless based on regional studies of individual aquaculture industries such as channel catfish in the United States (Boyd et al. 2000) and *Pangasius* in Vietnam (Bosma et al. 2009), aquaculture appears to be a relatively small contributor to water pollution and water use conflicts as compared to some other activities.

Water conservation in aquaculture

The most obvious way of lessening water use in aquaculture is through intensification. In an unaerated pond without water exchange, an increase in production from 1000 to 2000 kg/ha would not change consumptive water use but it would double the production per unit of water used. Usually the situation is not as simple as the preceding example because mechanical aeration and water exchange often are applied to ponds when production rates are increased. Mechanical aeration usually splashes water into the air to favor greater evaporation rates from ponds. There have been no studies of the influence of aeration on water loss from aquaculture ponds but experience suggests that the increase is rather minor. Water exchange typically is applied at 5–10% of pond volume per day—1.5–3 pond volumes per month. Water exchange does not increase evaporation from ponds, and unless groundwater is used for the process, it does not increase consumptive water use as defined by Boyd (2005). Verdegem and Bosma (2009) also concluded that intensification was a means for improving the efficiency of water use in aquaculture.

Some commonsense procedures can be used to conserve water in aquaculture and reduce both total water use and consumptive water use. Ponds should be designed and constructed with attention to reducing the potential for seepage, and old ponds that seep excessively should be repaired (Yoo and Boyd 1994). Of course, some of the water seeping from ponds becomes groundwater, and an abundance of ponds on the landscape is considered to positively affect groundwater recharge (Manson et al. 1968; Allred et al. 1971).

Evaporation rate is dependent upon the amount of water surface area exposed to the air. The ratio of evaporation to volume of water held in storage can be lessened by making ponds deeper. Two adjacent ponds, one 1 m deep and other 2 m deep, would hold 10 000 and 20 000 m^3, respectively, but both ponds would have essentially the same rates of evaporation. Assuming water levels are maintained in both ponds and the annual evaporation rate is 1 m, the water loss relative to storage is 100% in the shallow pond but only 50% in the deeper pond. Higher vegetation such as emergent plants, shrubs, and trees around the edges of ponds should be controlled to lessen water loss through transpiration.

Water in aquaculture can be reused as discussed in Chapter 1. Water reuse has the same effect as intensification by increasing the amount of production possible per unit of water use.

Aquaculture can sometimes be integrated with certain agricultural crops to improve water use. In Asia, considerable aquaculture is conducted in rice paddies allowing two crops—fish and rice—to be produced in the same water (Frei and Becker 2005). Ducks, pigs, and some other meat species may be combined with aquaculture, especially in Asia. The pond water is used for both aquacultural and agricultural species, and the waste from the agricultural species serves as fertilizer for aquacultural production (Pillay and Kutty 2005).

Water discharged from aquaculture production systems must be considered gray water. Nevertheless, as the degree of pollution is not usually great, the water is not impaired for most other uses. An effective way to use aquacultural wastewater is for irrigation of crops; the nutrient value of aquaculture effluent is not high, but despite

Figure 6.4 Integrated seaweed–abalone culture in Shandong Province, China. *Source:* David Cline.

it not having much value as a fertilizer, it is at least as useful as normal irrigation water (Ghate et al. 1993). A study of irrigation of melons in Brazil (Miranda et al. 2008) revealed that there was no difference in melon yield or quality between melons irrigated with river water and effluent from low-salinity shrimp ponds. Compared to river water, effluent irrigation affected soil quality by lowering pH and concentration of calcium and magnesium. It also increased the exchangeable sodium ratio. McIntosh and Fitzsimmons (2003) also demonstrated that effluent from low-salinity shrimp farms in Arizona (USA) could be used to irrigate wheat, supplying up to 30% of nitrogen fertilization required by the crop. Research and development on integration of agri-aquaculture systems worldwide has been reviewed (Gooley and Gavine 2003).

In some countries, fish cages may be installed in irrigation systems or irrigation water may be diverted through ponds or other fish culture systems and returned to the irrigation system. Studies by Chopin et al. (1999 and 2001) have suggested that seaweed culture can be integrated with salmon culture to remove nutrients contributed to seawater and allow production of a second crop. Studies also have revealed that abalone culture may be integrated with seaweed culture (Neori et al. 2000), and this practice apparently is done in China (Nie 1992; Cao et al. 2007). There is a project near Weihai in Shandong province in Northern China where huge amounts of seaweed are harvested from an area containing abalone farms (Fig. 6.4).

Conclusions

Global aquaculture production ($\approx$55.0 Mt) is rather small in comparison to total world agricultural production of 7670 Mt—only 0.72%. Aquaculture uses more water per unit of production than does most kinds of terrestrial agriculture. Based on the definition of consumptive water use in aquaculture proposed by Boyd (2005) that does not include green water, and deleting green water from water use in terrestrial agriculture, water use by aquaculture was estimated as 111.4 km^3/year and that of terrestrial agriculture is 1678 km^3/year. Thus consumptive use of water in aquaculture could be as much as 6.6% of total water use for agricultural production.

There appears to be plenty of freshwater for future expansion of aquaculture in many countries, but in water-scarce areas, it will be particularly important to intensify aquaculture in order to obtain as much production as possible from each unit of water.

Water conservation also should be practiced in aquaculture to reduce consumptive water use. Moreover, a reduction in both total and consumptive water use can lessen energy use and cost of pumping water into production facilities.

The eNGO perspective

The use of freshwater for aquaculture is much like the use of water for any other activity in the eyes of the eNGOs. The key impact areas of highest concern for the eNGOs are the changing of or depletion of habitat for wildlife. These circumstances are often a result of the damming of rivers for flood control, water supply, and hydroelectric power, or the draining of and conversion of wetland areas. There is also a growing concern among the eNGOs that the availability of freshwater to humans should be protected. Although not truly a direct threat to the environment, eNGOs believe that if there is a shortage of potable freshwater for humans, there is a potential for the removal of water from areas of high ecological importance to supplement this shortage. Of course, the degradation of water is of concern to the eNGOs as well, and water quality concerns are addressed in Chapter 10.

Because there is a tremendous amount of ocean and brackishwater of earth, the extraction of these waters is not of high concern to the eNGOs. This could change with advancements in desalinization but at this time there is little evidence that extraction of saline waters for human use is being addressed by eNGOs.

Drainage and conversion of wetlands for aquaculture activities was discussed in Chapter 5 primarily with respect to mangrove forests. This reallocation of water for aquaculture is essentially a form of habitat degradation and most eNGOs would consider any water reallocation from wetlands undesirable, both from the perspective of decreased habitat for wildlife and a diminished assimilative capacity for nutrient pollution destined for wetlands.

Although eNGOs generally have a strong stance against cage aquaculture, there is little discussion within the eNGO community about the amount of water used; rather the degradation of water quality that can result from this method of aquaculture is of highest concern. Water is seen as a medium for the culture of cage species. The water is not being removed from the system and does not constitute water extraction.

Pond aquaculture is often considered by the eNGOs to be using surface water and it is not viewed as a consumptive use of water. However, in relatively arid regions of the world where pond salinities can become too high for optimal growth of the culture species, some producers choose to add freshwater from aquifers to dilute pond waters and decrease the salinity. This is considered extraction of a rare resource that likely has human use implications. This practice has also been used to dilute brine water shipped from the sea to inland areas for the culture of marine shrimp. A 1998 moratorium on this practice in Thailand brought attention to the potential soil changes from the discharge of saline waters into freshwater areas (http://www.slu.se/Documents/externwebben/centrumbildningar-projekt/mkb-centrum%20dokument/Helpdesk/shrimps2.pdf). The implications were primarily socioeconomic—threatening crop lands and potentially making potable water unfit for human consumption.

In India, where the most abundant freshwater aquifers are near coasts, seepage from shrimp ponds has caused salination of some freshwater aquifers. This has had significant impacts on local communities. Shiva (1995) described a case study in Nellore district in the village of Kurru where the people's drinking water source became too saline for consumption as a result of shrimp pond water infiltrating the nearby aquifer. Protests erupted and the shrimp farm was required to ship in drinking water in tankers for the some 600 villagers. Although this example is pollution-related, it illustrates the delicacies of not only the sources of water used by the aquaculture industry, but also the potential impacts on other users of nearby groundwater sources.

Most eNGOs are supportive of closed aquaculture facilities where there is little-to-no effluent, and thus low reliance on water to maintain the culture medium except for seepage and evaporation. These systems represent a panacea of sorts for the eNGOs solving water pollution issues, escape issues, and water extraction issues. It also represents a possible solution to the eNGOs when water extraction issues are contentious or problematic.

WWF-US explored another less-reported water usage issue in the processing of aquaculture products. The use of polyphosphate chemicals in the processing of seafood is a common means to retain moisture (glazing) in the final product. However, there is a significant amount of abusive use of these products to increase weight gain. Products can be treated several times with polyphosphates and dipped in ice water several times in between freezings. It is likely that a product weight can be increased up to 15–20% using this process. In effect, water is being falsely represented as a seafood product. Further considering the volume of fisheries products traded globally, a significant amount of water is being used and transported simply as a disguise of greater fish product weight. The junior author discussed the reasoning for this process with a large Chinese tilapia producer. It was conveyed that when a retailer or wholesaler visits with the farm to discuss contracts, the buyer will request a certain amount of fish at a certain price. The buyer knows that the price that is being offered is less than what it costs to produce the product so it is implied or suggested that the producer make up that amount of money through glazing product to bring up the weights. Although a deceptive practice, the utilization of this water is typically a much smaller volume than that used in a processing facility. Probably, a greater resource issue related to the glazing practice is the amount of energy used to transport this water overseas for sale.

Water extraction for aquaculture is seldom an issue of interest to eNGOs. The most notable claim was by an eNGO stating that water extraction was lowering the water table near the Snake River (USA) because of trout farming activities. This was proven false as most of the water used in this region is from free-flowing springs that exit the ground in the Snake River gorge. Although the water table was decreasing, it was determined that the main reason for the decrease was a result of agriculture using wells to supply water for livestock and for irrigation.

There are few cases where eNGOs have become activists against aquaculture for water use and extraction issues. The eNGO community will likely make this issue a higher priority if habitats begin to be compromised because of water extraction.

References

Aladin, N. J. F. Crétaux, I. S. Plotnikov, A. V. Kouraev, A. O. Smurov, A. Cazenave, A. N. Egorov, and F. Papa. 2005. Modern hydro-biological state of the Small Aral Sea. *Environmetrics* 16:375–392.

Allred, E. R., P. W. Manson, G. M. Schwartz, P. Golany, and J. W. Reinke. 1971. Continuation of studies on the hydrology of ponds and small lakes. *Minnesota Agricultural Experiment Station Technical Bulletin 274*. Minneapolis, Minnesota: University of Minnesota.

Baumgartner, A. and E. Reichel. 1975. *The World Water Balance*. Amsterdam: Elsevier Scientific Publishing Company.

Bosma, R. H., C. T. T. Hanh, and J. Potting. 2009. Environmental impact assessment of the *Pangasius* sector in the Mekong Delta. Wageningen, Netherlands: Wageningen University.

Boyd, C. E. 1982. Hydrology of small experimental fish ponds at Auburn, Alabama. *Transactions of the American Fisheries Society* 111:638–644.

Boyd, C. E. 1986. Influence of evaporation excess on water requirements for fish farming. In *Conference on Climate and Water Management—A critical Era*, pp. 62–64. Boston: American Meteorological Society.

Boyd, C. E. 1987. Water conservation measures in fish farming. In *Fifth Conference on Applied Climatology*, pp. 88–91. Boston: American Meteorological Society.

Boyd, C. E. 1990. *Water Quality in Ponds for Aquaculture*. Auburn University, Alabama: Alabama Agricultural Experiment Station.

Boyd, C. E. 2005. Water use in aquaculture. *World Aquaculture* 36:12–15 and 70.

Boyd, C. E. and P. L. Chaney. 2013. Alabama fish ponds. In *Auburn Speaks, On Water 2013*, pp. 114–121. Auburn University, Alabama: Office of the Vice President for Research.

Boyd, C. E., L. Li, and R. Brummett. 2012. Relationship of freshwater aquaculture production to renewable freshwater resources. *Journal of Applied Aquaculture* 24:99–106.

Boyd, C. E., S. Soongsawang, E. W. Shell, and S. Fowler. 2009. Small impoundment complexes as a possible method to increase water supply in Alabama. *Proceedings 2009 Georgia Water Resources Conference*, April 27–29. Athens: University of Georgia.

Boyd, C. E., J. Queiroz, J. Lee, M. Rowan, G. N. Whitis, and A. Gross. 2000. Environmental assessment of channel catfish, *Ictalurus punctatus*, farming in Alabama. *Journal of the World Aquaculture Society* 31:511–544.

Boyd, C. E., C. S. Tucker, A. McNevin, K. Bostick, and J. Clay. 2007. Indicators of resource use efficiency and environmental performance in fish and crustacean aquaculture. *Reviews in Fisheries Science* 15:327–360.

Buchanan, R. C., R. R. Buddemeier, and B. B. Wilson. 2009. High Plains Aquifer. *Circular 18, Kansas Geological Survey*. Lawrence, Kansas: University of Kansas.

Cai, X., D. C. McKinney, and M. W. Rosegrant. 2003. Sustainability analysis for irrigation water management in the Aral Sea region. *Agricultural Systems* 76:1043–1066.

Cao, L., Wang, Y. Yang, C. Yang, Z. Yuan, S. Xiong, and J. Diana. 2007. Environmental impact of aquaculture and countermeasures to aquaculture pollution in China. *Environmental Science and Pollution Research* 14:452–462.

Chaney, P. L., C. E. Boyd, and M. Polioudakis. 2012. Number, size, distribution, and hydrologic role of small impoundments in Alabama. *Journal of Soil and Water Conservation* 67:111–121.

Changming, L., Y. Jingjie, and E. Kendy. 2001. Groundwater exploitation and its impact on the environment in the North China Plain. *Water International* 26:265–272.

Chopin, T., C. Yarish, R. Wilkes, E. Belyea, S. Lu, and A. Mathieson. 1999. Developing *Porphyra*/salmon integrated aquaculture for bioremediation and diversification of the aquaculture industry. *Journal of Applied Phycology* 11:463–472.

Chopin, T., A. H. Buschmann, C. Halling, M. Troell, N. Kautsky, A. Neori, G. Kraemer, J. Zertuche-Gonzalez, C. Yarish, and C. Neefus. 2001. Integrating seaweeds into aquaculture systems: a key towards sustainability. *Journal of Phycology* 37:975–986.

Frei, M. and K. Becker. 2005. Integrated rice-fish culture: coupled production saves resources. *Natural Resources Forum* 29:135–143.

Ghate, S. R., G. J. Burtle, and G. J. Gascho. 1993. Reuse of water from catfish ponds. *Proceedings 1993 Georgia Water Resources Conference*. Atlanta, Georgia: Georgia Institute of Technology.

Gleick, P. H. 2009. *The World's Water*. Washington: Island Press.

Gooley, G. J. and F. M. Gavine. (editors). 2003. *Integrated Agri-Aquaculture Systems*. Kingston: Rural Industries Research and Development Corporation.

Hoekstra, A. Y. 2008. The water footprint of food. In J. Förare, editor, *Water for Food*, pp. 49–60. Stockholm: The Swedish Research Council for Environment, Agricultural Sciences and Spatial Planning.

Hoekstra, A. Y. and A. K. Chapagain. 2007. Water footprints of nations: water use by people as a function of their consumption pattern. *Water Resource Management* 21:35–48.

Hoekstra, A. Y. and M. M. Mekonnen. 2012. The water footprint of humanity. *Proceedings of the National Academy of Sciences* 109:3232–3237.

Konikow, L. F. and E. Kendy. 2005. Groundwater depletion: a global problem. *Hydrogeology Journal* 13:317–320.

Leopold, L. B. 1974. *Water, a Primer*. San Francisco: W. H. Freeman and Company.

Manson, P. W., G. M. Schwartz, and E. R. Allred. 1968. Some aspects of the hydrology of ponds and small lakes. *Minnesota Agricultural Experiment Station, Technical Bulletin 257*. Minneapolis, Minnesota: University of Minnesota.

McIntosh, D. and K. Fitzsimmons. 2003. Characterization of effluent from an inland, low-salinity shrimp farm: what contribution could this water make if used for irrigation? *Aquacultural Engineering* 27:147–156.

Miranda, F. R., R. N. Lima, L. A. Crisóstomo, and M. G. S. Santana. 2008. Reuse of inland low-salinity shrimp farm effluent for melon irrigation. *Aquacultural Engineering* 39:1–5.

Neori, A., M. Shpigel, and D. Ben-Ezra. 2000. A sustainable integrated system for culture of fish, seaweed and abalone. *Aquaculture* 186:279–291.

Nie, Z. Q. 1992. A review of abalone culture in China. In S. A. Shepherd, M. J. Tegner, and S. A. Guzman del Proo, editors, *Abalone of the World: Biology, Fisheries and Culture*, pp. 592–602. Oxford: Blackwell.

Pillay, T. V. R. and M. N. Kutty. 2005. *Aquaculture: Principles and Practices*. New York: John Wiley and Sons.

Postel, S. 1996. *Dividing the Waters: Food Security, Ecosystem Health, and the New Politics of Scarcity*. Washington: Worldwatch Institute.

Schoof, R. R. and G. A. Gander. 1982. Computation of hydrograph reduction caused by farm ponds. *Water Resources Bulletin* 18:529–532.

Shah, T., A. D. Roy, A. S. Qureshi, and J. Wang. 2003. Sustaining Asia's groundwater boom: an overview of issues and evidence. *Natural Resources Forum* 27:130–141.

Shiva, V. 1995. Ethics, genetic engineering and biodiversity. In: H. Reinertsen and H. Haaland, editors, *Sustainable Fish Farming*, pp. 131–148. Rotterdam: A.A. Balkema.

Verdegem, M. C. J. and R. H. Bosma. 2009. Water withdrawal for brackish and inland aquaculture, and options to produce more fish in ponds with present water use. *Water Policy* 11:52–68.

Wetzel, R. G. 2001. *Limnology*, 3rd ed. New York: Academic Press.

White, W. R. 2010. *World Water: Resources, Usage, and Role of Man-made Reservoirs*. Bucks: Foundation for Water Research.

Yoo, K. H. and C. E. Boyd. 1994. *Hydrology and Water Supply for Aquaculture*. New York: Chapman and Hall.

Chapter 7

Energy use and atmospheric emissions

Nearly all energy sources used by humans have a direct or indirect relationship to solar radiation. Energy from sunlight allows green plants to photosynthetically reduce inorganic carbon in carbon dioxide (CO_2) to organic carbon in sugar ($C_6H_{12}O_6$) with release of molecular oxygen (O_2):

$$6CO_2 + 6H_2O \xrightarrow[\text{Mineral nutrients}]{\text{Light}} C_6H_{12}O_6 + 6O_2 \tag{7.1}$$

Photosynthesis is the origin of energy in wood, other biomass fuels, and fossil fuels. Solar radiation also drives the hydrologic cycle (Chapter 6) and assures stream flow for hydroelectric generation and cooling water for other types of electricity generation. Wind and solar power also originate from solar radiation.

Before the industrial revolution around 1750, energy use by humans was primarily manpower, animal power, wind power for sailing ships, and wood and other fuels for cooking, heating, and working with metals. Since then, there has been an increasing demand for fossil fuels and electricity generated mainly with fossil fuels to serve as energy sources for industrial processes and machinery, agricultural equipment, vehicles for passenger and cargo transport, and cold storage of perishable products as well as heating, lighting, and cooling of homes, offices, and many other facilities.

Fossil fuels resulted from massive deposition of organic carbon in dead biomass long ago and its gradual biological and geological transformation to oil, coal/peat, and natural gas. Human history has a much shorter time scale and fossil fuels are not renewable to us. There is concern over eventual depletion of fossil fuels and the resulting impacts on society. Of equal concern are the present and future effects of atmospheric pollution resulting from fuel production and use on climate, ecosystems, and humans. Increased concentrations of carbon dioxide, methane, and several other gases resulting from combustion of fuels can enhance the ability of the earth's

Aquaculture, Resource Use, and the Environment, First Edition. Claude E. Boyd and Aaron A. McNevin.
© 2015 John Wiley & Sons, Inc. Published 2015 by John Wiley & Sons, Inc.

atmosphere to trap and hold heat. This phenomenon is highly publicized as the cause of global warming, melting of polar ice and glaciers, and a rising sea level. Sulfur and nitrogen dioxides from combustion of fuels can cause acidification of rainfall, and carbon dioxide increase in the atmosphere leads to acidification of the ocean. In addition, some emissions contain organic chlorine and bromine compounds that can destroy ozone in the stratosphere reducing the ability of the ozone layer to screen out harmful ultraviolet (UV) radiation. Of course, air pollution at ground level also is harmful to human health.

Aquaculture uses energy for many purposes such as feed manufacturing, pond construction, supplying water to fill and maintain volumes of culture units, mechanical aeration, harvesting, processing, and transportation of supplies and products. Considerable energy also is used for producing liming materials, fertilizers, feed ingredients, and other products used in aquaculture. Aquaculture contributes to atmospheric emissions because it uses fossil fuels.

The purpose of this chapter is to provide a general discussion of world energy use, briefly explain the sources and effects of atmospheric emissions associated with use of fossil fuels to produce energy, and provide some information about energy use and emissions by fisheries and aquaculture.

Global energy use

Quantities of energy are reported in various units of measurement. In this chapter, amounts of energy usually will be given in joules, but for specific fuels, units of measurement common to each fuel industry will be used. Relationships among different units of measurement and some specialized terms related to fuels used in assessments of energy are provided (Table 7.1).

Primary fuels

Primary fuel use is the total amount of fuel used by society for all purposes. Estimates of energy contained in primary fuels made by different organizations vary because of differences in databases and procedures used to convert energy contents of primary fuels to common units of measurement. To illustrate, total global primary energy consumption in 2008 was estimated to be 533 EJ by the US Energy Information Agency (USEIA) (http://www.eia.gov/forecasts/ieo/world.cfm), 481 EJ by British Petroleum (BP 2012a), and 491 EJ by the Intergovernmental Panel on Climate Change (IPCC 2011).

Annual estimates of total primary fuel use (BP 2012a) from 1965 to 2011 are provided (Fig. 7.1). Primary energy use increased by a factor of 3.21 during this span because population rose 2.07 times and per capita energy use expanded by a factor of 1.55 (47.8–74.3 GJ/cap/year).

Total primary fuel use by source was estimated by the International Energy Agency (IEA 2012) for 1973 and 2010 (Table 7.2). Oil was the major fuel in both 1973 and 2010, and its total use increased from 6107 Mtoe in 1973 to 12 717 Mtoe in

Table 7.1 Orders of magnitude, abbreviations of units of measurement, definitions, and conversions of units for energy and fuels.

Orders of magnitude	Abbreviations of units
Thousand = 10^3 = kilo (K)	Joule = J
Million = 10^6 = mega (M)	Watt = W
Billion = 10^9 = giga (G)	Watt·hour = Wh
Trillion = 10^{12} = tera (T)	Kilowatt·hour = KWh
Quadrillion = 10^{15} = peta (P)	Tonne of oil equivalent = toe
Quintillion = 10^{18} = exa (E)	Million tonnes of oil equivalent = Mtoe
Sextillion = 10^{21} = zetta (Z)	Newton = N
	British thermal unit = BTU
Definitions	
1 N = force to accelerate 1 kg mass at 1 m/sec^2	Conversions
1 J = energy to apply 1 N through 1 m; 1 J = 1 N·m	1 toe = 7.33 barrels of oil
	1 billion m^3 natural gas = 0.9 toe
Conversions	1 t coal = 0.67 toe
1 J = 0.000277778 Wh	1 toe = 41.86 GJ
1 Wh = 3600 J = 3.6 KJ	1 toe = 11.63 MWh
1 W/m^2 = 1 J/m^2/sec	1 t coal = 8.141 MWh
1 KWh = 3.6 MJ	1 EJ = 23.88 Mtoe
1 BTU = 1055.87 J	
1 KWh = 3412 BTU	
1 TWh = 0.0036 EJ	

2011. But the percentage of oil use declined markedly during the 37-year period. This decline was offset by greater use of natural gas and nuclear power in particular and by small increases in use of coal/peat, hydroelectric power, and a few other sources. There was a slight decline in the percentage contribution of biofuels and wastes in spite of efforts to encourage the use of these fuels. Countries belonging to the Organization for Economic Cooperation and Development (OECD)—mainly developed countries—used 61.4% of total primary energy in 1973 but only 42.4% in 2010; this can be attributed to greater population growth rate and improving economies in rapidly developing, non-OECD countries—especially China and India.

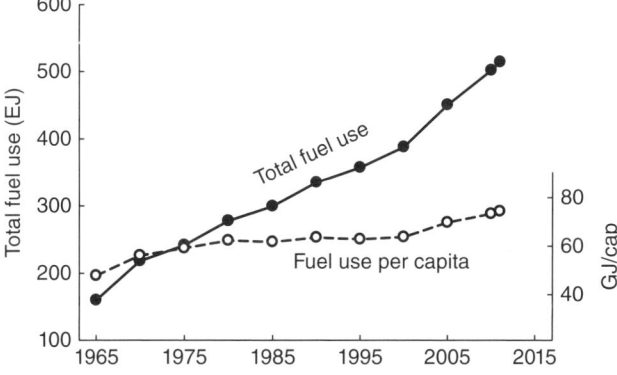

Figure 7.1 Annual estimates of global primary fuel use and energy use per capita. *Source:* Data for constructing figure from BP (2009, 2012).

Table 7.2 Global primary energy supply by source for 1973 and 2010.

	EJ/year	
Source	**1973**	**2010**
Oil	117.9	172.5
Natural gas	40.9	113.9
Coal/peat	62.9	145.4
Biofuels and wastes	26.8	53.2
Nuclear power	2.3	30.3
Hydropower	4.6	12.2
Other (geothermal, solar, wind, etc.)	0.3	4.8
Total	**255.7**	**532.3**

Source: Data for preparing table from www.eia.gov/countries/.

According to IEA (2012) about 81% of primary fuel use was from fossil fuels in 2010. A report by IPCC (2011) indicated that 89% of primary fuel use was from fossil fuels in 2008. Thus, the world is highly dependent upon nonrenewable fossil fuels for energy.

The 10 countries with greatest primary energy use are listed (Table 7.3). China has surpassed the United States as the world's leading user of energy. Nevertheless, an average person in the United States uses about four times as much energy as an average person in China. There is a huge gap between energy use in China and in the United States as compared with any of the other eight countries. Also, there is stark contrast between energy use per capita in India and in the other countries (Table 7.3). Other than for India and Brazil per capita energy consumption in the leading energy use countries is above the current world average of 74.3 GJ/cap/year.

The amount of energy used is projected to grow in the future. The USEIA predicted that energy use would increase from 533 EJ in 2008 to 812 EJ in 2035 with an average annual rate of increase of 0.6% in OECD countries and 2.3% in non-OECD countries (http://www.eia.gov/forecasts/ieo/world.cfm). The BP (2012b) energy forecast predicts an increase from 502 EJ in 2010 to 691 EJ in 2030, but with

Table 7.3 Top 10 energy consuming countries and their per capita primary fuel use.

	Primary energy use	
Country	**(EJ/year)**	**(GJ/cap)**
China	110.9	82
United States	93.2	299
India	31.8	26
Russia	30.4	213
Japan	19.6	153
Germany	13.3	163
Brazil	11.2	57
Canada	11.1	325
South Korea	10.8	217
France	10.8	171

Source: Data for constructing table from USEPA (2010); Aneja (1990); http://cdiac.ornl.gov/pns/current_ghg.html.

growth in global fuel use occurring almost entirely within non-OECD countries. An energy use forecast by IEA (2012) agreed more closely with the BP report than with the USEIA report.

Primary fuel reserves

According to the World Coal Association (http://www.worldcoal.org), coal reserve estimates range from 861 to 1004 Gt—112-year and 130-year supplies, respectively, at current use rates. Proven oil reserves are 1.482 trillion barrels or a 50-year supply at current use (http://en.wikipedia.org/wiki/List_of_countries_by_proven_oil-reserves). There is an estimated 300 trillion m^3 of proven natural gas reserves—a 100-year supply (http://en.wikipedia.org/wiki/List_of_countries_by_proven_natural_gas_reserves). Uranium reserve estimates are less certain than those of other fuels but OECD (2010) placed them at about 5.4 Mt in 2008. Global use of mined uranium is about 67 000 t/year; thus, there should be an 80-year supply (http://cfr.org/energy/global-uranium-supply-demand/p14705).

The actual amounts of primary fuel reserves are greater than proven reserves because deposits remain to be discovered. A recent study by the US Geological Survey (USGS 2012) gave an estimate of undiscovered oil and natural gas deposits of 565 000 Mboe (77 080 Mtoe) and 158.7 trillion m^3, respectively. These estimates represent about 38% and 52.9%, respectively of proven reserves of oil and natural gas. This suggests at least a 70-year supply of oil and a 150-year supply of natural gas at current use rates. Assuming similar undiscovered coal and uranium, there may be enough coal for another 150–175 years and uranium for more than 100 years. Although these estimates may look encouraging in the short run, the end of the supply of oil, natural gas, coal, and uranium nevertheless is in plain sight. The world obviously must wean itself from almost total dependence on fossil fuels because these resources ultimately will run out.

The cost of fuels will almost certainly rise as reserves decline. This intuitive fact is expressed more eloquently by the peak oil concept of Herbert (1949): when the maximum rate of oil extraction from the earth is reached in a country, the rate of production in that country will decline in the future. Peak oil was reached in the United States in 1970 resulting in massive imports of foreign oil and a rise in price.

Energy end use

A considerable amount of energy is used in refining or converting primary fuels to useful forms and then distributing the useful forms. The amount of energy applied for an intended application is called end-use energy. Electricity generation provides a good example of the conversion of primary fuel to end use. Electricity is generated primarily from energy released by burning fossil fuels (Table 7.4) and it is transmitted through power lines to the place of end use. In 2008, about 4399 Mtoe (36% of total primary fuels) were used to generate an estimated 20 185 TWh of electricity—39% energy recovery in electricity with the rest of the primary energy

Table 7.4 Shares of energy sources used for generating electricity.

Fuel	Shares (%) 1973	Shares (%) 2010
Coal/peat	38.3	40.6
Natural gas	12.1	22.2
Oil	24.7	4.6
Nuclear	3.3	12.9
Hydro	21.0	16.0
Other	0.6	3.7

Source: Adapted from IEA (2012).

converted to heat. About 3% of waste heat was put to a beneficial use. Of the electricity generated in 2008, only 16 430 TWh were applied at end use—the remainder was lost during transmission—an overall efficiency from primary fuel of 33% (http://en.wikipedia.org/wiki/Electricity_generation).

The end use of all primary fuels in 2010 was estimated to be 8877 Mtoe—68.2% of total primary fuel use (IEA 2012). A lower estimate of only 60% of primary fuel energy realized at final end use in 2008 was given by IPCC (2011).

The end use of oil is primarily for transportation while coal/peat is used mainly in industry. Electricity and natural gas are most important for agriculture, commercial and public services, and homes (Table 7.5). Most human energy uses cut across sectors, for example, a product is manufactured by industry, it is transported to markets worldwide, it is used by consumers, and waste resulting from it must be disposed.

Energy for the world food system

Estimates of energy consumption specifically by agriculture are approximate but FAO (2011a) presented an overview of energy use that places agriculture energy use in perspective with total world energy use. The FAO analysis also considered how energy use is proportioned within the world food system from farm to consumer. Information on energy use in world fisheries production is included but these data are less reliable than those for agriculture—especially those for aquaculture.

Table 7.5 Use of different energy sources by total amounts and by different use categories in 2010.

Amount and use categories	Oil	Natural gas	Coal/peat	Electricity	Other
Amount (EJ)	149.7	55.2	35.6	64.3	58.4
Shares of total (%)	41.2	15.2	9.8	17.7	16.1
Use category					
Nonenergy (%)	17.1	11.6	4.2	0.0	–
Industry (%)	9.0	35.2	79.5	41.5	–
Transport (%)	61.5	6.8	0.4	1.6	–
Other (agriculture, commercial and public services, residences, etc.) (%)	12.4	46.4	15.9	56.9	–

Source: http://www.epa.gov/climatechange/ghgemissions/gases.html.

Table 7.6 End use of energy in the world food system in 2008 in countries with high and low gross domestic products (GDP).

End use	Energy (EJ)		
	Global	High GDP countries	Low GDP countries
Plant crops	12.8	5.6	7.2
Livestock	5.1	4.4	0.7
Fisheries and aquaculture	2.4	1.8	0.6
Processing and distribution	40.9	24.0	16.9
Retail, preparation, and cooking	33.8	14.2	19.6
Total	**95**	**50**	**45**

Source: http://www.eia.gov/cfapps/ipdbproject/iedindex3.cfm?tid=90&pid=44&aid=8.

The end-use consumption of energy by all sectors was estimated to be 294 EJ globally in 2008 (FAO 2011a), and the world food system share was placed at 95 EJ/year—about 32% of the total. Energy estimates do not include human and animal energy, but indirect energy is included. Direct energy use at the farm level for pumping water, housing livestock, tilling and harvesting, heating, pest control, etc. was about 6 EJ/year. Indirect energy use for fertilizer manufacturing alone was estimated at 5 EJ/year, and there are many lesser indirect energy inputs.

Most of the end use of energy in the world food system is for processing, distributing, holding and selling, preparing, and cooking (Table 7.6). These activities consumed almost 80% of the total energy use in the global food system.

Countries with low gross domestic production (low GDP countries) consumed nearly the same amount of total energy as the high GDP countries for food production and use. Although FAO (2011a) did not give the population estimates for the two categories of countries, there is a considerably greater population in the low GDP category than in the high GDP one. Moreover, food export and imports occurred between the two categories of countries. Low GDP countries used more energy for plant crop production, while high GDP countries used more energy for livestock production and in the fisheries sector. Energy use for processing and distributing food was higher in high GDP countries, while the opposite applied for retail, preparation, and cooking.

Data from Smil (2008) revealed that in spite of the higher energy input to US corn production in 2007 as compared to 1945, the amount of energy used to produce corn has decreased from 2.7 to 2.2 GJ/t because of the much greater yield per hectare achieved in 2007. A similar relationship likely occurred with most other crops, but total energy use has increased drastically because of greater world production of crops.

Fisheries and aquaculture

A relatively small amount of energy was used for fisheries and aquaculture: 2 EJ/year for capture fisheries; 0.4 EJ/year for aquaculture (FAO 2011a). Capture fisheries energy use was based on the amount of fuel energy used to power the global fishing fleet (Tyedmers et al. 2005), while aquaculture energy use was based mainly on energy used to produce and deliver feeds (Smil 2008). Those making the FAO assessment apparently felt that fuel for fishing vessels and aquaculture feeds were the major energy uses, but energy is used for many other purposes in fisheries and aquaculture.

Approximately 16.6% of global animal protein intake is from fisheries products that are about equally divided between capture fisheries products and aquaculture products (FAO 2012). An estimated 2 EJ/year were used to capture about 8.3% of world animal protein intake (0.24 EJ/year for each percentage) while animal agriculture (excluding aquaculture) used 5.1 EJ/year to produce 83.4% of the animal protein intake (0.061 EJ for each percentage). The corresponding performance of aquaculture was 0.048 EJ for each percentage.

Life cycle assessments have been made on aquaculture species such as Galician mussels (Iribarren et al. 2010), tilapia (Pelletier and Tyedmers 2010), salmon (Pelletier et al. 2009; Winther et al. 2009), channel catfish (Boyd et al. 2011), Pacific white shrimp or whiteleg shrimp (Mungkung et al. 2012), and trout (Gronroos et al. 2006). The amounts of energy use per unit of production varied considerably, but the average was estimated to be roughly 26 GJ/t. In 2008, the year of the FAO (2011a) assessment, aquaculture production was 52.9 Mt so energy use could have been as great as 1.38 EJ or 0.166 EJ for each percentage of animal protein intake provided. Energy use in aquaculture, however, was likely less, because all species—except mussels—on which the life cycle analysis studies were performed had been produced using feed. Feed ingredients and manufacturing represent a large energy input, and ponds receiving feeds usually are mechanically aerated. Much aquaculture in developing countries uses no feed or aeration and has lower energy inputs as a result. Thus, aquaculture energy use probably is more than estimated by FAO (2011a) but still less than energy use for capture fisheries.

End use of renewable energy in agriculture was <1 EJ from a total of 7 EJ used directly on farms for crop production (FAO 2011a). This compares with 11 EJ of 87 EJ in industry, 12 EJ of 80 EJ in buildings, and 2 EJ of 94 EJ in transportation (IPCC 2011). Capture fisheries and aquaculture appear to be as highly dependent upon fossil fuels as are most other sectors.

Based on the FAO (2011a) report, the world food system consumed slightly over 30% of final end-use world energy and the estimate for fisheries and aquaculture was 2.53% of the energy for the food system and 0.82% of total global end-use energy—0.68% for capture fisheries and 0.14% for aquaculture. Aquaculture, no doubt, uses more energy than the FAO report suggests, but even at the maximum estimate of 1.38 EJ/year based on the LCA studies mentioned above, it used only 0.47% of global end-use energy.

Atmospheric emissions

Dependence on fossil fuels presents a serious resource use issue because their supply is finite and diminishing. Moreover, the use of fossil fuels contaminates the atmosphere with combustion by-products. Fossil fuels are organic matter comprised mainly of carbon, hydrogen, and oxygen, but they also contain sulfur and nitrogen. Combustion of fossil fuels releases carbon dioxide, water vapor (H_2O), nitrogen dioxide

(NO_2, but usually symbolized as NO_X), nitrous oxide (N_2O), sulfur dioxide (SO_2), methane (CH_4), and incompletely oxidized organic carbon (soot or black carbon) and other particles. Combustion of fossil fuels may be expressed by the following unbalanced equation:

$$\text{Fossil fuel} + O_2 \rightarrow CO_2 + H_2O + CH_4 + N_2O + NO_x + SO_2 \\ + \text{soot and other particles} + \text{heat.} \tag{7.2}$$

There are other natural and anthropogenic sources of the products of fossil fuel combustion listed above. Respiration by plants, animals, and organisms of decay release huge amounts of carbon dioxide globally. In respiration, organisms oxidize organic carbon to carbon dioxide to obtain energy:

$$C_6H_{12}O_6 + 6O_2 \rightarrow 6CO_2 + 6H_2O + \text{energy.} \tag{7.3}$$

Wildfires release all of the products shown in Equation 7.2. Animals discharge methane and decomposition of manures releases nitrous oxide, methane, and sulfur dioxide. Microbial activity in soils treated with nitrogen fertilizers is a source of nitrous oxide. Methane and nitrous oxide are produced in anaerobic sediment of wetlands, lakes, and other water bodies. Sulfur dioxide and hydrogen sulfide (H_2S) are released by microbial activity in wetlands, oceans, lakes, and waterlogged soils as well as by volcanos.

Natural atmospheric emissions have been relatively constant over most of human history, and in the past, could be balanced by natural sinks. For example, photosynthesis in the ocean and on land fixes about 200–235 Gt carbon (734–862 Gt carbon dioxide) annually. Respiration by the earth's terrestrial and aquatic biota releases roughly the same amount of carbon dioxide back into the atmosphere each year (Post et al. 1990). Before the industrial age, the carbon cycle (Fig. 7.2) remained in balance, but since the beginning of the industrial revolution, vegetative growth has not been sufficient to remove all of the carbon dioxide that has entered the atmosphere from combustion of fossil fuels and as a result of deforestation.

Annual anthropogenic emissions of carbon dioxide have been increasing and reached 33.5 Gt (12.5 Gt carbon) in 2010 (Fig. 7.3). Since 1850, an estimated 330 Gt of carbon have been released into the atmosphere by anthropogenic activities, but only about one-third of this carbon was removed by natural processes—mainly sequestration in the ocean. Atmospheric carbon dioxide concentration has risen from an estimated 280 ppm in 1850 to 394 ppm in 2010 (Fig. 7.3). Television news worldwide announced on May 9, 2013 that the atmospheric carbon dioxide concentration measured at Mauna Loa, Hawaii reached 400.04 ppm for the first time since monitoring began there in 1958.

It can be easily shown that the increase in atmospheric carbon dioxide is of anthropogenic origin. The atmosphere currently contains about 750 Gt carbon and the carbon dioxide concentration is around 400 ppm (1.88 ppm CO_2 per Gt carbon). Since 1850, about 330 Gt of anthropogenic carbon have been released into the atmosphere (Canadell et al. 2007), and about one-third of this has been sequestered—mainly in the ocean. The increase of 221 Gt carbon should correspond to a rise in atmospheric

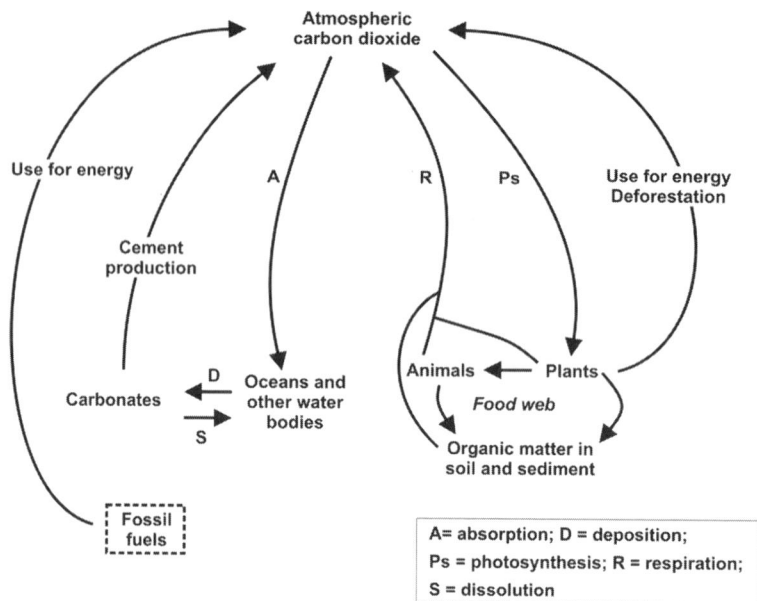

Figure 7.2 The world carbon budget.

carbon dioxide concentration of 118 ppm (221 Gt ÷ 1.88 ppm CO_2/Gt)—the measured increase has been about 120 ppm.

Anthropogenic sources of methane, nitrous oxide, sulfur dioxide, and nitrogen
dioxide also are substantially relative to natural sources (Table 7.7). This has led
to considerable increases in the concentrations of methane and nitrous oxide in the
atmosphere. Sulfur dioxide, hydrogen sulfide, and nitrogen dioxide have a short life
span in the atmosphere because they are quickly oxidized. Their concentrations vary
greatly from place to place, but the annual inputs to the atmosphere are much greater
today than in the past as illustrated for sulfur emissions in Fig. 7.4.

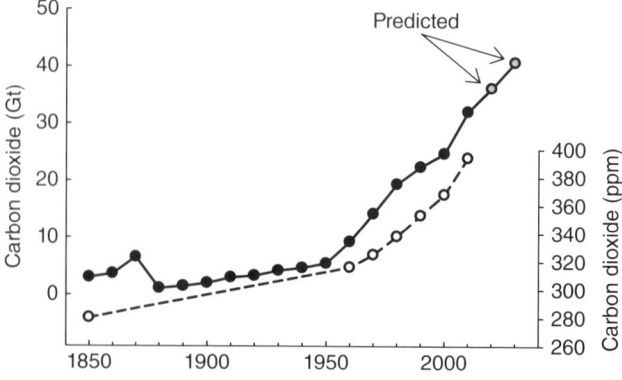

Figure 7.3 Annual estimates of global, anthropogenic carbon dioxide emissions at 10-year intervals
(dots) (1850–2030) and changes in atmospheric carbon dioxide concentrations measured at Mauna
Loa, Hawaii from 1958 to 2012 (circles). *Source:* http://cdiac.ornl.gov/#.

Table 7.7 Contribution of anthropogenic sources to global emissions of greenhouse gases to the atmosphere and atmospheric concentrations.

	Total	Anthropogenic	1750 (Estimated)	2010 (Measured)
Carbon dioxide	790 Gt	33.9 Gt	280 ppm	393 ppm
Methane	566 Mt	258 Mt	700 ppb	1816
Nitrous oxide	18.1 Mt	6.7 Mt	270 ppb	324 ppb
Sulfur dioxide	146 Mt	103 Mt	–	–
Nitrogen dioxide (as N)	59.1 Mt	32.6 Mt	–	–
Ozone	–	–	25 ppb	34 ppb
F-gases (12 compounds combined)	–		0 ppb	1263 ppt

Sources: Adapted from USEPA (2010); Aneja (1990), http://cdiac.ornl.gov/pns/current_ghg.html; http://simple.wikipedia.org/wiki/carbon_cycle.

Other gases of anthropogenic origin of concern are halogenated compounds that destroy ozone (O_3). Destruction of ozone in the stratosphere is harmful because it allows more UV light to reach the earth's surface, and excessive exposure to UV light causes skin damage that can lead to cancer in humans. Chlorofluorocarbon (CFC) compounds were used as refrigerants and solvents but have been phased out almost entirely in favor of hydrochlorofluorocarbon (HCFC) compounds. There now is a plan to phase out HCFC compounds by 2030. Nevertheless these compounds remain for many years in the atmosphere and all fluorinated gases (F-gases) in the atmosphere are of anthropogenic origin. Concentrations of the various F-gases range from about 3 to 500 ppt (http://cdiac.ornl.gov/pns/current_ghg.html).

An assessment by Bond et al. (2013) found that black carbon emissions were likely around 7.5 Mt globally in 2000. Black carbon absorbs sunlight and generates heat in the atmosphere. It also may be deposited on snow or ice where it absorbs sunlight and accelerates melting. The lifetime of black carbon in the atmosphere is less than 1 month, and its concentration in the atmosphere varies greatly with location.

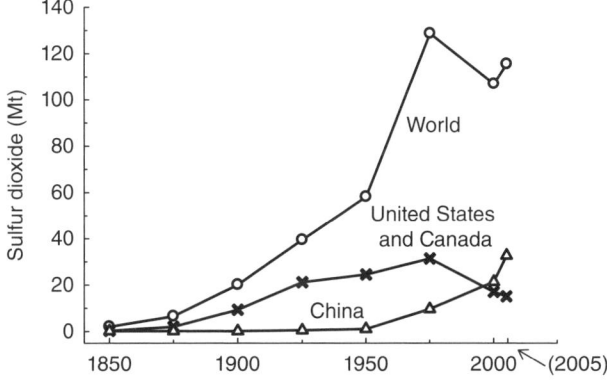

Figure 7.4 Annual estimates of anthropogenic sulfur dioxide emissions at 25-year intervals (1850–2005) for China, United States (US) and Canada, and the world. *Source:* Data for constructing figure from Smith et al. (2011).

Solar radiation, atmospheric gases, and temperature

The sun

The sun is a star comprised mainly of hydrogen (73.46%) and helium (24.58%). As a result of the continuous transmutation of hydrogen to helium in the sun, it radiates a tremendous amount of electromagnetic radiation into space (Fig. 7.5). The earth intercepts a minute portion of the sun's radiation that consists of about 40% visible light (0.38–0.76 µm wave length), 10% UV radiation (0.40–0.10 µm), and 50% infrared or IR radiation (0.7–5 µm). Solar radiation also contains X-rays and gamma rays shorter than UV radiation and microwaves and radio waves longer than IR radiation (Fig. 7.5, Table 7.8). However, most solar radiation is centered between 0.2 and 2 µm. Radiation often is classified as short-wave or long-wave radiation—most long-wave radiation has wave lengths between 4 and 100 µm (http://en.wikipedia.org/wiki/sun).

   The amount of solar radiation received by a plane perpendicular to the sun's rays at the top of the earth's atmosphere is called the solar constant despite the quantity

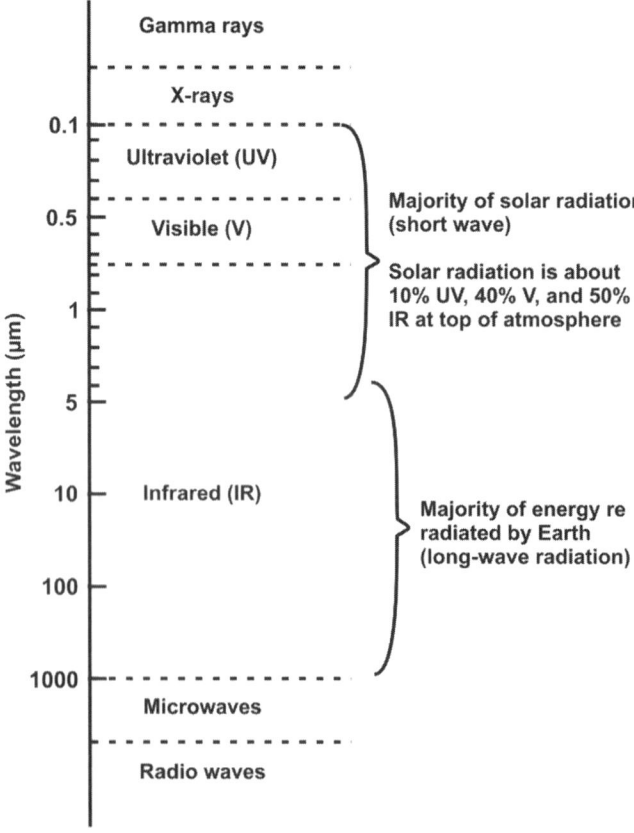

Figure 7.5 Electromagnetic spectrum.

Table 7.8 Comments on radiation of different wavelengths.

Wavelength (μm)	Comments
$<10^{-5}$	Gamma rays—kill cells, but used to treat cancer
10^{-5} to 10^{-2}	X-rays—diagnostic use in medicine; excess can cause cancer
0.280–0.380	Ultraviolet
0.280–0.315	Damages skin (skin cancer)
0.315–0.380	Improves immune system; suntan
0.380–0.760	Visible
0.380–0.500 ⎫ 0.650–0.700 ⎭	Most efficient wavelengths for photosynthesis
0.700–1000	Infrared—many applications in industry, science, and medicine. Night vision most well-known application. Long wave radiation is reradiated by earth and can be trapped in atmosphere to warm earth.
1000	Microwaves and radio waves—important for communication and power applications.

not being exactly constant. The solar constant varies over time according to known cycles but for unknown reasons. The amount of incoming radiation changes during the year—the time of the earth's travel around the sun—as a result of the earth's orbit being elliptical and causing the distance of the earth from the sun to vary. Averaged over several years, the solar constant is 1368 W/m^2 or 82.08 KJ/m^2/min (http://earthobservatory.nasa.gov/Features/SORCE/).

The amount of solar radiation at the limit of the earth's atmosphere is affected by the sunspot cycle. Sunspots are dark areas that appear for days to weeks on the sun's surface. They are the result of magnetic activity resulting in a vortex of gases beneath the sunspot—similar to hurricanes in the earth's atmosphere. The dark area is cooler because of less convection of energy to the sun's surface, but more energy than usual is emitted in the areas surrounding sunspots (http://solarscience.msfc.nasa.gov/feature1.shtml). Solar radiation is greatest when the number of sunspots is maximum, and it decreases when sunspots are minimum. The period between minimum and maximum sunspot activity is 11 years, and the resulting variation in the solar constant averages 1.4 W/m^2. There possibly are longer cycles in the sunspot activity that are undiscovered. The difference in the solar constant resulting from the sunspot cycle is not thought by most climatic scientists to be great enough to appreciably affect the earth's temperature. However, the cycle is not consistent. During the period 1645–1715—known as the Maunder Minimum—almost no sunspot activity was observed (Fig. 7.6). This period corresponded to a period of unusually low temperatures known as the little ice age, and the lower solar irradiance during this time likely caused the phenomenon (Bard et al. 2000). A shorter period of diminished solar irradiance centered around 1810 is known as the Dalton Minimum. However, strong volcanic eruptions and resulting atmospheric contamination during the Dalton Minimum possibly had a greater effect on solar irradiance than resulted from less sunspot activity (Wagner and Zarita 2005). Solar irradiance has been increasing since 1900 and the period 1950–2000 had greater solar irradiance than any half-century period from 1600 to 2000 (Fig. 7.6). Some scientists feel that this increased irradiance may have a major effect on global warming, and it is a natural phenomenon.

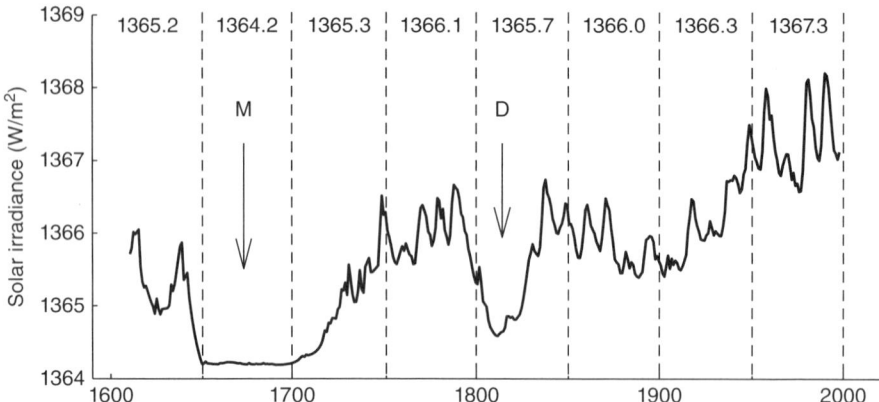

Figure 7.6 Solar irradiance from 1600 to 2000 showing the Maunder Minimum (M) and the Dalton Minimum (D). *Source:* Bard, E., G. Reinbeck, F. Yiou, and J. Jouzel. Solar irradiance during the last 1200 years based on cosmogenic nuclides. *Tellus* 52B:985–992. Copyright © 2000, John Wiley & Sons, Inc.

There also are cycles in the earth's tilt. Although the tilt averages 23.5° from perpendicular with the plane of earth's orbit around the sun, the tilt is thought to vary from 22.5° to 24.0° on a 40 000-year cycle. When tilt is maximum, more sunlight strikes higher latitudes. The beginnings and ends of ice ages are thought to have coincided with minimum and maximum tilt (Hays et al. 1976).Other cycles that can affect the quantity of incoming radiation are the earth's wobble on a 20 000-year cycle, and variation in the shape of the earth's orbit around the sun (eccentricity) on a 100 000-year cycle (http://www.livescience.com/6937-ice-ages-blamed-tilted-earth.html).

The atmosphere

Sunlight must pass through the earth's atmosphere to reach its surface. The atmosphere consists of several layers: troposphere (surface to 10 or 12 km); stratosphere (top of troposphere to 48–50 km); mesosphere (top of stratosphere to around 80 km); and thermosphere (top of mesosphere to 100 km). The air that has a density of 1.2 kg/m³ at the earth's surface becomes more rarified (lower density) with distance from the earth's surface—beyond 30–35 km up, the density of air is <0.1 kg/m³. The ozone layer important for filtering out harmful UV radiation is at about 30 km in the stratosphere (http://www.eoearth.org/article/Atmosphere-layers).

The main gases in dry tropospheric air are nitrogen, oxygen, argon, carbon dioxide, and neon, but air contains water vapor in variable amounts ranging from near 0 to 5% (Table 7.9). The air usually contains suspended particles, soot from incomplete combustion of fuels, sulfur dioxide, salts from sea spray, and other substances not listed in Table 7.9. These substances may be short-lived in the atmosphere and vary greatly in concentration with location.

The gases in the atmosphere—especially water vapor in clouds—reflect or absorb the incoming radiation preferentially by wavelength. The earth's surface also reflects

Table 7.9 Composition by volume of dry air in the lower troposphere.

Gas	Concentration	Gas	Concentration
Nitrogen (N$_2$)	78.084%	Hydrogen (H$_2$)	550 ppb
Oxygen (O$_2$)	20.946%	Nitrous oxide (N$_2$O)	325 ppb
Argon (Ar)	0.934%	Carbon monoxide (CO)	100 ppb
Carbon dioxide (CO$_2$)	394 ppm	Xenon (Xe)	90 ppb
Neon (Ne)	18.18 ppm	Ozone (O$_3$)	70 ppb
Helium (He)	5.24 ppm	Nitrogen dioxide (NO$_2$)	20 ppb
Methane (CH$_4$)	1.79 ppm	Iodine (I$_2$)	10 ppb
Krypton (Kr)	1.14 ppm	Ammonia (NH$_3$)	Trace

Source: Adapted from http://wikipedia.org/wiki/Atmosphere-of-earth).
Water vapor ≈ 0.25% (varies from trace to 5%).

some incoming radiation. The proportion of radiation reflected by the earth and its atmosphere is called the albedo, and the earth's albedo is about 30%. The light received at the earth's surface is about 44% visible, 3% UV, and 53% IR—much less UV than at the top of the atmosphere.

The earth and its atmosphere absorb about 70% of incoming radiation converting it to heat but the heated earth and atmosphere reradiate long-wave radiation with an average wavelength of around 10 μm (Fig. 7.5). The radiation budget of the earth is complex (Fig. 7.7). Of incoming solar radiation, about 30% is reflected by the atmosphere, clouds, and earth surface. The remaining radiation is absorbed mainly by land and water but the atmosphere and clouds also absorb a portion. Part of the radiation absorbed by land and water enters the air as latent heat of evaporation, some is contained in rising air (convection) and some is reradiated as long waves

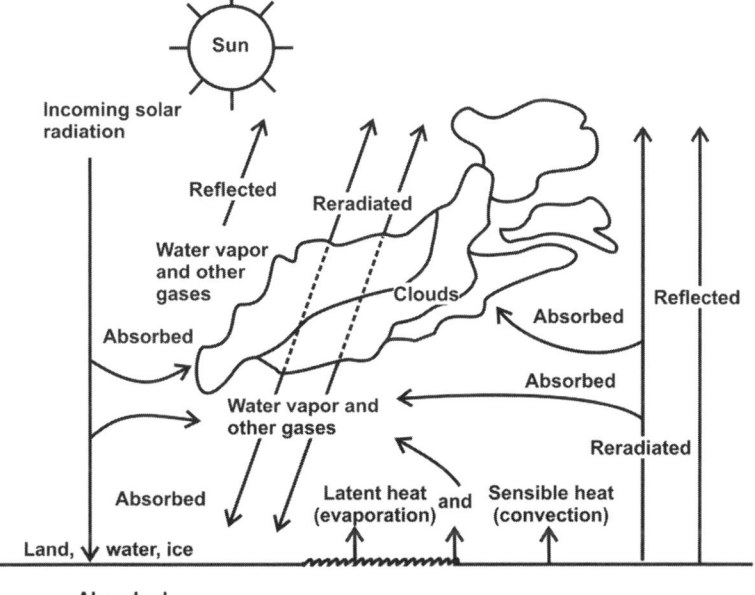

Figure 7.7 Schematic of earth's radiation budget.

and absorbed by the atmosphere and clouds or lost directly to space. Of course, the atmosphere and clouds reradiate long waves to space. When the amount of energy reradiated back into space equals the amount of solar radiation absorbed by the earth and its atmosphere, the average temperature remains constant.

The amount of radiation reaching the earth's cross-section is about 70% of the solar constant or 957.6 W/m^2 (57.46 KJ/m^2/min). The earth has a diameter of 6368 km^2, so its cross-sectional area exposed to sunlight is 127 400 000 km^2. The incoming solar radiation is equal to about 7.32 EJ/min. This is a tremendous amount of energy—the total annual energy use by the world's population was only 551 EJ in 2011. Of course, the earth is a sphere and the surface of a sphere is four times that of a disk of the same diameter. The sunlight falling on the earth would be about 239.4 W/m^2 if it were spread equally over the surface. The amount varies, however, with latitude, time of day, and season because the earth is a rotating sphere tilted on its axis of rotation causing the sun's rays to strike at different angles from the vertical at different places and times.

Nitrogen, oxygen, argon, and most other gases in the atmosphere do not impede the escape of long-wave radiation back into space. However, water vapor, carbon dioxide, methane, nitrous oxide, and F-gases absorb part of the reradiated long-wave radiation slowing its passage back into space and increasing the heat content of the atmosphere. This process depicted in Fig. 7.8 is known as the greenhouse effect of the earth's atmosphere; the gases responsible for this effect are called greenhouse gases. The common observation that the air at night is often warmer when there is cloud cover than when the sky is clear provides tangible evidence of the greenhouse effect. The greenhouse effect warms the earth, and without it the earth would have an average air temperature more than 30°C less. Life on earth, as we know, depends upon the natural greenhouse effect of the atmosphere.

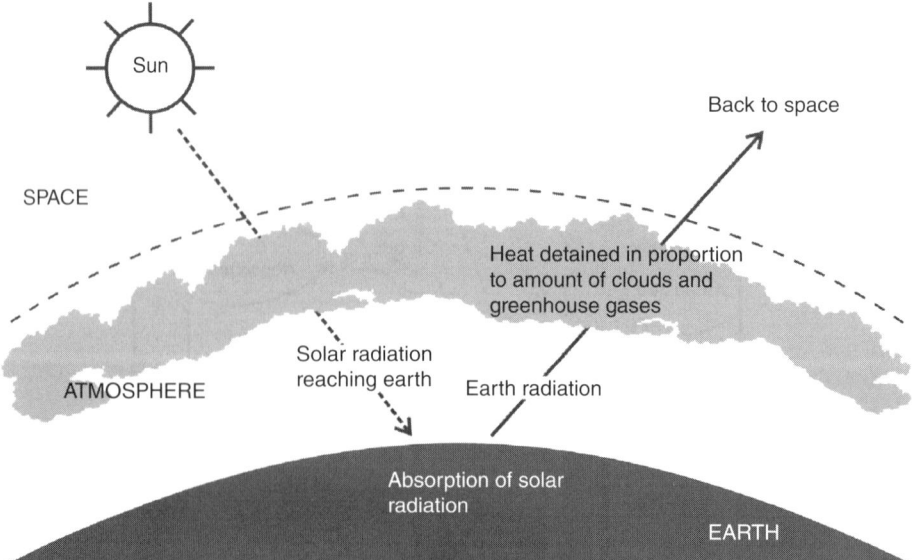

Figure 7.8 Schematic of greenhouse effect of earth's atmosphere.

Greenhouse gases

Records suggest that the global average surface air temperature has been rising. The temperature in 2010 appears to be about 0.8°C greater than in 1910, and the increase has been particularly rapid since the late 1970s (Fig. 7.9). There has been an increase in solar irradiance since 1900 of about 2 W/m² (Fig. 7.6). Nevertheless many atmospheric scientists do not feel that the observed increase in global surface temperature could be solely the result of increasing solar irradiance (http://earthobservatory.nasa.gov/Features/SORCE/). The prevailing conclusion is that the main reason for the increased surface temperature is an enhanced atmospheric greenhouse effect caused by emissions of greenhouse gases from fossil fuel combustion.

Greenhouse gases are not equal with their potential to trap long-wave radiation and increase the heat content of the atmosphere. The atmospheric warming potential of a given greenhouse gas depends upon how much long-wave radiation it can absorb and upon how long this gas remains in the atmosphere. Carbon dioxide—the gas emitted in the largest quantity by human activities—is taken as the standard (global warming potential = 1.0) for ranking greenhouse gases according to their potential for increasing the heat content of the atmosphere (Table 7.10). Ozone that occurs in the troposphere as a result of pollution also can act as a greenhouse gas.

Another way of expressing the relative importance of anthropogenic increases in different greenhouse gases is to rank them according to radiative forcing—the reduction in earth's reradiation of energy into space. The Carbon Dioxide Information Analysis Center (http://cdiac.ornl.gov/pns/current_ghg.html) gave the following radiative forcing values for greenhouse gases: carbon dioxide, 1.85 W/m²; methane, 0.51 W/m²; nitrous oxide, 0.18 W/m²; tropospheric ozone, 0.35 W/m²; F-gases combined, 0.32 W/m². The total radiative forcing of these gases is 3.21 W/m². Moreover, black carbon also contributes to atmospheric heat content, and Bond et al. (2013) concluded that the radiative forcing of black carbon is 0.71 W/m²—second only to carbon dioxide.

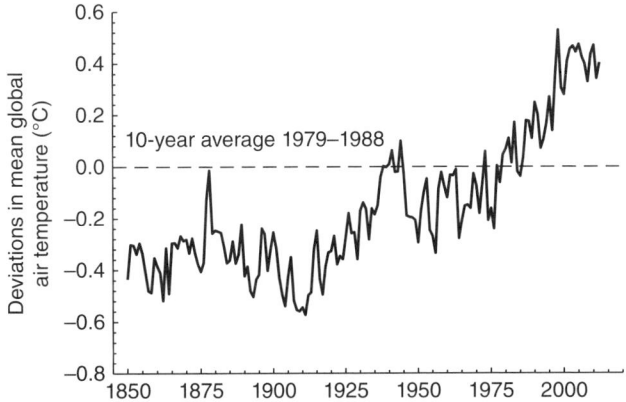

Figure 7.9 Deviations from the global, 10-year average air temperature (1979–1988) from 1850 to 2012. *Source:* http://www.climate4you.com/GlobalTemperatures.htm.

Table 7.10 Global warming potential of greenhouse gases.

Gas	Lifetime (year)	Global warming potential ($CO_2 = 1$)
Carbon dioxide (CO_2)	1	1
Methane (CH_4)	12	21
Nitrous oxide (N_2O)	120	310
F-gases		
HFCs (hydrofluorocarbons)	1—270	140–11 700
PFCs (perfluorocarbons)	800–50 000	6500–9200
Sulfur hexafluoride (SF6)	3200	23 900

Source: Adapted from http://www.epa.gov/climatechange/ghgemissions/gases.html.

Presentation of data on greenhouse gases may be combined by converting the quantities of each gas to a carbon dioxide equivalent (CO_2-e) and summing. This practice also can lead to confusion because in many reports, it is not clear whether information is for carbon dioxide alone or for carbon dioxide equivalent of all greenhouse gases. The most commonly reported statistic related to global warming is carbon dioxide emissions because carbon dioxide is the major greenhouse gas released by fuel combustion. The shares of anthropogenic greenhouse gas were given as carbon dioxide (energy related), 61%; carbon dioxide (nonenergy related), 20%; methane, 13%; nitrous oxide, 5%; F-gases, 1% (http://www.epa.gov/climatechange/ghgemissions/global.html).

Carbon dioxide emissions from anthropogenic sources slowly increased from 1850 to 1950, but since then, they have increased more rapidly (Fig. 7.3). The 10 major countries with respect to greenhouse gas emissions are listed in Table 7.11. China and the United States have much greater emissions than other countries. In 2010, about 32 000 Mt of carbon dioxide were released—the carbon dioxide equivalent of all greenhouse gases released into the atmosphere in 2010 was considerably more than 40 000 Mt. The annual quantity of global emissions is expected to increase until at least 2030 as illustrated in Fig. 7.3 for carbon dioxide emissions.

Table 7.11 Top 10 countries for emissions of greenhouse gases in 2008.

Country	Carbon dioxide (Mt)	Estimated total greenhouse gases (Mt CO_2-e)[a]
China	7032	8681
United States	5461	6742
India	1743	2152
Russia	1709	2110
Japan	1208	1491
Germany	786	970
Canada	544	672
Iran	538	664
United Kingdom	523	646
South Korea	509	628

[a] Assuming that carbon dioxide is 81% of carbon dioxide equivalent (CO_2-e) for all gases.
Source: Adapted from http://en.wikipedia.org/wiki/List_of_countries_by_carbon-dioxide_emissions.

Table 7.12 Greenhouse gas emissions by fisheries and aquaculture in 2008.

| Source | Emissions (Mt CO_2-e) | |
	Capture fishery	Aquaculture
Fishing vessels	40–90	–
Aquaculture	–	35–40
Postharvest and processing[a]	5–7.5	5–7.5
Transport[a]		
Air	1.5–2.0	1.5–2.0
Non-air	170	170
Total	**216.5–269.5**	**211.5–219.5**

[a]Assumed a 50% split between fishery and aquaculture products.
Source: Adapted from FAO (2011a).

The source of greenhouse gases by sector were given as energy supply, 26%; industry, 19%; forestry, 17%; agriculture, 14%; transportation, 13%; buildings, 8%; waste and wastewater, 3% (http://www.epa.gov/climatechange/ghgemissions/global.html). In this assessment, the agriculture emissions apparently included only those accrued in crop production, and other emissions related to the world food system were included in other categories. FAO (2011a) concluded that greenhouse gas emissions in 2006 from the entire world food system totaled 9.7 Gt or about 22% of global emissions of 44.17 Gt CO_2-e. The shares of greenhouse gas emissions were separated by sector as follows: plant crops, 28.2%; livestock, 34.9%; fisheries (including aquaculture), 2.1%; processing and distribution, 21.7%; retail, preparation, and cooking, 13.1%.

An attempt was made to separate fisheries and aquaculture with respect to greenhouse gas emissions (Table 7.12) based on the information provided by FAO (2011a). Assuming the maximum estimate of greenhouse gas emissions made by FAO, fisheries production would have been responsible for 487 Mt of greenhouse gases (CO_2-e)— 269.5 Mt for fisheries and 219.5 Mt for aquaculture. These values are 2.8% and 2.2% of total food system emissions, and 0.61% and 0.49% of total global greenhouse gas emissions, respectively.

The FAO (2011a) estimates of greenhouse gases from aquaculture are undoubtedly low because fossil fuel is used in aquaculture for many purposes besides feed production (Boyd and Tucker 1998). Carbon footprints available for a few aquaculture species usually are between 2 and 6 kg CO_2/kg product (Pelletier et al. 2009; Winther et al. 2009; Mungkung et al. 2012; Boyd et al. 2011). Typical carbon footprints for wild-caught fish are 1–3 kg CO_2/kg product (average = 2.47 kg CO_2/kg) (Winther et al. 2009). Carbon footprints for other common meats are 12–16 kg CO_2/kg beef, 4–8 kg CO_2/kg pork, and 3–4 kg CO_2/kg chicken (Pelletier et al. 2010a, 2010b).

Using a conservative estimate of 5 kg CO_2/kg of production, a live weight to useful product ratio of 1:0.75, and the 2008 FAO estimate of 52.9 Mt of aquaculture production, the carbon dioxide emissions could be around 198 Mt or possibly about 245 Mt CO_2-e of total greenhouse gases. This would represent about 2.5% of carbon dioxide emissions by the world food system and around 0.55% of total global emissions.

Aquaculture ponds sequester organic carbon in sediment (Boyd et al. 2010). The average sequestration rate for 233 ponds representing the culture of eight common species or species group in nine countries was 148.9 C/m^2/year (546 g CO_2/m^2/year). The total amount of organic carbon sequestered globally was estimated to be 16.6 Mt/year (60.8 Mt CO_2/year). This amount of carbon dioxide would offset about 23% of estimated greenhouse gas emissions by the global aquaculture industry. However, it is likely that at the farm level, many types of aquaculture are essentially carbon emissions neutral or are net carbon sequesters.

Effects of anthropogenic greenhouse gas increase

Concentrations of greenhouse gases are predicted to continue to increase, and average global surface temperature that has increased about 0.8°C in the past century is forecast to increase by 1.7–2.7°C by 2050 (US Global Change Research Program 2009). Not all scientists agree with this viewpoint (Goreham 2013). Some feel that the increase in temperature may be more the result of natural cycles and events or a combination of natural cycles and anthropogenic activities rather than of increased concentrations of greenhouse gases alone. The climate of the world has changed many times over the millennia and as discussed earlier, solar irradiance also has fluctuated. Variations in irradiance over the past 1200 years have been reported by various researchers to vary from 2 to 5 W/m^2 over periods of several decades (Bard et al. 2000). This is within the range of radiative forcing attributed to greenhouse gases and black carbon, and solar irradiance has been generally increasing since 1900 (Bard et al. 2000). Moreover, van Geel et al. (2002) emphasized that the global climate system is more sensitive to small variations in solar activity than generally believed. It is wise to keep an open mind about the causes of climate change because to quote Mark Twain, "a consensus is always wrong." Fossil fuel use and air pollution need to be reduced, but increasing greenhouse gas concentrations are not the only factor contributing to global warming. Even ICPP recognizes that natural phenomena increase radiative forcing.

The predicted influence of increased greenhouse gas concentrations and global warming on the earth's ecosystems and plant, animal, and human inhabitants are considerable. The most serious forecasts are extinction of many plant and animal species, changes in precipitation patterns, increases in temperature and carbon dioxide concentration leading to an expansion of water volume (sea level rise) and acidification, melting of glaciers and sea ice, also causing changes in habitat and contributing to sea level rise, and more severe weather events.

There is already evidence of more severe droughts in many regions of the world and more rainfall and flooding in others. It is predicted that river flow will increase by 10–40% at high latitudes and in some wet tropical areas, while a similar decrease in river flow is in some dry, midlatitude regions and in the subtropics (http://www.epa.gov/climatechange/impacts-adaptation/international.html). The impacts on natural ecosystems, water supply, agriculture, and flood damage associated with such predicted changes in rainfall and river flow changes would be sizeable.

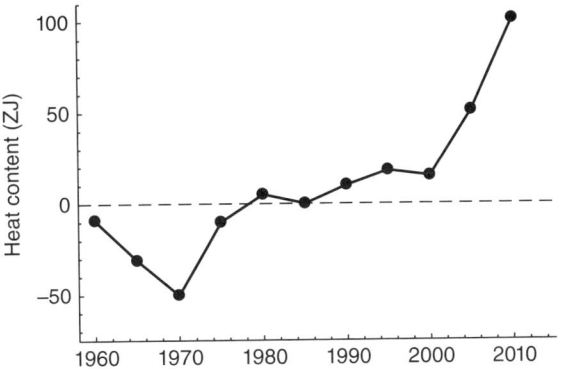

Figure 7.10 Heat content of the 0–700 m layer of the ocean. The baseline (0 ZJ) is heat content for 1985. *Source:* http://www.rodc.noaa.gov/OC5/3M_HEAT_CONTENT/.

The heat content of the upper ocean layer (0–700 m) has steadily increased since the mid-1980s (Fig. 7.10). This results in thermal expansion of water because warmer water is less dense than cooler water. Both the Arctic and Antarctic polar ice is being lost at an accelerated rate (Polyakov et al. 2010; Velicogna 2009; Chen et al. 2009). Glacier volume also is decreasing and northern hemisphere snow cover is retreating (http://www.ncdc.noaa.gov/indicators/).

The combined effect of melting ice and expanding volume of ocean water is sea level rise (Fig. 7.11). Globally mean sea level has been rising at a rate of about 1.7 mm/year for the past 100 years, but the rate of rise averaged over the past 20 years is 3.5 mm/year (http://www.ncdc.noaa.gov/indicators/). Predictions for the future rise in global mean sea level vary from 10 to 48 cm by 2050 and 18 to 200 cm by 2010 (IPCC 2007; Rahmstorf 2007; Pfeffer et al. 2008). Sea level rise would inundate low-lying areas and many of the world's major cities are located in such places. In addition, sea level rise will increase coastal erosion and make the impacts of storms more severe.

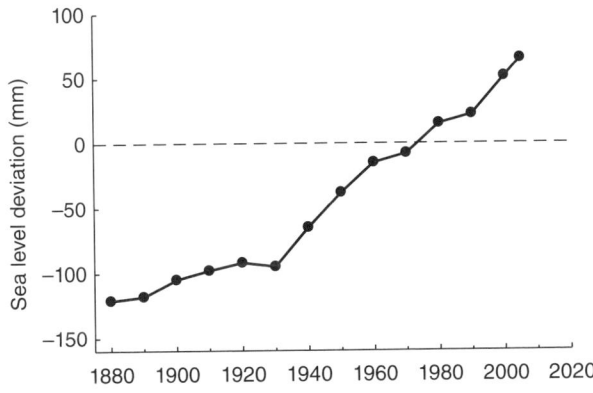

Figure 7.11 Sea level rise since 1860. The baseline (0-mm deviation) was measured at sea level in 1880. *Source:* NOAA (2010).

At greater carbon dioxide concentration in the atmosphere, the ocean absorbs more carbon dioxide. About one-third of anthropogenic carbon emissions are absorbed by the ocean (http://www.eoearth.org/article/Ocean-acidification). Carbon dioxide has an acidic effect in water:

$$CO_2 + H_2O = HCO_3^- + H^+. \tag{7.4}$$

Increasing the carbon dioxide concentration in the ocean will decrease pH. The average pH of the ocean decreased from about 8.12 in the late 1980s to 8.09 in 2008. The carbon dioxide concentration was 0.58 mg/L in 1990; it is expected to increase to 0.88 mg/L by 2065 and to 1.08 mg/L by 2100. This could lead to a much greater rate of pH decline in the future. A decrease in pH in the ocean would increase the solubility of calcium carbonate minerals such as aragonite and calcite that comprise the shells of many marine organisms and coral (Orr et al. 2005; Langer et al. 2006). This process of acidification could have drastic effects on biodiversity in the ocean.

In addition to causing changes in surface temperature and rainfall patterns, global warming is expected to increase the severity of droughts, hurricanes, and other severe weather events. The annual climate extremes index assesses the frequency of extreme temperature, precipitation, and storm intensity. This index fluctuated from 1920 until the mid-1970s in the United States with no trend of increase or decrease over time; since then there has been a clear, trend of increase for the index (http://www.ncdc.noaa.gov/indicators/). Of course, the reader should realize that the reporting of severe weather events has improved greatly over the years and most of the increase in the index may be the result of better reporting. The same may also be said for many of the other data presented in this chapter.

The negative effects of air pollution on human health are considerable by contributing to respiratory infections, heart disease, and lung cancer. According to the World Health Organization (WHO) urban outdoor air pollution is estimated to cause 1.3 million deaths annually (http://www.who.int/mediacentre/factsheets/fs313/en/). The WHO has set air quality guidelines as follows:

Particulate matter
Particles <2.5 μm: 10 μg/m^3 annual mean; 25 μg/m^3 24-hour mean
Particles <10 μm: 20 μg/m^3 annual mean; 50 μg/m^3 24-hour mean
Ozone 100 μg/m^3 8-hour mean
Nitrogen dioxide 40 μg/m^3 annual mean; 200 μg/m^3 1-hour mean
Sulfur dioxide 20 μg/m^3 24-hour mean; 500 μg/m^3 10-minute mean

Effects of sulfur and nitrogen dioxide

Rainwater is naturally acidic because it becomes saturated with carbon dioxide while falling through the atmosphere. The pH of rainfall is normally around 5.6 (Boyd 2000). Increasing the carbon dioxide concentration in the atmosphere has little effect on the pH of rain, but in many regions worldwide, the pH of rain is well below that possible from carbon dioxide alone. For example, rainfall in the northeastern

United States and Canada has an annual, average pH of 4.2–4.4 and rain from individual storms has lower pH (Haines 1981). This phenomenon known as acid rain, results mainly from sulfur (SO_2 and H_2S) and nitrogen (NO_x).

Hydrogen sulfide and other sulfides emitted with sulfur dioxide are quickly oxidized to sulfur dioxide in the atmosphere. In the gas phase of the atmosphere, sulfur dioxide and nitrogen oxides (NO_x) form strong acids (Boyd 1990) as illustrated below:

$$SO_2 + OH^- = HOSO_2 \tag{7.5}$$

$$HOSO_2 + O_2 \rightarrow HO_2^- + SO_3 \tag{7.6}$$

$$SO_3 + H_2O \rightarrow H_2SO_4 \text{ (sulfuric acid)} \tag{7.7}$$

$$NO_2 + OH^- \rightarrow HNO_3 \tag{7.8}$$

$$\text{(or } NO_x + xOH^- \rightarrow HNO_3 \text{ (nitric acid).} \tag{7.9}$$

In cloud droplets, sulfur dioxide reacts in a slightly different manner:

$$SO_2 + H_2O = SO_2 \cdot H_2O \tag{7.10}$$

$$SO_2 \cdot H_2O = H^+ + HSO_3^- \tag{7.11}$$

$$HSO_3^- = H^+ + SO_3^{2-}. \tag{7.12}$$

But the sulfite ion (SO_3^{2-}) can react with water forming sulfuric acid.

Burning of fuels is the main source of acid-forming substances in the atmosphere. All sectors that burn fuel contribute such emissions roughly in proportion to energy use. The world food production system represents about 30% of the world's end-use energy and therefore a similar proportion of emissions. World capture fisheries production uses around 0.68% of global end-use energy consumption based on the analysis by FAO (2011a). Examination of energy use reported in life cycle assessments of the production of several common species suggest that aquaculture may use as much as 0.47% of global end-use energy and as a result contribute a similar proportion of emissions.

Burning of crop residues after harvest and other agricultural wastes often cause much smoke and haze at times in developing countries. These emissions, however, contribute only about 0.1% of sulfur emissions (Smith et al. 2011).

It is interesting to note that global sulfur dioxide emissions that had been increasing rapidly declined following an effort that began in the 1970s in developed countries to remove sulfur from emissions (Fig. 7.4); data for 2005 suggest that emissions are increasing again. The emissions for the United States and Canada (Fig. 7.4) show how emissions increased from 1850 until widespread concern in the 1970s led to regulations and a decline in emissions. Emissions by China remained low until the 1970s, but they have rapidly increased as the country's industrial sector and economy expanded in the absence of effective regulations.

The combined discharge of United States and Canada is about half as much sulfur dioxide as China (Fig. 7.4). According to Smith et al. (2011) other countries and regions with large sulfur dioxide emissions are India (6275 Mt), Russia (5975 Mt), Western Europe (6242 Mt), and Central Europe (4832 Mt). International shipping was responsible for 12 078 Mt sulfur dioxide in 2008—more than any single country other than China.

Smith et al. (2005) predicted that global sulfur emissions will rise slightly until 2020 and then steadily decline with 2050 levels being only 50% of 2020 levels. Presumably this will be the result of more effective regulation of sulfur emissions in China and other emerging economies.

Acids formed in the atmosphere are washed out and delivered to the earth's surface in rainfall; they also may be deposited directly from the air onto vegetation, the ground, or other surfaces—wet deposition versus dry deposition. The obvious effect of acidic deposition is to depress the pH of soils and surface waters. The effects of acidic rain are more severe in areas with highly leached, acidic soils where surface waters typically are of low pH and weakly buffered, such as the eastern United States and Canada and Northern Europe (Boyd 1990). Because the acid-forming gases and their products are free to move in the atmospheric circulation, emissions from highly populated, industrial regions may cause acidic rain far away in rural areas where there are few emissions.

In some water bodies, pH has declined by 1–2 units since the 1940s (Seip and Tollan 1978; Cowling 1982). Many lakes and streams in the eastern United States and Canada have a pH below 5 (Beamish and Harvey 1972; Schofield 1976; Haines and Akielaszek 1983). Acidification has led to adverse effects at all trophic levels; abundance, production, growth, and species diversity have diminished. Fish have suffered acute mortality, reduced growth, reproductive failures, skeletal deformities, and increased accumulation of heavy metals (Haines 1981).

Mitigation of acidity by liming can be effective but the expense prohibits the use of this practice to highly prized water bodies (Flick et al. 1982). Acid rain usually is not an important issue in aquaculture because acidity resulting from the oxidation of ammonia, from fertilizers and feeds, greatly exceeds acidity from acid rain (Boyd and Tucker 1998). Liming materials to offset the acidity from nitrification also counteracts the effects of acid rain.

Acid rain also has negative effects on terrestrial ecosystems (Hutchinson and Havas 1980; Feng et al. 2002; Likens et al. 1979). These effects include soil acidification, increased leaching of nutrients, negative effects on vegetative growth, and loss of sensitive species of plants and animals. Mass mortalities of large trees have been observed at some sites (Johnson and Siccama 1983).

Elevated carbon dioxide concentration and mineral acids resulting from sulfur and nitrogen dioxide emissions increase limestone solubility. Using calcium carbonate to represent limestone, the reactions are

$$CaCO_3 + H_2O + CO_2 = Ca^{2+} + 2HCO_3^-. \tag{7.13}$$

The reaction depicted in Equation 7.13 is of the equilibrium type, and increasing carbon dioxide concentration forces it to the right—more calcium carbonate dissolves (Boyd 2000). Mineral acidity dissolves limestone by direct action of hydrogen ion:

$$CaCO_3 + H_2SO_4 = Ca^{2+} + SO_4^{2-} + H_2O \qquad (7.14)$$

$$CaCO_3 + 2HNO_3 = Ca^{2+} + NO_3^- + H_2O. \qquad (7.15)$$

Monuments and buildings often are made from limestone. They may be severely damaged by acid rain.

Actions to combat atmospheric emissions

The first broad international agreement related to acid rain was the Convention on Long Range Transboundary Air Pollution that was signed in 1979 by all nations of Eastern and Western Europe and North America (http://www.jstor.org/stable/2201373). The countries pledged to reduce emissions of sulfur and nitrogen dioxides and other air pollutants. The United States and Canada also signed an Air Quality Agreement in 1991 to reduce emissions of air pollutants in both countries. Japan has been the leading country in Asia seeking to develop agreements among countries to reduce air pollution (http://easts.dukejournals.org/content/1/2/263.extract). The United Nations Framework Convention on Climate Change (UNFCCC) was initiated in 1992 and most countries now participate in this program that has the objective of stabilizing greenhouse gas concentrations in the atmosphere at a level where ecosystems can adapt naturally to climate change, food production is not threatened, and economic development can proceed in a sustainable fashion. Countries participating in the UNFCCC are required to report greenhouse gas emissions and to pledge to reduce emissions (http://unfccc.int/2860.php). The IPCC was established in 1988 by the UN Environmental Program and the WMO to provide a scientific analysis of the current state of knowledge in climate change and its effects (http://ipcc.ch/); IPCC and UNFCCC work together closely.

As a result of these international accords, countries have enacted regulations to meet their reporting and emissions control commitments. In addition, many government and nongovernment organizations have initiated education programs to improve public understanding of the importance of reducing emissions and of the effects of climate change. Most governments even provide incentives—often as tax breaks—to encourage adoption of new technology for reducing atmospheric emissions.

There is considerable effort to control emissions through particle removal by electrostatic precipitation and fabric filtration, selective and nonselective catalytic reduction of nitrogen dioxide, and wet/dry scrubbing for sulfur dioxide control (http://www.southernresearch.org/environment-energy/air-pollution-control/particulate-so2-and-nox-control). Cars in most countries must be equipped with positive crankcase ventilation to avoid release of unburned hydrocarbons in exhaust fumes, and with catalytic converters for final treatment of exhaust fumes (http://en.wikipedia.org/wiki/vehicle_emissions_control). These efforts combined with regulations to increase energy efficiency have been successful as demonstrated by data

Table 7.13 Energy content of common fuels and their life cycle carbon dioxide emissions.

Fuel	Energy content (MJ/kg)	Life cycle carbon dioxide emissions (kg/GJ)
Coal	29	134
Natural gas	38	75
Liquefied petroleum gas (LPG)	46	90
Gasoline	44	81
Diesel	43	87
Fuel oil	42	97
Biodiesel	37	13–41
Bioethanol	27	24–34
Wood chip	14	4–5
Wood pellets	17	4–7
Grass/straw	14.5	1.5–4
Grid electricity	–	150

Source: Adapted from http://www.biomassenergycentre.org.uk/portal/page?_pageid=75,163182&_dad=portal&_schema=PORTAL.

showing that in the United States between 1970 and 2005 annual atmospheric emissions declined as follows: carbon monoxide, 197.3–89 Mt; nitrogen dioxide, 26.9–19 Mt; particulate matter, 12.2–2.0 Mt; sulfur dioxide, 31.2–15 Mt; volatile organic compounds, 33.7–16 Mt; lead, 0.221–0.003 Mt (http://www.epa.gov/airtrends/2006/emissions_summary_2005.html). Although carbon dioxide emission in the United States increased from 4328 Mt in 1970 to 5826 Mt in 2005, it now appears to be decreasing. The 2009 carbon dioxide estimate was 5299 Mt.

With respect to aquaculture, there has been an effort to develop better practices to minimize negative environmental impacts of production. These include measures such as more efficient use of feeds, water, and aeration that would lessen energy use, and this will lessen atmospheric emissions (see Chapter 13). The FAO guidelines for aquaculture ecolabel certification programs also recommend development of standards for energy efficiency and reduction in carbon dioxide and other greenhouse gas emissions (FAO 2011b).

Energy sources differ in their heat contents and carbon dioxide emissions upon combustion (Table 7.13). Electricity has the greatest carbon dioxide emissions per unit of energy because emissions are calculated based on the quantity of fuel used to generate the electricity. The same procedure is used to estimate carbon dioxide emissions of other fuels, but each kilowatt·hour of electricity generated by burning fuel is delivered—minus transmission losses—at the point of use. Other fuels typically are burned at the point of use and a portion of the energy released is lost as heat before final use. In terms of carbon dioxide emissions per unit of useful energy, electricity does not differ from other fuels as much as the data in Table 7.13 might suggest.

Some fuels burn cleaner than others—produce less carbon emissions per unit of energy released. Therefore, cleaner fuels such as natural gas are encouraged as a means of reducing carbon dioxide emissions. However, fossil fuels are not a renewable resource and switching to cleaner fuels cannot be considered a long-term solution.

Limits have been imposed on carbon emissions in many countries. An upper limit may be placed on total carbon dioxide emissions and allowable quantities of emissions allocated among sources or industries. In this way, an entity that is able to reduce its emissions below its allotment may sell the unused part of its allotment (called a carbon credit) to another entity that is unable to comply with its allotment. This arrangement is known as carbon cap and trade.

Carbon sequestration projects in which carbon dioxide is sequestered in soil, sediment, or living biomass or removed chemically can be established and certified. These projects allow the operators to acquire carbon credits that can be used or traded. Carbon exchanges that operate much like the stock market have been established to facilitate trading in carbon credits.

Some governments have put taxes on carbon dioxide emissions. Presumably, the income from these taxes is used in programs to reduce carbon emissions.

The use of biomass for fuel is ecologically appealing because it could result in sequestration of carbon emissions through photosynthesis. Carbon dioxide released from biofuels would be used in photosynthesis to produce more biofuels; in essence, fuel recycling would occur. This is much different than the prevailing situation in which carbon dioxide sequestered in biomass by photosynthesis long ago and gradually transformed to fossil fuels is released into the atmosphere to disrupt the carbon cycle and increase atmospheric carbon dioxide levels.

A disadvantage of biomass fuels such as bioethanol and biodiesel is that they usually come from crops used as human and livestock food. Moreover, there is not an unlimited supply of additional land that could be brought into cultivation to produce biomass for fuels. There is much interest in cultivation of algae for producing biofuels as more agricultural land would not be necessary. The idea of algal culture for food and fuel is not new (Burlew 1953), but so far there has been little success in commercializing algal production for these purposes.

The most likely scenario for assuring a continuing energy supply is to conserve existing fossil fuels, use novel sources such as wind, solar, and geothermal energy, continue to develop biofuels, use energy from waste heat and biomass, and attempt to find new energy sources. Mankind always has adapted to change, and as history tends to repeat itself, humans will likely adapt to a new, but yet unknown paradigm related to energy and also to climate.

Conclusions

Total primary energy use in 2008 was an estimated 493 EJ with end use of 294 EJ. End use of energy by the world food system for producing, processing, distributing, preparing, and cooking food was 95 EJ or 32% of the total. Capture fisheries and aquaculture sectors consumed about 2 EJ and 0.4 EJ of end-use energy, respectively, or 0.68% and 0.14% of total end-use energy, respectively (FAO 2011a). However, the energy use for aquaculture obtained from LCA studies suggests a greater energy use that possibly is as high as 1.38 EJ/year or 0.47% of the current total global end-use energy.

Greenhouse gas emissions by the world food system in 2008 were about 9700 Mt CO_2-e or about 22% of global emissions of 44.17 Gt of CO_2-e reported by FAO (2011a). Data from the FAO report suggest that 269.5 Mt CO_2-e were from capture fisheries and 219.5 Mt CO_2-e were from aquaculture. Extrapolation from life cycle analysis studies indicated that aquaculture might produce 198 Mt of carbon dioxide alone. This equates to 245 Mt CO_2-e of greenhouse gas emissions—slightly higher than the FAO estimate. Nevertheless, the greater estimate is only 0.55% of total global emissions.

Greenhouse gas emissions are increasing at a rapid rate, and sulfur and nitrogen dioxide emissions that were in decline are increasing again because of greater emissions by China, India, and a few other rapidly developing countries. These emissions are causing acid rain and acidification of the ocean. Air temperature increase (global warming) that is causing melting of polar ice and glaciers and warming and expansion of ocean water with resulting sea level rise is attributed mainly to greenhouse gas emissions by many scientists. However, increased solar irradiance may also be a major factor in global warming.

The world must decrease its dependence on fossil fuels because these nonrenewable resources are diminishing. Uranium for nuclear power also is nonrenewable and in limited supply. In order to assure a continuing supply of energy will likely involve a combination of several measures to include conservation of fossil fuels, greater utilization of wind, solar, and geothermal energy, continued development of biofuels, use of waste heat and biomass, and research to find new energy sources. The effort to find a more sustainable energy paradigm also will include efforts to lessen greenhouse gas emissions. Mankind is highly adaptable and innovative; it is likely we will find solutions to the energy dilemma.

The eNGO perspective

The use and sources of energy are of paramount concern to the eNGO community. Regardless of the region, industry sector or consumer, energy utilization is a driving force behind all eNGOs because its implications are global in nature. There is a fundamental conviction that humans, through the burning of fossil fuels, are a major contributor to global warming and climate change. The eNGOs strategy on energy and climate change is quite simplistic in principle—reduce emissions, conserve plant biomass, and identify and migrate to cleaner forms of energy (i.e., solar and wind power).

As stated in this chapter, the forecast of greater greenhouse gas emissions will have a dramatic impact on our global ecosystem. From an eNGO perspective promoting the acknowledgement of this is a critical first step toward reducing the impacts of climate change. If this aspect is accepted by citizens, businesses, and governments, stakeholders can move forward toward a strategy for reducing impacts.

Although aquaculture production at the farm level is a relatively small contributor to greenhouse gas emissions, the supply chain for aquaculture including processing, feed ingredient production and shipping, and final product shipping can contribute a sizeable amount to total greenhouse gas emissions by the food production sector. Thus the primary eNGO strategy for reducing aquaculture's impact on climate change is by working through the entire supply chain. This strategy is best accomplished by working at the point of greatest aggregation. Retail companies selling aquaculture products to end-consumers are the most logical point of focus for the eNGOs. However, the challenge with this level of engagement is prioritizing emission cuts from the production and shipment of a variety of products, not necessarily just aquaculture. To this end, many of the eNGOs have either a climate change program or a broader business engagement program that addresses more cross-cutting issues related to the macro-level use of energy by specific retailers and their suppliers. Because the eNGOs have taken on a strategic structure that allows them to work with any type of business to address climate change impacts, aquaculture impacts related to climate change may not be of highest priority. Nevertheless, every eNGO engaged in aquaculture issues has some expectations for the aquaculture industry which can be as benign as quantifying emissions of greenhouse gases and exploring other forms of renewable energy to actual emission reduction targets.

The eNGOs contend that there is no future for those businesses that intend to rely on fossil fuels as a means to bring products to market. To the extent possible eNGOs are trying to educate aquaculture businesses but at the same time the eNGOs are learning from many of the most innovative businesses in the world. For example, one tilapia farm located in Honduras informed a forum of eNGOs about the ability to convert tilapia wastes to biodiesel to fuel trucks used to ship tilapia from their processing plant to ports for export. From an investment and growth perspective, businesses will have to become more efficient with their energy use. This is a cost-saving measure and a means to forecast farther into the future based on rising energy costs.

There is likely no "acceptable" level of fossil fuel energy use at aquaculture facilities from an eNGO perspective, although there is an understanding that most aquaculture production activities rely on fossil fuels. Further what is considered "tolerable" by the eNGOs is not clear. Primarily, eNGOs will look toward percentage reductions in total carbon emissions as a way to show responsible farming practices. What these percentages should be is relative to the specific eNGO and the specific company—5% reduction of greenhouse gases in 5 years is good, but 10% is better, and 20% better still.

The conservation of forests to protect habitat of threatened or endangered species is well understood but eNGOs are also adamant that the presence of forests will aid in mitigating greenhouse gas emissions by sequestering, in particular, carbon in the form of carbon dioxide. With this in mind, eNGOs lobby heavily for the protection and conservation of forests as a means to mitigate climate change. Thus, aquaculture operations that remove vegetation to construct farms are typically encouraged to identify ways in which they can offset that removal. The larger the farming operation, typically the larger investment in carbon offsets. For example, some farms undertake forest replanting programs as part of the terms of an eNGO partnership. WWF-Indonesia has a partner shrimp farm in Borneo that has developed a large mangrove preservation area in response to the concerns associated with the need for the farm to offset carbon emissions. Of course, the size of the mangrove preservation area was dependent on the availability of appropriate land, but also on the amount of funds the farm could allocate to procure this area.

The promotion of and fostering the adoption of Reduced Emission from Forest Deforestation and Degradation (REDD) has become a major push for the eNGOs during the past decade. Although not prevalent in the aquaculture industry, REDD is an effort to maintain forested regions around the world. The principle is to build up a region's forest management capabilities to conserve forests. In return, companies can purchase carbon offset credits from these "sustainably managed" forests. Conservation International, through a partnership with Disney, facilitated the investment of $US 10 million into REDD projects in Peru and the Democratic Republic of Congo. It is likely that aquaculture facilities cannot develop REDD projects themselves, but there are more opportunities in developing countries now to purchase carbon credits to support efforts to sustain forest cover and its carbon sequestration function.

Changing energy sources is often a challenge for aquaculture operations because most production occurs in regions of the world where energy sources are unstable or unreliable. Although an eNGO would prefer that a farm use wind power to sustain the operations, hydroelectric and coal-fired plants may be the only available sources for the farm to obtain energy. It is possible for some farms to generate their own power through capture of solar radiation or constructing windmills on site. However, requesting this of farms would likely only be possible for large farms with an integrated production and processing system.

Fundamentally, the exploration of energy use and sources is a fairly new endeavor for the eNGOs working on aquaculture projects. Most eNGOs are requesting farms to monitor and quantify energy use. The intent is to build enough information to begin to identify more efficient farms and practices and bring other farms toward the adoption of these practices. Identifying alternative energy sources for aquaculture operations still remains a challenge for both the industry and the eNGOS, although most progressive farms are working toward energy alternatives to maintain production into the future.

References

Aneja, V. P. 1990. Natural sulfur emissions into the atmosphere. *Journal of the Air and Waste Management Association* 40:469–476.

Bard, E., G. Reinbeck, F. Yiou, and J. Jouzel. 2000. Solar irradiance during the last 1200 years based on cosmogenic nuclides. *Tellus* 52B:985–992.

Beamish, R. and H. Harvey. 1972. Acidification of the LaCloche Mountain Lakes, and resulting fish mortalities. *Journal of the Fisheries Research Board of Canada* 29:1131–1143.

Bond, T. C., S. J. Doherty, D. W. Fahey, P. M. Forster, T. Berntsen, B. J. DeAngelo, M. G. Flanner, S. Ghan, B. Kärcher, D. Koch, S. Kinne, Y. Kondo, P. K. Quinn, M. C. Sarofim, M. G. Schultz, M. Schulz, C. Venkataraman, H. Zhang, S. Zhang, N. Bellouin, S. K. Guttikunda, P. K. Hopke, M. Z. Jacobson, J. W. Kaiser, Z. Klimont, U. Lohmann, J. P. Schwarz, D. Shindell, T. Storelvmo, S. G. Warren, and C. S. Zender. 2013. Bounding the role of black carbon in the climate system: a scientific assessment. *American Geophysical Union*: 118:5380–5552.

Boyd, C. E. 1990. *Water Quality in Ponds for Aquaculture.* Auburn University: Alabama Agricultural Experiment Station.

Boyd, C. E. 2000. *Water Quality, an Introduction.* Boston: Kluwer Academic Publishers.

Boyd, C. E. and C. S. Tucker. 1998. *Pond Aquaculture Water Quality Management.* Boston: Kluwer Academic Publishers.

Boyd, C. E., M. Polioudakis, and T. Hanson. 2011. Carbon footprint of US farm-reared catfish. Report to US Catfish Farmers Association, Jackson, Mississippi.

Boyd, C. E., C. W. Wood, P. L. Chaney, and J. F. Queiroz. 2010. Role of aquaculture pond sediments in sequestration of annual global carbon emissions. *Environmental Pollution* 158:2537–2540.

BP. 2009. *BP Statistical Review of World Energy.* London: BP.

BP. 2012a. *BP Statistical Review of World Energy.* London: BP.

BP. 2012b. *BP Energy Outlook 2030.* London: BP.

Burlew, J. S. (editor). 1953. *Algal Culture from Laboratory to Pilot Plant.* Washington: Carnegie Institute.

Canadell, J. G., C. Le Quéré, M. R. Raupach, C. B. Field, E. T. Buitenhuis, P. Ciais, T. J. Conway, N. P. Gillett, R. A. Houghton, and G. Marland. 2007. Contributions to accelerating atmospheric CO_2 growth from economic activity, carbon intensity, and efficiency of natural sinks. *Proceedings of the National Academy of Sciences* 104:18865–18870.

Chen, J. L., C. R. Wilson, D. Blankenship, and B. D. Tapley. 2009. Accelerated Antarctic ice loss from satellite gravity measurements. *Nature Geoscience* 2:859–862.

Cowling, E. B. 1982. Acid precipitation in historical perspective. *Environmental Science and Technology* 16:110–123.

FAO (Food and Agriculture Organization). 2011a. *"Energy-smart" Food for People and Climate.* Rome: FAO.

FAO (Food and Agriculture Organization). 2011b. *Technical Guidelines on Aquaculture Certification.* Rome: FAO.

FAO (Food and Agriculture Organization). 2012. *The State of World Fisheries and Aquaculture.* Rome: FAO Fisheries and Aquaculture Department.

Feng, Z., H. Miao, F. Zhang, and Y. Huang. 2002. Effects of acid deposition on terrestrial ecosystems and their rehabilitation in China. *Journal of Environmental Sciences* 14:227–233.

Flick, W. A., C. Schofield, and D. A. Webster. 1982. Remedial actions for interim maintenance of fish stocks in acidified waters. In R. E. Johnson, editor, *Acid Rain/Fisheries*, pp. 287–306. Bethesda: American Fisheries Society.

Goreham, S. 2013. *The Mad, Mad, Mad World of Climatism.* New Lenox: New Lenox Books.

Gronroos, J., J. Seppala, F. Silvenius, and T. Makinen. 2006. Life cycle assessment of Finnish cultivated rainbow trout. *Boreal Environmental Research* 11:401–414.

Haines, T. A. 1981. Acid precipitation and its consequences for aquatic ecosystems: a review. *Transactions of the American Fisheries Society* 110:669–707.

Haines, T. A. and J. J. Akielaszek. 1983. Acidification of headwater lakes and streams in New England. In *Lake Restoration, Protection and Management.* pp. 83–87. Washington: United States Environmental Protection Agency.

Hays, J. D., J. Imbrie, and N. J. Shackleton. 1976. Variations in the earth's orbit: pacemaker of the Ice Ages. *Science* 194:1121–1132.

Herbert, M. K. 1949. Energy from fossil fuels. *Science* 109:103–109.

Hutchinson, T. C. and M. Havas, editors. 1980. *Effects of Acid Rain on Terrestrial Ecosystems.* New York: Plenum Press.

IEA (International Energy Agency). 2012. *Key World Energy Statistics.* Paris: IEA.

IPCC (Intergovernmental Panel on Climate Change). 2007. Introduction. In R. K. Pachauri and A. Reisinger, editors, *Climate Change 2007: Synthesis*. pp. 12–15. Geneva: IPCC.

IPCC (Intergovernmental Panel on Climate Change). 2011. *Special Report on Renewable Energy and Climate Change Mitigation*. In O. Edenhofer, R. Pichs-Madruga, Y. Sokona, K. Seyboth, P. Matschoss, S. Kadner, T. Zwickel, P. Eickemeier, G. Hansen, S. Schlömer, and C. von Stechow, editors. Cambridge, UK: Cambridge University Press.

Iribarren, D., A. Hospido, M. T. Moreira, and G. Feijoo. 2010. Carbon footprint of canned mussels from a business-to-consumer approach. A starting point for mussel processors and policy makers. *Environmental Science and Policy* 13:509–521.

Johnson, A. H. and T. G. Siccama. 1983. Acid deposition and forest decline. *Journal of Environmental Science and Technology* 17:294A–305A.

Langer, M. R., M. Geisen, K. H. Baumann, J. Klas, U. Riebesell, and S. Thoms. 2006. Species-specific responses of calcifying algae to changing seawater carbonate chemistry. *Geochemistry, Geophysics, and Geosystems* 7:Q09006.

Likens, G. E., R. F. Wright, J. N. Galloway, and T. J. Butler. 1979. Acid rain. *Scientific American* 241:43–51.

Mungkung, R., S. H. Gheewala, and A. Tomnantong. 2012. Carbon footprint of IQF peeled tail-on breaded shrimp *Litopenaeus vannamei*. How big is it compared to other aquatic products? *Environment and Natural Resources Journal* 10:31–36.

NOAA (National Oceanographic and Atmospheric Administration). 2010. Technical Considerations for use of geospatial data in sea level change mapping and assessment. NOAA Technical Report NO 2010-01, US Department of Commerce, Washington, DC.

OECD (Organization for Economic Cooperation and Development). 2010. Total population. In *OECD Factbook 2010: Economic, Environmental and Social Statistics*. pp. 12–15. Paris: OECD Publishing.

Orr, J. C., V. J. Fabry, O. Aumont, L. Bopp, S. C. Doney, R. A. Feely, A. Gnanadesikan, N. Gruber, A. Ishida, F. Joos, R. M. Key, K. Lindsay, E. Maier-Reimer, R. Matear, P. Monfray, A. Mouchet, R. G. Najjar. G. K. Plattner, K. B. Rodgers, C. L. Sabine, J. L. Sarmiento, R. Schlitzer, R. D. Slater, I. J. Totterdell, M. F. Weirig, Y. Yamanaka, and A. Yool. 2005. Anthropogenic ocean acidification over the twenty-first century and its impact on calcifying organisms. *Nature* 437:681–686.

Pelletier, N. and P. Tyedmers. 2010. Life cycle assessment of frozen tilapia fillets from Indonesian lake-based and pond-based intensive aquaculture systems. *Journal of Industrial Ecology* 14:467–481.

Pelletier, N., R. Pirog, and R. Rasmussen. 2010a. Comparative life cycle environmental impacts of three beef production strategies in the upper Midwestern United States. *Agricultural Systems* 103:380–389.

Pelletier, N., P. Lammers, D. Stender, and R. Pirog. 2010b. Life cycle assessment of high- and low-profitability commodity and deep-bedded niche swine production systems in the upper Midwestern United States. *Agricultural Systems* 103:599–608.

Pelletier, N., P. Tyedmers, U. Sonesson, A. Scholz, F. Ziegler, A. Flysjo, S. Kruse, B. Cancino, and H. Silverman. 2009. Not all salmon are created equal: Life cycle assessment (LCA) of global salmon farming systems. *Environmental Science and Technology* 43:8730–8736.

Pfeffer, W. T., J. T. Harper, and S. O'Neal. 2008. Kinematic constraints on glacier contributions to 21st-century sea-level rise. *Science* 321:1340–1343.

Polyakov, I. V., L. A. Timokhov, V. A. Alexeev, S. Bacon, I. A. Dmitrenko, L. Fortier, I. E. Frolov, J. C. Gascard, E. Hansen, V. V. Ivanov, S. Laxon, C. Mauritzen, D. Perovich, K. Shimada, H. L. Simmons, V. T. Sokolov, M. Steele, and J. Toole. 2010. Arctic Ocean warming contributes to reduced polar ice cap. *Journal of Physical Oceanography* 40:2743–2756.

Post, W. M., T. H. Peng, W. R. Emanuel, A. W. King, V. H. Dale, and D. L. DeAngelis. 1990. The global carbon cycle. *American Scientist* 78:310–326.

Rahmstorf, S. 2007. A semi-empirical approach to predicting future sea-level rise. *Science* 315:368–370.

Schofield, C. 1976. Acid precipitation: effects on fish. *Ambio* 5:228–230.

Seip, H. and A. Tollan. 1978. Acid precipitation and other possible sources of acidification of rivers and lakes. *Science of the Total Environment* 10:253–270.

Smil, V. 2008. *Energy in Nature and Society—General Energetic of Complex Systems*. Cambridge: MIT Press.

Smith, S. J., H. Pitcher, and T. M. L. Wigley. 2005. Future sulfur emissions. *Climatic Change* 73:267–318.

Smith, S. J., J. van Aardenne, Z. Kilmont, R. J. Andres, A. Volke, and S. D. Arias. 2011. Anthropogenic sulfur dioxide emissions: 1850–2005. *Atmospheric Chemistry and Physics* 11:1101–1116.

Tyedmers, P. H., R. Watson, and D. Pauly. 2005. Fueling global fishing fleets. *Ambio* 34:635–638.

USEPA (United States Environmental Protection Agency). 2010. *Methane and Nitrous Oxide Emissions from Natural Sources*. Washington: EPA (430-R-10-001).

USGS (United States Geological Survey). 2012. An estimate of undiscovered conventional oil and gas resources of the World, 2012. *USGS Factsheet 2012–3028*. Denver, Colorado: USGS.

United States Global Change Research Program. 2009. *Global Climate Change Impacts in the United States*. New York: Cambridge University Press.

van Geel, B., O. M. Raspopov, H. Renssen, J. van der Plicht, V. A. Dergachev, and H. A. J. Meijer. 2002. The role of solar forcing upon climate change. In F. Chambers and M. Ogle, editors, *Climate Change*, pp. 331–338. New York: Routledge.

Velicogna, I. 2009. Increasing rates of ice mass loss from the Greenland and Antarctic ice sheets revealed by GRACE. *Geophysical Research Letters* 36:L19503–L19506.

Wagner, S. and E. Zorita. 2005. The influence of volcanic, solar and CO_2 forcing on the temperatures in the Dalton Minimum (1790–1830): A model study. *Climate Dynamics* 25:205–218.

Winther, U., F. Ziegler, E. S. Hognes, A. Emanuelsson, V. Sund, and H. Ellingsen. 2009. Carbon footprint and energy use of Norwegian seafood products. SINTEF Fisheries and Aquaculture Report SFH80A096068, Trondheim, Norway.

Chapter 8

Protein conversion and the fish meal and oil issue

The whole bodies of humans and most other animals contain about 15% protein—usually more than any other component except water. Proteins are an essential and major component of animal and human diets. According to WHO recommendations (FAO/WHO/UNU 2002), the typical adult man needs 56 g protein daily and the typical adult woman requires 46 g protein daily. Human protein requirements can be satisfied with an all-plant diet but animal products tend to contain more and higher-quality protein than contained in plant materials (de Man 1999). Most people get their protein from both plants and animals, and large amounts of plant protein and some animal protein are fed to animals to produce meat protein for human use. Because of the expense and relative scarcity of proteins, a high efficiency of feed protein conversion is desirable in animal production.

Aquaculture products are an important source of animal protein for humans as discussed in Chapter 3. Some aquaculture species require a high percentage of protein in their diet, and marine fish meal often is included in aquaculture feeds as a protein source. Fish oil is also added to diets for some species. The quantity of small, pelagic fish that can be captured from the oceans for making fish meal and oil is limited. Overexploitation of these fish also would have negative impacts on marine food chains (Naylor et al. 2009). In this chapter we will discuss the controversy about the use of fish meal in aquaculture feeds. In addition, we will assess protein conversion by aquaculture species and compare it with that of some other meat animals.

Amino acids and proteins

Proteins are made from amino acids. An amino acid is an organic acid with an amino group (NH_2) and a carboxyl group (COOH) attached to an organic moiety (R)

Aquaculture, Resource Use, and the Environment, First Edition. Claude E. Boyd and Aaron A. McNevin.
© 2015 John Wiley & Sons, Inc. Published 2015 by John Wiley & Sons, Inc.

as shown below:

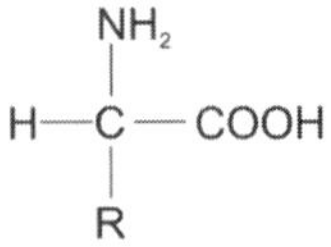

$$
\begin{array}{c}
NH_2 \\
| \\
H-C-COOH \\
| \\
R
\end{array}
\tag{8.1}
$$

There are over 20 different amino acids.

Amino acids join together by the amino group on one amino acid reacting with the carboxyl group of another amino acid and splitting off water:

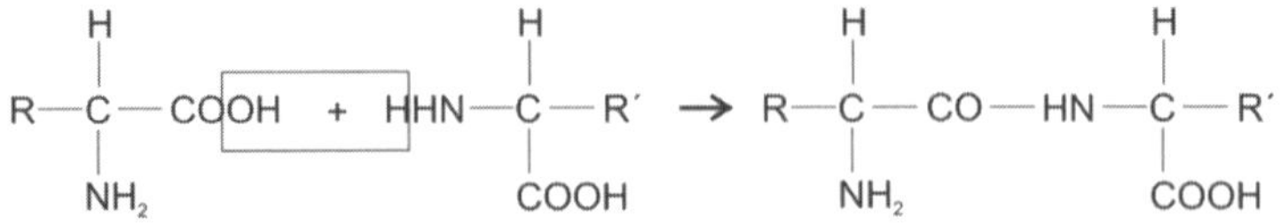

$$
\begin{array}{ccc}
\underset{NH_2}{\overset{H}{R-C-COOH}} \ + \ \underset{COOH}{\overset{H}{HHN-C-R'}} & \rightarrow & \underset{NH_2}{\overset{H}{R-C-CO-HN-\overset{H}{\underset{COOH}{C}}-R'}}
\end{array}
\tag{8.2}
$$

The bond between two amino acids is called a peptide bond, and two amino acids so bonded are called dipeptides. A dipeptide can react with an amino acid to form a tripeptide. Proteins are polypeptides that are no longer straight chains but ring systems, and there are many different kinds of protein molecules.

The ultimate source of amino acids is plants. Plants use sugar produced in photosynthesis and ammonia nitrogen to make amino acids from which they synthesize proteins. Animals can synthesize some amino acids from nutrients in their diet, while other amino acids—known as essential amino acids—must be present in adequate amounts in their diet. The diet of animals should have an adequate amount of protein and sufficient quantities of essential amino acids. Different species differ considerably in protein requirement and even in suites of essential amino acids.

Plant products tend to be low in protein content as compared to animal products (de Man 1999). Certain essential amino acids also are typically at lower concentration in plant biomass than in animal biomass. Although humans can choose their food carefully and obtain an adequate intake of protein and essential animal acids from an all-plant diet, human diets usually contain a certain amount of animal protein. Of course, animal production requires feed sources that have adequate amounts of protein and essential amino acids.

Protein conversion in aquaculture

Fish and other aquatic animals are capable of highly efficient protein conversion. Tilapia feed usually contains about 32% crude protein, and tilapia have a protein content (live weight basis) of around 14% (Boyd et al. 2007). At a feed conversion of 1.8, 0.576 kg protein in feed (1.8 × 0.32) yields 0.14 kg protein in tilapia biomass. Thus 24.3% of the feed protein is recovered in tilapia biomass at harvest.

Percentage recoveries of feed protein in harvest biomass of several other species groups are provided in Table 8.1. The percentages range from 25.7 for shrimp to 43.0 for salmon with an average of 30.9%—a ratio of feed protein:fish protein slightly

Table 8.1 Conversion of protein by five aquaculture species groups.

| | FCR[a] | Feed crude protein (%)[a] | Dressout as fillets (%)[b] | Crude protein (%) | | Protein recovery (%) | |
				Fillet[c]	Whole fish[a]	Fillet	Whole
Catfish	1.8	32	45	15.55	14.9	12.1	25.9
Tilapia	1.7	32	35	18.50	14.0	11.9	25.7
Shrimp	1.5	38	37	20.31	17.8	13.0	31.2
Trout	1.2	45	69	20.87	15.6	26.7	28.9
Salmon	1.0	42	72	19.90	18.5	33.3	43.0

The feed conversion ratio (FCR) is based on results with high-quality feed and good feed management.
Sources: Prepared with data from [a]Boyd et al. (2007); [b]http://www.commerce.state.ak.us/ded/dev/seafood/recoveries; [c]http://www.highproteinfoods.net/fish-shelfish.

greater than three. Although protein recovery from feed by salmon is higher than for the other species, the protein for salmon comes from fish meal, while the majority of protein in catfish and tilapia feed is from plant sources.

The original method of producing meat animals was to rear them in grazing areas where food was from natural sources. Almost two-thirds of the agricultural area in the world still is used as grazing area for livestock (Eswaran et al. 1999). The traditional, open-range grazing system for meat animal production does not differ much from the practice of capturing fish and other aquatic food organisms from the sea or inland waters.

Of course as the human population density increased it became necessary in many regions to confine animals within fenced pastures or lots and to provide supplemental feed to increase production. Using fences to confine animals and providing feeds to supplement natural food has an analogy in extensive and semi-intensive aquaculture where fertilizers and supplemental feeding are used to increase production.

Because of the rising demand for meat by the human population, meat production is becoming more intensive. The major meat animals increasingly are produced at high density in rearing houses or feed lots. This approach has also been adopted in intensive feed-based aquaculture in ponds, cages, raceways, and other systems. Like in intensive production of terrestrial meat animals, aquatic animals in intensive culture receive their nutrients almost exclusively from feed.

The efficiency of conversion of plant protein to animal protein on a global basis was estimated at 8.3% (van der Hoek 1998). But there are many kinds of animals and meat production systems, and the ratio varies considerably—just as it does in aquaculture. The conversion of dietary protein to whole animal is about 34% for chicken, 20% for pork, and 10% for beef (van der Hoek 1998). Thus aquaculture production (Table 8.1) appears to be about as protein efficient as chicken production and much more efficient than pork and beef production.

The efficiency of protein conversion to edible meat is lower than for whole animals. The conversion of feed protein to edible meat protein is: 20% for chicken; 10% for pork; 4% for beef (Oenema and Tamminga 2005)—the same is true for fisheries products. In the case of tilapia fillet yield is around 35%, and fillets contain around 18.5% crude protein. The protein content of fillets from 1 kg live tilapia equals 0.065 kg or 11.9% of the feed protein. Protein recoveries in fillets of the five species groups (Table 8.1) ranged from 11.9 to 33.3% with an average of 19.4% comparing favorably with chicken and better than pork and beef.

Carnivorous species such as salmon and trout have a higher protein conversion efficiency than omnivorous species such as catfish, tilapia, and shrimp (Boyd et al. 2007). However carnivorous species require a higher protein content in their diet, their feeds have a greater fish meal and fish oil content, and the FIFO ratio is greater than for omnivorous or filter-feeding species (Tacon and Metian 2008).

The most ecologically efficient species are those that can be reared in ponds utilizing natural food organisms. Production of these species nevertheless will be low unless nitrogen and phosphorus fertilizers are applied, and even with fertilization, production will be much less than in feed-based culture. Tilapia production may reach 1500 kg/ha in ponds treated with inorganic fertilizers, 3000 kg/ha in ponds receiving animal manures, but over 5000 kg/ha in ponds with feeding (Boyd and Tucker 1998). Moreover, feed provides a more uniform taste and most large-scale retailers and food purveyors prefer this consistency. Thus feed-based aquaculture is a necessity if aquaculture is to continue to fill the gap between capture fishery production and the demand for fisheries products by the rapidly growing human population.

Fish meal and fish oil

Fish meal is an excellent source of protein for inclusion in aquaculture feeds because it has a high protein content (Table 8.2) and a balanced amount of essential amino acids. It also contains phosphorus, calcium, other minerals and essential omega-3 fatty acids, that is, docosahexaenoic acid (DHA) and eicosapentaenoic acid (EPA). Fish meal may be included in feeds for many aquaculture species, but it is particularly

Table 8.2 Proximate composition and mineral nutrient concentrations in four mechanically extracted fish meals.

Variable	Herring	Menhaden	White fish	Anchovy
Dry matter (%)	92	92	92	92
Crude protein (%)	72	64.5	62.3	65.4
Ether extract[a] (%)	8.4	9.6	5.0	7.6
Ash (%)	10.4	19.0	21.3	14.3
Phosphorus (%)	1.67	2.88	3.81	2.43
Calcium (%)	2.20	5.19	7.31	3.73
Magnesium (%)	0.14	0.15	0.18	0.24
Potassium (%)	1.08	0.70	0.83	0.90
Iron (mg/kg)	114	544	355	220
Manganese (mg/kg)	4.8	37	8.4	9.5
Zinc (mg/kg)	125	144	211	103
Copper (mg/kg)	5.6	10.3	5.9	9.03

Source: Modified from National Research Council (1993).
[a]Crude fat.

important in diets for marine shrimp and carnivorous fish such as salmon and trout (Table 5.4).

Fish meal is made from menhaden, herring, anchovies, sardines, and other small, pelagic fish captured from the ocean. This fishery is referred to by various names: the fish meal fishery, the pelagic fishery, the feed fish fishery, the reduction fish fishery, or the industrial fish fishery. This pelagic fishery usually is found in areas of the ocean where upwelling brings nutrients into surface water to sustain high phytoplankton productivity and fish production such as occurs along the coast of northern Chile, Peru, and Ecuador.

The fish are pulverized and the oil and water pressed out. The solids are cooked and pulverized further to make a high-protein-content meal that is used primarily as a protein source in animal feed (FAO 1986). The liquid that is extruded during fish meal manufacturing is further processed to make fish oil. This product can be used in aquaculture feeds to supply energy and essential fatty acids.

The yield of fish meal varies with the species of fish or other fisheries products or by-products from which it is made (Shepherd et al. 2005). In 2008 16.5 Mt of whole fish caught by reduction fishery fleets and 5.5 Mt of trimmings and rejected food fish from processing were used to produce 4.82 Mt of fish meal from a total of 22 Mt of raw material. Each kilogram of fish meal required 4.56 kg of live fish as raw material (www.seafish.org). We will use a value of 4.5 for the fish input:fish meal ratio. Fish meal production fluctuated but exhibited an overall upward trend from 1960 to a peak of 7.44 Mt in 1994. Since then it has continued to fluctuate annually but the trend in production is downward (Fig. 8.1). Fish oil production has shown a similar change as observed for fish meal, because it is a by-product of fish meal production (Fig. 8.1). In 2008 fish meal and fish oil production was approximately 5 Mt and 1 Mt, respectively. It does not seem likely that fish meal and fish oil production from the capture fishery will increase in the future.

Fish processing offal can be rendered to make fish offal meal that is lower in protein and higher in ash content than fish meal. Nevertheless fish offal meal is suitable for use in animal feeds including aquaculture feeds. The estimated production of fish

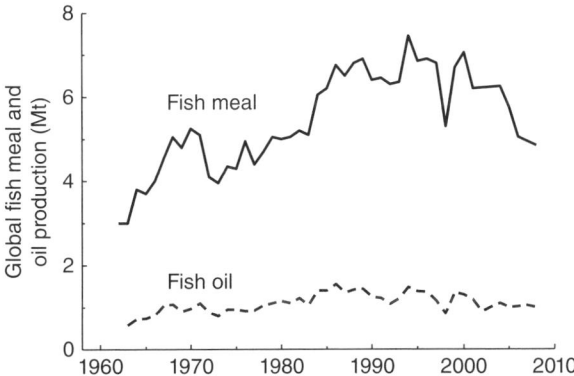

Figure 8.1 Annual production of fish meal and fish oil: 1962–2009. *Source:* www.iffo.net.

Table 8.3 Proximate composition and mineral concentrations in meals made from fish, shrimp, and crab processing wastes.

Variable	Tilapia[a]	Catfish[b]	Shrimp[b]	Crab[b]
Dry matter (%)	92.6	92.0	88.0	92.0
Crude protein (%)	54.8	50.8	39.5	32.0
Ether extract[c] (%)	11.7	9.6	3.2	2.5
Ash (%)	26.3	18.0	27.2	41.0
Phosphorus (%)	4.1	–	–	1.6
Calcium (%)	8.4	–	–	14.6
Magnesium (%)	0.15	–	–	0.94
Potassium (%)	0.38	–	–	0.45
Iron (mg/kg)	187	–	–	4356
Manganese (mg/kg)	13.9	–	–	133
Zinc (mg/kg)	67.5	–	–	–
Copper (mg/kg)	9.0	–	–	32.7

Sources: Compiled with data taken from [a]Dale et al. (2004); [b]National Research Council (1993); [c]Crude fat.

offal meal in 2008 was 1.23 Mt—about 25% of total global fish meal production of 6.05 Mt (www.iffo.net).

The oil content of species used for making fish meal differs greatly resulting in a variable fish oil yield (Boyd et al. 2007). The world supply of fish oil in 2008 was 1.02 Mt (www.iffo.net). If we multiply the ratio, annual fish meal production:annual fish oil production by 4.5, we get an average estimate of fish oil yield of 1 kg per 21.3 kg of live fish. This agrees well with the assumption that fish oil yield is around 5% of live fish input for fish meal manufacturing (Tacon and Metian 2008). Fish offal oil also can be made from fish processing wastes but good estimates of annual production of fish offal oil are not available.

Wastes from processing aquaculture species also can be used to make relatively good-quality fish offal meal (Table 8.3). Oil also is produced as a by-product of fish offal production. Most processing plants for aquaculture products send their waste to renderers but records of amounts of offal meal and oil are incomplete (Tacon and Metian 2008). Tacon and Metian did list an anonymous report stating that 600 000 t of salmon produced by aquaculture in Chile resulted in 270 000 t of processing waste and mortalities that were converted to 48 600 t salmon offal oil and 43 200 t of salmon offal meal.

Fish meal and fish oil are used in the diets for many animals but inclusion in aquaculture feeds is a major use. In 2009 an estimated 63% of global fish meal consumption was used in aquaculture feeds; of the rest, 25% went into pig feed, 8% into poultry feed, and 4% was used for other purposes (www.iffo.net). With respect to fish meal inclusion in aquaculture feed in 2009, the breakdown follows: salmon and trout feed, 27%; crustacean (mainly shrimp) feed, 26%; marine fish feed, 26%; tilapia feed, 6%; eel feed, 5%; cyprinid feed, 5%; other feeds, 6%. An estimated 81% of the world's fish oil is used in aquaculture feeds, while 13% is consumed by humans and 6% is used for industrial purposes (www.iffo.net). Salmon and trout feeds make up 68% of the fish oil use in aquaculture. Around 19% of fish oil used in aquaculture is for marine fish feed and about 6% is included in crustacean feed.

Fish in–fish out ratio

Environmentalists argue—and rightfully so—that unless the amount of aquaculture production achieved using feed exceeds the quantity of live fish needed to make the fish meal and fish oil included in the feed, then that particular type of aquaculture actually detracts from world fisheries production (Naylor et al. 2000, 2009). Thus the FIFO ratio is an important variable related to the efficiency of fish meal and fish oil use in feed (Tacon and Metian 2008). For example, assuming salmon feed is 24% fish meal and 16% fish oil (all made from live marine fish) and the feed conversion ratio (FCR) is 1.2, then 0.288 kg fish meal (1.2 kg × 0.24) and 0.192 kg fish oil (1.2 kg × 0.16) would be included in the feed for 1 kg salmon. An estimated 1.30 kg live fish would be needed to make the fish meal (0.288 kg × 4.5). The oil yield from making the fish meal would be 0.065 kg (1.30 kg × 0.05); thus another 0.127 kg fish oil would be obtained from 2.54 kg live fish (0.127 kg ÷ 0.05). Thus a total of 3.84 kg live fish would be expended to produce 1 kg live weight of salmon.

In calculating FIFO ratio, only fish meal and fish oil made from live marine fish should be included. Any fish offal meal or oil should be subtracted from the total fish meal and fish oil included in the feed. Although meal and oil from offal can be subtracted from marine fish meal and oil included in the feed, targeted fisheries for fish meal and oil are not ecologically equivalent to farm salmon and meal and oil made from their processing waste.

It can be assumed that 0.225 kg of fish oil results from the production of 1 kg of fish meal because 1 kg fish meal requires 4.5 kg live fish (a fish meal:live fish ratio of 0.222) and yields 5% of their weight as oil (a fish meal:fish oil ratio of 4.44). If the fish meal content of feed is 4.44 or more times greater than the fish oil content, all of the fish oil in a feed can be compensated for by fish oil resulting from production of the included fish meal. In this case the FIFO ratio can be estimated as:

$$\text{FIFO} = \text{FCR} \times \left[\frac{\text{FM}}{(100)(0.222)} \right] \tag{8.3}$$

where FM, fish meal (%). But if the fish meal content is less than 4.44 times fish oil content, the appropriate equation is:

$$\text{FIFO} = \text{FCR} \times \left[\frac{\left(\dfrac{\text{FO}}{100} - \dfrac{\text{FM}}{(100)(4.44)} \right)}{0.05} + \left[\frac{(\text{FM})(4.5)}{100} \right] \right] \tag{8.4}$$

where FM, fish meal (%).

Fortunately most species groups have lower FIFO ratios than salmon. Moreover, according to Tacon and Metian (2008) there was a reduction in the FIFO ratio between 1995 and 2006 for major species groups as follows: salmon, 7.5–4.9; trout, 6.0–3.4; eel, 5.2–3.5; marine fish, 3.0–2.2; shrimp, 1.9–1.4. They also reported that several species groups with large global production totals have FIFO ratios below 1.0: Chinese carp, 0.2; milkfish, 0.2; tilapia, 0.4; catfish, 0.5; freshwater crustaceans, 0.6.

Nevertheless aquaculture production has been steadily increasing since 1995 and the total amount of fish meal and oil used in aquaculture feeds is much greater than it was in 1995 regardless of the improved efficiency of use in feeds (Tacon and Metian 2008).

It is only fair to mention that the FIFO ratios for wild-caught fish species would likely be similar to those of aquaculture species. However, in an ecologically balanced system this would be the natural situation and considered sustainable. The problem is that natural fisheries are being overfished and cannot sustain a production great enough to meet human needs. Thus small pelagic fish are captured and rendered in fish meal and oil for use in aquaculture feeds to increase global aquaculture production to offset the deficit of wild-caught fisheries products. There is an opinion that exploitation of the reduction fisheries does not detract from fisheries production for direct human consumption because fish meal is not widely used as human food and relatively little fish oil is used directly as a dietary supplement for humans. Those of this opinion hold that the main problem with excessive capture of small pelagic fish for fish meal and oil production is disruption of the natural food supply of carnivorous, marine fish. This opinion however is not exactly true because 10–20% of global fisheries landings since 1961 have been consumed by humans (Alder et al. 2008).

Many in the aquaculture industry are concerned about excessive inclusion of fish meal and fish oil in feeds because inclusion in aquaculture feeds is already using over 60% of the fish meal and 80% of fish oil supplies (www.iffo.net). Fish oil usage is particularly problematic because fish oil is a by-product of fish meal manufacturing, and it appears that fish meal production from the capture fishery cannot be increased in the future. Unless aquaculture can reduce fish meal and oil use, it will not be able to continue to expand to meet the growing gap between capture fisheries production and demand for fisheries products. This concern has led to much research on the replacement of fish meal and oil with plant meals and oils (Watanabe 2002; Naylor et al. 2009). Most aquaculture feeds that contain fish meal and oil also contain plant meals and oils from barley, canola, corn, cottonseed, peas, lupins, soybeans, and wheat. It is possible to greatly reduce fish meal and oil content or replace them completely by including a greater amount of one or more plant meals and oils or a combination of plant meals and oils and animal by-products such as feather, meat, bone, blood, or fish offal meals (Carter and Hauler 2000; El-Sayed 1998; Regost et al. 2003; Kaushik et al. 1995; Robinson and Li 1994; Amaya et al. 2007; Enterria et al. 2011). Some examples of reductions in fish meal and fish oil inclusion in diets of major aquaculture species groups between 1995 and 2007 are provided (Table 8.4).

Rapid growth of aquaculture production nevertheless has caused a drastic increase in fish meal and fish oil use for aquaculture feed during the period 1995–2007 (Naylor et al. 2009). Growth of aquaculture is expected to continue as the world's population increases, and the demand for fish meal and fish oil for aquaculture feed likely will continue. In addition, regardless of the type of plant meal or oil included in an aquaculture feed, it still is important to strive for efficient conversion of feed protein to animal protein. Protein in particular is more scarce and expensive than other major feedstuffs. Protein sources spared from aquaculture feeds can be used for other purposes.

Table 8.4 Feed conversion ratio (FCR) and inclusion of fish meal and fish oil in feed for several aquaculture species groups.

Species group	Typical FCR	Average % fish meal in feed	Average % fish oil in feed
Shrimp			
1995	2.0	28	2
2007	1.7	18	2
Salmon			
1995	1.5	45	25
2007	1.3	24	16
Marine fish			
1995	2.0	50	15
2007	1.9	30	7
Chinese carp			
1995	2.0	10	0
2007	1.7	5	0
Tilapia			
1995	2.0	14	1
2007	1.7	5	0
Channel catfish[a]			
1995	2.2	6	2.0
2007	2.2	2	1.3

Source: Naylor, R. L., R. W. Hardy, D. P. Bureau, A. Chiu, M. Elliott, A. P. Farrell, I. Forster, D. M. Gatlin, R. J. Goldburg, K. Hua, and P. D. Nichols. 2009. Feeding aquaculture in an era of finite resources. *Proceedings National Academy of Science* 106:15, 103–15, 110.
[a]Catfish data from discussions with feed mill operators in Alabama and Mississippi.

Although fish oil can be replaced with vegetable oil in aquaculture feeds, at vegetable oil inclusions above 50% of supplementary lipid, there is a reduction in the ratio of omega-3 to omega-6 fatty acids in fish tissue (Bell et al. 2001). This finding is both positive and negative; less fish oil use is desirable, but a high ratio of omega-3:omega-6 fatty acids is considered beneficial to human health mainly by protecting against cardiovascular diseases (Adarme-Vega et al. 2012).

There also is an argument that wild fish contain more omega-3 fatty acids than aquacultured fish. Some studies have shown a lower ratio of omega-3 to omega-6 fatty acids in farmed fish (Alasalvar et al. 2002; Lenas and Nathanailides 2011), while Nettleton (2000) and Hardy (2003) presented data that contradict the idea that farmed fish are lower in omega-3 fatty acids.

The possible future shortage of fish oil, the possible human health benefits of omega-3 fatty acids, and the increasing proportion of aquacultured to wild-caught seafood is leading to efforts to find new sources of omega-3 fatty acids to use in aquaculture feeds. One promising source is mass production of marine microalgae rich in DHA and EPA (Adarme-Vega et al. 2012). Other possibilities are vegetable oils that contain biosynthetic precursors of omega-3 fatty acids, and genetically modified oil-seed crops that contain omega-3 fatty acids (Miller et al. 2008).

Conclusions

Aquaculture species are at least as efficient and possibly more efficient in converting feed protein to edible protein for humans as are traditional agricultural sources of meat. Nevertheless aquaculture feeds typically have a higher protein content than feeds for agricultural animals. The protein in aquaculture feeds is primarily from plant meals, animal processing waters, and marine fish meal. Oil must be included in aquaculture feed, and in many instances, fish oil is used.

The fish meal and oil issue is a concern for aquaculture because it could limit the amount of global production possible. However, great strides have been made in reducing the amount of fish meal and fish oil in feeds, and noticeable improvements in the FCR have occurred. The efficiency of protein conversion also is increasing. These are positive events but the aquaculture industry must continue to find ways of reducing fish meal and oil use and improving protein conversion.

It is not certain that a reduction in capture of small pelagic fish for use in aquaculture feeds would increase world capture fisheries by an amount greater than the aquaculture production lost by reducing fish meal and fish oil use. Of course the ideal situation would be to find replacements for fish meal and fish oil that would allow aquaculture to continue to grow without the necessity for harvesting fish for making feed ingredients. This would possibly lead to greater production of marine fish that forage on small pelagic fish and greater capture fisheries production.

Some planktivorous and omnivorous aquatic animal species do not require any fish meal and oil, and others require much less of these resources than do carnivorous species. Humans have specific ideas about what they want to eat and there is a market for the carnivorous species and shrimp that tend to use the greatest amount of fish meal and oil. As long as this market exists there will be producers interested in supplying products for it. There is an effort by many eNGOs to educate consumers about the ecological impacts of different products.

The eNGO perspective

The use of wild fish as a component in aquafeeds probably is the most important justification for eNGOs to engage with the aquaculture industry. The eNGOs work on aquaculture was an off-shoot of their work on ocean-related issues. The original thought was that aquaculture could take pressure off of wild fisheries. The eNGOs saw aquaculture as a part of the solution to dwindling stocks of fish in the ocean, because much of the early aquaculture, for example, carp, tilapia, catfishes, molluscs, and seaweed, had a relatively low trophic status. Upon closer examination eNGOs realized that a substantial amount of wild fish was being harvested to make feed for cultured species. For example Atlantic salmon, marine shrimp, groupers, cod, and tunas are more carnivorous and occupy a higher trophic state. It is not surprising that the eNGOs sought to address the wild fish dependency using the most widely consumed of these species—shrimp and salmon. Moreover, wild fish dependency has become the key criterion for the acceptance or activism against novel species development in aquaculture.

There are two main aspects of wild fish dependency that eNGOs are most concerned—the actual fish stock being utilized and the level of efficiency with which the wild fish component is fed to the culture species. The status of the fish stock is dependent on numerous management aspects that relate to the general stock "health." These issues include effective policies governing their harvest, the current stock status (whether overfished or not), the forecasts of sustaining stock biomass, the ecological role of the species in the greater marine food web, and the resilience of the species.

There is a natural balance in the exploitation and recovery of wild fish. This is a fundamental aspect of the ecology of living organisms. Exploitation of fish beyond the capability of the fish to reproduce and recruit new offspring into the population will result in a decline of biomass. It is this aspect of ecology that is stressed by the eNGOs. They maintain that overexploitation of fish stocks is occurring at an alarming rate and most of the governments and industries have not proven they have the ability or will to manage these natural resources. Much of this sentiment is supported by the current data suggesting 85% of world fish stocks are either fully exploited, overexploited, depleted, or recovering (FAO 2010). The question posed by eNGOs is, "What evidence is there that global fisheries can be well-managed, and why should reduction fisheries be seen any differently?" Further if current fishing practices for food fish are not allowing fisheries to sustain themselves, any recovery would require the presence of prey fish. These prey fish are those targeted for production of fish meal and oil.

The nine most economically important species captured for fish meal and oil are depicted in Fig. 8.2. Although many of these fisheries are subject to quotas set by governments or fisheries management groups, there is little evidence of a clear trend in landings. Thus the eNGOs consider the harvests and consequently the fish stocks as highly variable. This would suggest that the fisheries may or may not be managed in a responsible manner. For the sustainability of the fish stocks, the eNGOs will likely invoke the "precautionary principle" which allows them to err on the side of high conservatism when it comes to harvests.

It should be noted that the eNGOs, to varying degrees dependent on the particular organization, do have some faith in the management of fish stocks that are certified by the Marine Stewardship Council (MSC). The MSC certifies fish stocks that are managed in a responsible manner. The certification program was initiated by WWF and Unilever in the late 1990s. The MSC had considered adopting aquaculture certification (as will be discussed in Chapter 14), but later decided against this direction. However, there is considerable interest by eNGOs for MSC to identify and certify reduction fisheries such that aquaculture of carnivores can be achieved without the potential for degradation of these fisheries. Of the main fisheries targeted for fish meal and oil, the MSC has certified or has begun assessments of the following: Atlantic herring, Pacific thread herring, and European pilchard (specific fishing regions omitted) (http://www.msc.org/track-a-fishery/fisheries-in-the-program/fisheries-by-species).

Many eNGOs find some solace in the notion that certain reduction fisheries will be MSC certified. Of course, there are others that distrust or have been critical of the MSC and their certifications (Jacquet et al. 2010).

Another issue that is important to the eNGOs is the development of new markets for carnivorous fish species when lessons on current commoditized aquaculture products have shown a level of unsustainability in the utilization of reduction fisheries. There is a degree of tolerance for these current and new markets by eNGOs. Their tolerance is likely a result of their formal partnerships with retailers, food service

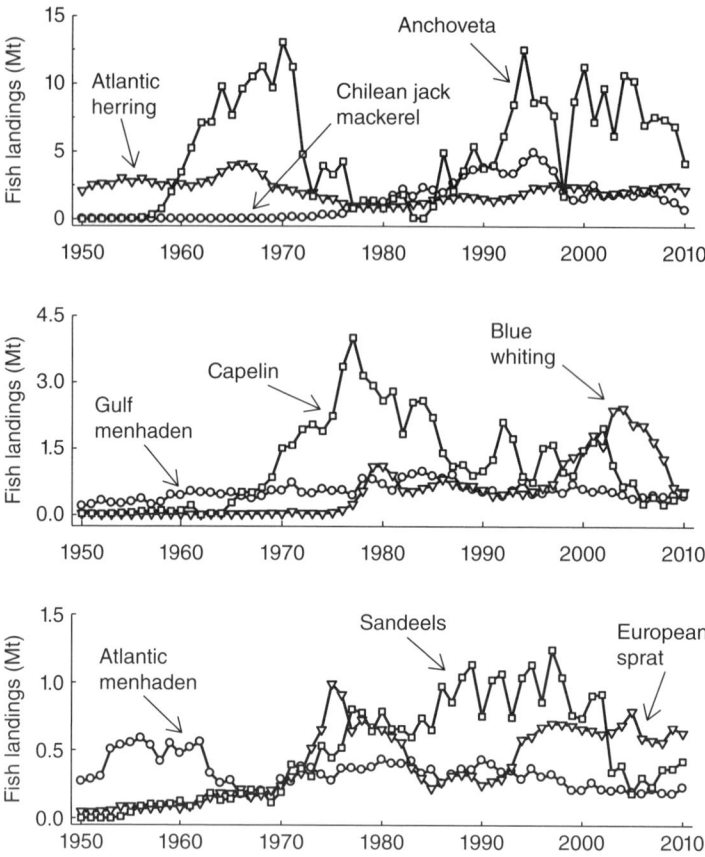

Figure 8.2 Landing of most economically important species of fish for fish meal and oil production. *Source:* FAO (2011).

companies, or restaurants that sell these products. It is challenging for eNGOs to walk the tightrope between attempting to transform a business stepwise and being ostracized by industry because they are viewed as too radical. The latter negates future efforts to work closely with businesses to improve practices and performance and realize some environmental betterment. Of course there are those that see eNGO-business partnerships as a form of "green-washing" (MacDonald 2008) which makes the balance of environmental change and business engagement even more delicate.

The eNGOs will often seek a higher level of efficiency in the conversion of wild fish to cultured product (i.e., the FIFO ratio) in addition to lobbying for less carnivorous species. At a minimum, the eNGOs contend that the FIFO ratio should not be above 1.0. For species such as tilapia or catfish, they seek even lower ratios somewhere between 0.5 and 0.8.

There are those that will surmise that species such as tilapia should be raised without any manufactured feed because it is an omnivore that can grow on a completely vegetarian diet in extensive or semi-intensive systems (with only local feed additives)(Belton et al. 2009, 2010). While tilapia can be raised in ponds using manures and other local feedstuffs, there is apparently little consumer desire in the United States, Europe, and Japan (regions importing the largest amounts of farmed tilapia) to purchase such products. The large scale change of fish such as tilapia (traded internationally) from a feed-based diet to an extensive primary production-based diet is unlikely, and most eNGOs recognize this.

One of the most novel and potentially industry-changing discoveries in fish meal and oil utilization came from DuPont, which developed a genetically modified yeast by introducing algal genes that can produce long chain omega-3 fatty acids such as EPA and DHA. This is a remarkable invention. Further, Verlasso (http://www.verlasso.com/), a salmon-farming operation in southern Chile is currently raising salmon with these ingredients and has achieved a 75% reduction in the dependency on wild fish (Personal communication, Scott Nichols). The use of wild fish to culture aquatic organisms is likely the most important issue to the environmentalists, and although many of the eNGOs recognize the achievements of Verlasso, there is little evidence of eNGOs lobbying retail partners to move procurement toward products such as this. Further, some eNGOs that are staunch supporters of organic production will reject innovations such as Verlasso's because of the utilization of genetically modified organisms to produce the omega-3 fatty acids.

The high rate of growth in the aquaculture sector coupled with a general trend toward introduction of more carnivorous fish species to markets (groupers, tunas, cod, etc.) generates a high degree of eNGO criticism. Although the majority of aquaculture (seaweeds, bivalves, and carps) are not highly carnivorous and do not have a burden on wild fisheries, the eNGOs tend to paint a picture of aquaculture with carnivorous species transposed over all aquaculture production. However, eNGOs find industry claims that carnivore aquaculture is needed to fight poverty and feed the world similarly misguided and disingenuous.

The perception of the eNGOs is that for aquaculture to sustain itself, it will require a directional change toward producing less carnivorous fish or finding synthetic or manufactured alternatives to fish meal and oil. There will be a need for current commoditized carnivores (such as shrimp and salmon) to be raised at higher levels of efficiency with regards to the utilization of wild fish. The eNGOs will soon only tolerate fish that are raised with wild fish certified by the MSC.

References

Adarme-Vega, T. C., D. K. Y. Lim, M. Timmins, F. Vernen, Y. Li, and P. M. Schenk. 2012. Microalgal biofactories: a promising approach towards sustainable omega-3 fatty acid production. *Microbial Cell Factories* 11:96.

Alasalvar, C., K. D. A. Taylor, E. Zubcov, F. Shahidi, and M. Alexis. 2002. Differentiation of cultured and wild sea bass (*Dicentrarchus labrax*): total lipid content, fatty acid and trace mineral composition. *Food Chemistry* 79:145–150.

Alder, J., B. Campbell, V. Karpouzi, K. Kaschner, and D. Pauly. 2008. Forage fish: from ecosystems to markets. *Annual Review of Environment and Resources* 33:153–166.

Amaya, E. A., D. A. Davis, and D. B. Rouse. 2007. Replacement of fish meal in practical diets for the Pacific white shrimp (*Litopenaeus vannamei*) reared under pond conditions. *Aquaculture* 262:393–401.

Bell, J. G., J. McEvoy, D. R. Tocher, F. McGhee, P. J. Campbell, and J. R. Sargent. 2001. Replacement of fish oil with rapeseed oil in diets of Atlantic salmon (Salmo salar) affects tissue lipid compositions and hepatocyte fatty acid metabolism. *Journal of Nutrition* 131:1535–1543.

Belton, B., F. Murray, J. Young, T. Telfer, and D. Little. 2010. Passing the Panda standard: a TAD off the mark? *Ambio*:39:2–13.

Belton, B., D. Little, and K. Grady. 2009. Is responsible aquaculture sustainable aquaculture? WWF and the eco-certification of Tilapia. *Society and Natural Resources* 22:840–855.

Boyd, C. E. and C. S. Tucker. 1998. *Pond Aquaculture Water Quality Management*. Boston: Kluwer Academic Publishers.

Boyd, C. E., C. S. Tucker, A. McNevin, K. Bostick, and J. Clay. 2007. Indicators of resource use efficiency and environmental performance in fish and crustacean aquaculture. *Reviews in Fisheries Science* 15:327–360.

Carter, C. G. and R. C. Hauler. 2000. Fish meal replacement by plant meals in extruded feeds for Atlantic salmon, *Salmo salar* L. *Aquaculture* 185:299–311.

Dale, N. M., M. Zumbado, A. G. Gernat, and G. Romo. 2004. Nutrient value of tilapia meal. *Journal of Applied Poultry Research* 13:370–372.

de Man, J. M. 1999. *Principles of Food Chemistry*, 3rd ed. New York: Spring Science and Business Media, Inc.

El-Sayed, A-F. M. 1998. Total replacement of fish meal with animal protein sources in Nile tilapia, *Oreochromis niloticus* (L.), feeds. *Aquaculture Research* 29:275–280.

Enterria, A., M. Slocum, D. A. Bengston, P. D. Karayannakidis, and C. M. Lee. 2011. Partial replacement of fish meal with plant protein sources singly and in combination in diets for summer flounder, *Paralichthys dentatus*. *Journal of the World Aquaculture Society* 42:753–765.

Eswaran, H., F. Beinroth, and P. Reich. 1999. Global land resources and population supporting capacity. *American Journal of Alternative Agriculture* 14:129–136.

FAO (Food and Agriculture Organization). 1986. The production of fish meal and oil. FAO Fisheries Technical Paper, T142, Rome, Italy.

FAO (Food and Agriculture Organization). 2010. *The State of World Fisheries and Aquaculture*. Rome: FAO Fisheries and Aquaculture Department.

FAO (Food and Agriculture Organization), WHO (World Health Organization), and UNU (United Nations University). 2002. Protein and amino acid requirements in human nutrition. WHO Technical Report Series 935, Geneva, Switzerland.

Hardy, R. W. 2003. Farmed fish and omega-3 fatty acids. *Aquaculture Magazine* 29:63–65

Jacquet, J., D. Pauly, D. Ainley, S. Holt, P. Dayton, and J. Jackson. 2010. Seafood stewardship in crisis. *Nature* 467:28–29.

Kaushik, S. J., J. P. Cravedi, J. P. Lalles, J. Sumpter, B. Fauconneau, and M. Laroche. 1995. Partial or total replacement of fish meal by soybean protein on growth, protein utilization, potential estrogenic or antigenic effects, cholesterolemia and flesh quality in rainbow trout, *Oncorhynchus mykiss*. *Aquaculture* 133:257–274.

Lenas, D. S. and C. Nathanailides. 2011. High content of n-6 fatty acids in the flesh of farmed fish sea bass *Dicentrarchus labrax* L. and gilthead sea bream *Sparus aurata* L.: implications for the human health. *International Aquatic Research* 3:181–187.

MacDonald, C. 2008. *Green, Inc*. Guilford: The Lyons Press.

Miller, M. R., P. D. Nichols, and C. G. Carter. 2008. *n-3* oil sources for use in aquaculture—alternatives to the unsustainable harvest of wild fish. *Nutrition Research Reviews* 21:85–96.

National Research Council. 1993. *Nutrient requirements of fish*. Washington, D.C: The National Academy of Sciences.

Naylor, R. L., R. J. Goldburg, J. H. Primavera, N. Kautsky, M. C. M. Beveridge, J. Clay, C. Folks, J. Lubchenco, H. Mooney, and M. Troell. 2000. Effect of aquaculture on world fish supplies. *Nature* 405:1017–1024.

Naylor, R. L., R. W. Hardy, D. P. Bureau, A. Chiu, M. Elliott, A. P. Farrell, I. Forster, D. M. Gatlin, R. J. Goldburg, K. Hua, and P. D. Nichols. 2009. Feeding aquaculture in an era of finite resources. *Proceedings National Academy of Science* 106:15103–15110.

Nettleton, J. A. 2000. Fatty acids in cultivated and wild fish. Proceedings of International Institute of Fisheries Economics and Trade, 10th Conference, Corvalis, Oregon.

Oenema, O. and S. Tamminga. 2005. Nitrogen in global animal production and management options for improving nitrogen use efficiency. *Science in China* 48:871–887.

Regost, C., J. Arzel, J. Robin, G. Rosenlund, and S. J. Kaushik. 2003. Total replacement of fish oil by soybean or linseed oil with a return to fish oil in turbot (*Psetta maxima*) 1. Growth performance, flesh fatty acid profile, and lipid metabolism. *Aquaculture* 217:465–482.

Robinson, E. H. and M. H. Li. 1994. Use of plant proteins in catfish feeds: replacement of soybean meal with cottonseed meal and replacement of fish meal with soybean meal and cottonseed meal. *Journal of the World Aquaculture Society* 25:271–276.

Shepherd, C. J., I. H. Pike, and S. M. Barlow. 2005. Sustainable feed resources of marine origin, pp. 59–66. European Aquaculture Society Special Publication No. 35, European Aquaculture Society, Oostende, Belgium.

Tacon, A. G. J. and M. Metian. 2008. Global overview on the use of fish meal and fish oil in industrially compounded aquafeeds: trends and future prospects. *Aquaculture* 285:146–158.

van der Hoek, K. W. 1998. Nitrogen efficiency in global animal production. *Environmental Pollution* 102:127–132.

Watanabe, T. 2002. Strategies for further development of aquatic feeds. *Fisheries Science* 68:242–252.

Chapter 9

Chemicals in aquaculture

The chemicals used most widely in aquaculture are liming materials and fertilizers. Liming materials are mainly agricultural limestone and lime. The most common chemical fertilizers are superphosphate, triple superphosphate, and urea but several other compounds are used (Boyd and Tucker 1998). An array of other chemicals including oxidants, coagulants, osmoregulators, algicides, herbicides, fish toxicants, antifoulants, therapeutants, disinfectants, anesthetics, agricultural pesticides, and hormones also are used in aquaculture (Boyd and Massaut 1999; Arthur et al. 2000; Schnick 2001).

Chemicals used in aquaculture can enter natural waters to cause water pollution. Some chemicals pose a danger to farm workers through potential toxicity or as fire and explosion hazards, and residues of certain chemicals may contaminate aquaculture products and present a food safety concern. The FAO (1997) urged countries to regulate the use of chemicals in aquaculture that are hazardous to human health and to the environment. In many countries governmental regulations on chemical use in aquaculture have been promulgated (Schnick 1988, 2001; Cabello 2006). For example, in the United States the USDA and the United States Food and Drug Administration (USFDA) must clear all chemical uses on food fish (http:// www.fda.gov/AnimalVeterinary/DevelopmentApprovalProcess/ Aquaculture/default.htm). Most importing countries also have food safety regulations that involve inspection of incoming shipments of seafood products for specific chemical residues. The US program requires importers to comply with Hazard Analysis and Critical Control Points (HACCP) rules and inspections including microbial and chemical analyses for specific contaminants are made (http://seafood.ucdavis. edu/seafoodhaccp.html).

Aquaculture, Resource Use, and the Environment, First Edition. Claude E. Boyd and Aaron A. McNevin.
© 2015 John Wiley & Sons, Inc. Published 2015 by John Wiley & Sons, Inc.

This chapter presents information on most chemicals that are used in aquaculture. It also provides some perspective on the contribution of aquaculture to chemical pollution and other adverse impacts of chemical use by humans.

Chemical use

There has been much discussion of the negative impacts of chemicals and many environmentally concerned citizens seem aghast at the mere mention of chemical use in food production. This concern seems particularly strong when aquaculture products are mentioned because of lack of knowledge about aquaculture and the considerable publicity about the negative impact of this endeavor. However, our modern lifestyle is dependent upon the use of chemicals in industry, sanitation, medicine, food production, transportation, and most other undertakings. The old DuPont advertising slogan "better things for better living ... through chemistry" that originated in the mid-1930s was quite accurate. The widespread concern over chemical uses probably prompted DuPont to remove "through chemistry" from the slogan in the early 1980s, and change the slogan to "miracles of science" in recent years. Without the use of chemicals infectious human diseases would be rampant because of lack of sanitation, disease vector control, and effective medicines resulting in human suffering and shorter life expectancy. Crop production would be lower and pest damage to crops would be much greater resulting in more food shortages than now exist. Simply put, peoples' lives would be a lot more difficult without chemicals. People often wish for the proverbial simpler life of times gone by but the senior author remembers the response of his physician grandfather to this lament: "Yes, the good old days, the good old salted pork and yellow fever days."

Without intent to minimize the hazards associated with chemical use in modern society for they are serious and many, chemicals are ubiquitous and necessary for many purposes. The goals should be to strive to use chemicals conservatively and responsibly, and to find alternatives to chemicals and chemical uses that are particularly hazardous or ineffective. Drastic restrictions on the use of many chemicals are not realistic and in many cases would be counterproductive.

Chemicals are used in aquaculture but hardly any of them were developed specifically for aquaculture. The majority were adopted from agriculture while others were taken from sanitary engineering, human medicine, and industry. Space considerations will not allow a discussion of the precautions that should be taken for safe use of individual aquacultural chemicals. But in general, chemicals should be stored securely to avoid unauthorized use and handled carefully to avoid spills. Workers should be provided instructions on how to use chemicals and of the hazards of each. Workers also should be provided protective clothing and other gear and equipment needed to apply chemicals safely. The discharge of chemicals to the environment should be minimized or avoided entirely where possible. Moreover, chemicals such as therapeutants, algicides, herbicides, and disinfectants, which may be highly toxic, accumulate in tissues of the culture species. Most chemicals—especially the potentially most hazardous—are accompanied by material safety data (MSD) sheets. The instructions on MSD sheets should be read and heeded. We will, however, discuss some especially hazardous characteristics of certain compounds or groups of compounds.

Liming materials

The common liming agent for aquaculture (and traditional agriculture) is agricultural limestone. This mineral consists of either calcium carbonate or a mixture of calcium and magnesium carbonates. Pure calcium carbonate ($CaCO_3$) is known as calcite while a 1:1 mixture of calcium and magnesium carbonate ($CaCO_3 \cdot MgCO_3$) is called dolomite. Most limestone is neither pure calcite nor pure dolomite, but a mixture containing more calcium carbonate than magnesium carbonate. Agricultural limestone is made by finely pulverizing limestone; the particle size of the product will vary, but nearly all of the particles normally will be 60 mesh or smaller (Boyd and Tucker 1998).

Agricultural limestone when applied to acidic soil or water neutralizes acidity (H^+) and raises pH:

$$CaCO_3 + 2H^+ = Ca^{2+} + H_2O + CO_2 \tag{9.1}$$
$$CaCO_3 \cdot MgCO_3 + 4H^+ = Ca^{2+} + 2H_2O + 2CO_2. \tag{9.2}$$

It also reacts with carbon dioxide (CO_2) in water to increase bicarbonate ($HCO_3{}^-$) concentration that is the main source of alkalinity in most waters. Typical reactions may be illustrated as follows:

$$CaCO_3 + H_2O + CO_2 = Ca^{2+} + 2HCO_3^- \tag{9.3}$$
$$CaCO_3 \cdot MgCO_3 + 2H_2O + 2CO_2 = Ca^{2+} + Mg^{2+} + 4HCO_3^-. \tag{9.4}$$

The pH of water in equilibrium with atmospheric carbon dioxide and solid phase limestone is 8.3 (Boyd and Tucker 1998). Agricultural limestone treatment does not increase pH high enough in aquaculture systems to harm fish or other aquatic life.

Lime is made by heating limestone at high temperature in a kiln to drive off carbon dioxide as illustrated for calcitic limestone:

$$CaCO_3 \rightarrow CO_2 \uparrow + CaO. \tag{9.5}$$

The resulting product (CaO) is called burnt lime, quick lime, or unslaked lime, and it may be slaked (treated with water) to form calcium hydroxide [$Ca(OH)_2$]. Of course lime usually is not made from pure calcitic limestone and therefore contains magnesium oxide or hydroxide also.

When put in water burnt lime and hydrated lime both react to yield calcium ion and hydroxyl ion:

$$CaO + H_2O = Ca(OH)_2 \tag{9.6}$$
$$Ca(OH)_2 = Ca^{2+} + 2OH^-. \tag{9.7}$$

Hydroxyl ion is basic and can raise pH to 10 or 12 following treatment resulting in toxicity to aquatic organisms. Lime applications greater than 50 kg/ha should be avoided in ponds containing fish or shrimp because of possible pH toxicity. Hydroxide reacts with carbon dioxide to form bicarbonate:

$$CO_2 + OH^- = HCO_3^- \qquad (9.8)$$

and the high pH effect of lime only lasts a few hours or days. Water should not be discharged from ponds for 3 or 4 days after lime application, and lime treatment should be postponed if heavy rain that could cause overflow is expected.

Lime is such a common aquaculture chemical that workers tend to be careless with it. Lime-water slurries often are made to facilitate the application of lime. These slurries have a pH of 13 to 14 and accidental contact can cause serious burns to the skin and damage to eyes including blindness.

Both types of lime are used in agriculture and aquaculture. According to West and McBride (2005) the use of lime in agriculture in the United States has declined drastically because it is much less expensive to manufacture agricultural limestone. The use of lime in agriculture is still common in many developing countries, and lime is used widely in aquaculture ponds as a liming material and for disinfecting pond bottom soils.

Reliable data on the global use of liming materials in agriculture could not be found, but in the United States, about 20 Mt were used in 2003 (West and McBride 2005). The United States uses about 10% of the world's chemical fertilizer nutrients. Assuming a similar relationship for liming materials, world liming material use may exceed 200 Mt.

There are about 11 000 000 ha of aquaculture ponds (Verdegem and Bosma 2009) and where lime materials are applied, rates are typically between 1000 and 3000 kg/ha (Boyd and Tucker 1998). Assuming that 25% of all aquaculture ponds are limed at 2000 kg/ha/year, the annual use of liming material could be as much as 5.5 Mt/year—about 2.75% of the estimated amount used in agriculture.

The global reserves of limestone are huge and no references of impending shortages could be found. Nevertheless, there is much concern over the adverse environmental impacts of limestone quarrying (Parise and Pascali 2003; Langer and Arbogast 2003; van Beynen and Townsend 2005; Darwish et al. 2011). Agricultural limestone and lime production requires use of fossil fuels with release of greenhouse gases. Carbon dioxide is released when limestone is burned to make lime, and carbon dioxide that was sequestered in limestone is released to the environment when agricultural limestone reacts with acidity.

Fertilizers

Fertilizers include manufactured chemicals that dissolve in water to release nutrients or organic wastes or by-products from agriculture and food processing that are decomposed microbially with the release of nutrients. The primary plant nutrients in fertilizers are nitrogen, phosphorus, and potassium. Fertilizer nutrients are not directly used by the culture species; they supplement nutrients naturally present in the bottom soil and water to stimulate phytoplankton productivity that in turn increases the abundance of natural food organisms allowing greater production of the aquaculture species.

Organic fertilizers

The most common organic fertilizers used in aquaculture are animal manures, cut grass, and crop residues (Boyd and Tucker 1998). Application rates often are in the range of 25 to 100 kg/ha/d of dry matter, and in a growing season, many tons of organic matter are applied per hectare.

Based on data from Verdegem and Bosma (2009) about 28.4% of pond aquaculture production was feed based—a lower estimate than 67% given by FAO (2012). Many ponds, however, receive small applications of feed merely to supplement the natural food resulting from fertilization; thus, to be conservative it will be assumed that all ponds are fertilized. No estimate of the global area of chemically fertilized versus organically fertilized ponds is available but possibly 25% or less of fertilized ponds receive chemical fertilizers. Using these assumptions, around 8 250 000 ha of ponds are treated with organic fertilizers and 2 750 000 ha receive chemical fertilizers. Livestock manure is likely the most common organic fertilizer used in aquaculture, and application rates vary widely. Assuming an average input of around 200 kg N and 75 kg P/ha/year in manure worldwide, as much as 1.65 Mt N and 0.62 Mt P may be applied annually to ponds in organic fertilizers. Of course, these estimates are too high because some ponds that receive feeds are not fertilized.

Amounts of nitrogen and phosphorus contained in livestock manure globally have been estimated as 17.34 Mt/year (Liu et al. 2010) and 16–20 Mt/year (Smil 2000), respectively. Thus a relatively small proportion of the nutrients in manures and other agricultural wastes—10% of nitrogen and 3.5% of phosphorus at most—are applied to aquaculture ponds. The recycling of nutrients from agricultural waste products by using them as pond fertilizers is a wise use of resources. This practice conserves chemical fertilizers and extracts value from waste.

Human wastes also are sometimes used to fertilize aquaculture ponds (Edwards 1980; 1988). No reliable information on the extent of human waste fertilization of ponds could be found. Moreover, this practice is not advisable because of sanitation and possible human health issues.

Chemical fertilizers

Nutrients in chemical fertilizers (and sometimes in organic fertilizers) are reported as percentages of N, P_2O_5, and K_2O. The actual forms of nutrients in chemical fertilizers are urea [$(NH_2)_2CO$], ammonium (NH_4^+), nitrate (NO_3^-), phosphate ($H_2PO_4^-$), and potassium (K^+). Urea quickly oxidizes to ammonia (NH_3):

$$(NH_2)CO + H_2O \rightarrow 2NH_3 + CO_2. \tag{9.9}$$

Ammonia is in equilibrium with ammonium:

$$NH_3 + H^+ \rightarrow NH_4^+. \tag{9.10}$$

At the pH of most soils and waters the majority of the ammoniacal nitrogen is in NH_4^+ form.

Nitrogenous fertilizer compounds usually are made by the reduction of atmospheric nitrogen (N_2) to ammonia by an industrial process. The source of hydrogen for reducing atmospheric nitrogen usually is natural gas. About 1230 m^3 of natural gas is required to produce 1 t of ammonia nitrogen but less than 5% of world natural gas consumption is expended to manufacture ammonia (Fixen and Johnston 2012). Ammonia is used as the nitrogen source for making most other nitrogen fertilizer compounds including the largest proportion of nitrate fertilizers. Some sodium nitrate fertilizer is extracted from the ore caliche that is rich in this compound (Ericksen 1983).

The source of fertilizer phosphorus is rock phosphate, a mineral known as apatite. This mineral is mined and the ore is treated with sulfuric acid to make superphosphate fertilizer and to produce phosphoric acid to manufacture triple superphosphate fertilizer. The phosphorus compound in both fertilizers is calcium phosphate $[Ca(H_2PO_4)_2]$. Calcium phosphate can be reacted with ammonium to make ammonium phosphates (Boyd and Tucker 1998).

Potassium for fertilizers is extracted from ores mined from deposits of sylvite (KCl), sylvinite (KCl and NaCl), hartsalz (K_2SO_4), and langbeinite [$K_2Mg_2(SO_4)_3$] (Boyd and Tucker 1998). Potassium fertilizers also can be produced from brine solutions obtained from closed-basin lakes such as the Dead Sea. The most common potassium fertilizer is potassium chloride (muriate of potash).

Sulfur is a plant nutrient and it is sometimes included incidentally or intentionally in fertilizer. Sulfur is indispensable in phosphate fertilizer manufacturing because it is used to convert phosphates in apatite to water-soluble form. About 80–85% of the world's sulfur production is used to make sulfuric acid, and about half of the world's sulfuric acid production is used in fertilizer manufacturing (IFDC 2008). There is a variety of sources of sulfur but most sulfur on the world market is extracted as a by-product from natural gas and crude oil.

The most common chemical fertilizers and their nutrient concentrations are provided (Table 9.1). Two or more of these basic fertilizers often are blended together to provide mixed fertilizers of various nutrient ratios. A common mixed fertilizer used in aquaculture contains 20% N, 20% P_2O_5, and 5% K_2O. It can be made from urea, triple superphosphate, ammonium phosphate, and muriate of potash or certain other combinations of basic fertilizer compounds.

The annual use of fertilizers in agriculture is presented in Fig. 9.1. Quantities of fertilizer nutrients used in agriculture have increased drastically since 1961—about ninefold for N, fourfold for P_2O_5, and threefold for K_2O. Nitrogen use per year still seems to be increasing and this probably is because of the large nitrogen input to grain crops. Phosphorus and potassium use, however, have tended to level off and have not increased greatly since the 1980s. Global quantities of fertilizer nutrients in 2009 were 105 Mt N, 37.9 Mt P_2O_5, and 27.5 Mt K_2O (http://faostat.fao.org).

Table 9.1 Approximate nutrient contents of common commercial fertilizers.

Fertilizer	Nutrient (%)		
	N	P$_2$O$_5$	K$_2$O
Urea	45	0	0
Calcium nitrate	15	0	0
Sodium nitrate	16	0	0
Ammonium nitrate	33–35	0	0
Ammonium sulfate	20–21	0	0
Superphosphate	0	18–20	0
Triple superphosphate	0	44–54	0
Monoammonium phosphate	11	48	0
Diammonium phosphate	18	48	0
Calcium metaphosphate	0	62–64	0
Potassium nitrate	13	0	44
Potassium sulfate	0	0	50
Potassium chloride (muriate of potash)	0	0	60

Source: Boyd (1990).

These quantities equate to about 16 kg N, 7 kg P$_2$O$_5$, and 4 kg K$_2$O per capita per year for the world's population. Greater fertilizer use has been a major factor allowing agricultural production to increase.

There have been reports of impending shortages of raw material for fertilizer manufacturing (Bumb and Baanante 1996; Dery and Anderson 2007; Vaccari 2009). A recent and thorough assessment of world fertilizer nutrient reserves (Fixen and Johnston 2012) concluded, however, that reserves and resources for nitrogen, phosphorus, potassium, and sulfur appear adequate for the foreseeable future. Nevertheless, these authors stressed that nutrient costs will rise over time, and wise use of non-renewable nutrient resources should be a critical concern in agriculture.

The amount of fertilizer used in aquaculture ponds varies greatly among culture species, basic fertility of pond soil and water, and managerial opinions.

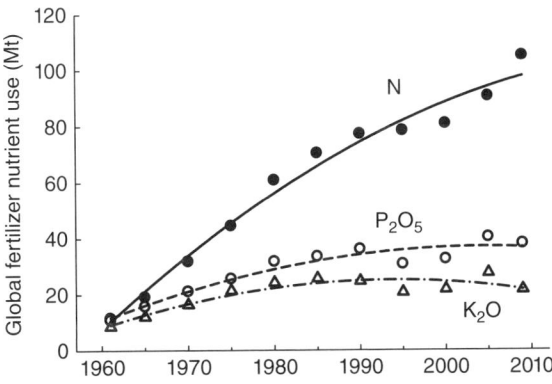

Figure 9.1 Global fertilizer nutrient use from 1961–2009. *Source:* http://faostat3.fao.org/faostat-gateway/go/to/download/R/RF/E.

Applications typically range from about 2 kg/ha N and 1 kg P_2O_5/ha to 10 kg/ha N and 5 kg P_2O_5/ha (Boyd and Tucker 1998). Assuming that about 2 750 000 ha of ponds receive about 15 applications of chemical fertilizers per year at input rates of 5 kg N/ha and 2.5 kg P_2O_5/ha per application, global annual fertilizer use in aquaculture might be as much as 206 250 t of N and 103 125 t of P_2O_5. These amounts represent about 0.20% and 0.27% of annual world N and P_2O_5 use in agriculture. Potassium fertilizer is seldom used in pond fertilizers, but when it is, application rates are about 2 kg/ha of K_2O. Thus a maximum of 82 500 t or 0.30% of world K_2O use in agriculture could be applied to aquaculture ponds. The actual amount is much less. Aquaculture definitely is not a major factor affecting world fertilizer nutrient supplies and markets.

Fertilizers increase concentrations of ammoniacal nitrogen, nitrate, and phosphate in water and pond effluents can contribute to eutrophication of natural waters. Nutrient concentrations are relatively low, seldom exceeding 1 mg/L total ammonia nitrogen, 0.5 mg/L nitrate nitrogen, or 0.3 mg/L phosphate phosphorus, and fertilized ponds are not considered to be major sources of nutrient pollution (Boyd and Li 2012). Toxic concentrations of ammonia nitrogen do not occur in fertilized ponds unless unusually large nitrogen applications are made (Boyd et al. 2006). Nevertheless, avoiding excessive fertilizer use will prevent overly dense phytoplankton blooms in ponds, reduce production costs, lessen the potential for water pollution, and conserve resources.

A few nitrogen fertilizers—sodium, potassium, ammonium, and calcium nitrates in particular—are explosion hazards. These materials can be mixed with diesel fuel and used as a substitute for dynamite. They should not be stored near petroleum products and sparks or open flames.

Animal dung and other agricultural waste products can create an odor and filth problem when stockpiled in large quantities on a farm. It is possible that antibiotics fed to livestock could pass from manures to fish or shrimp in ponds. This possibility can be minimized by composting manure before using it in ponds. There is not a human pathogen concern about using livestock manure in ponds, but some people might not want to learn that their fish or shrimp had been produced in ponds receiving animal dung. Although not proven, we believe that the use of human wastes in aquaculture ponds could be a source of pathogens for workers or even those consuming the fish. Certainly many people even in poor rural areas probably would not prefer fish produced in human wastes.

Therapeutants

Antibiotics

Antibiotics are chemicals produced by various species of microorganisms and fungi; these compounds can inhibit the growth of or kill bacteria and other microorganisms. Antibiotics are employed widely in human and veterinary medicine to treat infectious diseases. The two most well-known antibiotics probably are penicillin and streptomycin, but there are many others.

There is concern about the discharge of antibiotics into the environment because they can be toxic to natural organisms and repeated exposure of bacteria to an antibiotic can lead to development of resistance making the antibiotic less effective for its intended purpose. Studies have revealed that antibiotics can be toxic to algae and microcrustaceans (Holten-Lützhøft et al. 1999; Christensen et al. 2006; Goncalves-Ferreira et al. 2007), and of course antibiotics are toxic to non-target bacteria. The release of antibiotics into the environment can negatively impact the biodiversity of planktonic and benthic communities.

Repeated exposure of microorganisms to antibiotics in the water environment can lead to the selection of spontaneous mutants that exhibit antibiotic resistance (Zhang et al. 2009). This phenomenon has been demonstrated for bacteria from shrimp ponds (Tendencia and de la Peña 2001), trout farms (Spanggaard et al. 1993; Schmidt et al. 2000), *Pangasius* farms (Sarter et al. 2007), and in waters and marine sediment near salmon farms (Nygaard et al. 1992; Miranda and Zemelman 2002). Development of antibiotic resistance in target, pathogenic organisms could render an antibiotic worthless for its intended use in aquaculture. However, a broader concern is the horizontal transmission of antibiotic resistance from bacteria at aquaculture sites to bacteria responsible for human diseases as discussed by Rhodes et al. (2000) and Benbrook (2002). The transfer of antibiotic resistance in aquaculture to human pathogens has not been clearly established (Burridge et al. 2010), but there are reasons to suspect that such transfers have occurred (Webber et al. 1994).

The usual means of administering antibiotics in aquaculture is with medicated feed. It has been estimated that 70–80% of the antibiotic in medicated feed enters the water of the culture system via urinary and fecal excretion and in uneaten feed (Hektoen et al. 1995). The antibiotic remains in the water until degraded by natural processes or is adsorbed by sediment. Some antibiotics degrade rapidly, but most are persistent (Kemper 2008).

Aquaculture is only one source of antibiotic contamination. Many studies have shown that antibiotics occur at measurable concentrations in waters, sediments, and terrestrial soil near livestock production facilities or in fields where livestock manures have been used as fertilizers. Antibiotics also occur in domestic wastewater. Natural water bodies are contaminated with antibiotics from municipal wastewater and in runoff from animal production facilities. According to Kemper (2008) little is known about the actual hazard of antibiotic pollution to either humans or the environment, but the increasing resistance to antibiotics by bacteria and the diminishing effectiveness of therapeutic drugs is of global concern. A proper risk management approach is to strive to reduce the discharge of antibiotics for human, veterinary, and aquaculture use to the environment.

A wide variety of antibiotics are used in aquaculture (Schnick 1988; Primavera et al. 1993; Supriyadi and Rukyani 2000; Gräslund et al. 2003; Costello et al. 2001). The antibiotics are the same or closely related to those used for treating human, livestock, and pet diseases. Some of the most common antibiotics used in aquaculture are listed in Table 9.2. The greatest use of antibiotics is in the culture of high-value species such as rainbow trout, salmon, and shrimp, but they also are used in culture of *Pangasius* for export, and the culture of channel catfish in the United States and a few other fish species in developed countries. Antibiotics are not used in

Table 9.2 Antibiotics commonly used in aquaculture.

Class	Compound
Aminoglycosides	Neomycin[*]
	Gentamycin[*]
	Streptomycin (obsolete)
β-Lactams; penicillins	Amoxicillin[*]
	Ampicillin[*]
Fenicoles	Chloramphenicol
	Florfenicol
Fluoroquinolones	Ciprofloxacin[*]
	Enrofloxacin[*]
	Flumequine[*]
Macrolides	Erythromycin[*]
Non-fluorinated quinolones	Oxolinic acid
	Sarafloxin
Sulfonamides	Sulfamethazine
	Sulfamerazine
	Sulfadimethoxine
Trimethoprim	Trimethoprim[*]
Tetracyclines	Chlortetracycline
	Oxytetracycline[*]
	Tetracycline[*]

Those with an asterisk also are commonly used in human medicine.

molluscan bivalve culture except in hatcheries, and there probably is little antibiotic use in aquaculture for domestic consumption in developing countries.

It is difficult to find reliable information on the quantities of antibiotics used in aquaculture. A study of US aquaculture (Benbrook 2002) gave high and low estimates of antibiotic use for the late 1990s. The averages of these high and low estimates and data on average annual production in the late 1990s are given in Table 9.3. Antibiotic use was highest for trout (920 g/t production) and lowest for miscellaneous species (250 g/t production)—average antibiotic use was 380 g/t. Benbrook (2002) concluded that antibiotic use per unit of production was similar between aquaculture and beef cattle, but hog production used about fivefold more antibiotics than aquaculture per unit of production.

Table 9.3 Antibiotic use in US aquaculture in the late 1990s.

Species	Annual production (t)	Antibiotic use	
		(kg/year)	(g/t of fish)
Channel catfish	249 400	85 700	340
Rainbow trout	24 910	23 000	920
Salmon	15 000	9 500	630
Other species	68 900	17 200	250
Total	**358 210**	**135 400**	**380**

Source: Prepared from information presented by Benbrook (2002).

Table 9.4 Antibiotic use in salmon aquaculture in the four leading producer countries in 2007.

Country	Production (t)	Antibiotic use	
		(kg)	(g/t of fish)
Norway	821 997	649	0.79
Chile	380 391	385 600	1014
Scotland	132 528	1553	11.7
Canada	102 509	21 330	208

Source: Prepared from information presented by Burridge et al. (2010).

Results of an evaluation of antibiotic use in salmon culture in the four leading salmon-farming countries (Burridge et al. 2010) are summarized (Table 9.4). As of 2007, Chile used antibiotics in salmon culture at a greater rate relative to production than reported for US salmon culture in the late 1990s. The other three countries, Norway, Scotland, and Canada, used much less antibiotics per unit of production than did Chile. Although there is no evidence of reduced antibiotic use in Chile, there is for the other three countries. Between 2006 and 2008 total antibiotic use in salmon farming declined by factors of 2.27 in Norway, 2.66 in Canada, and 63.6 in Scotland (Burridge et al. 2010).

No estimate of world antibiotic use was found, but in 1999 the countries in the European Union (EU) used 13 288 t and in 2006 the United States used 16 200 t. In the EU, 29% were used in veterinary medicine, 65% in human medicine, and 6% as growth promoters, while in the United States, 70% were used in livestock farming and 30% for other purposes (Kemper 2008). The estimate of antibiotic use in US aquaculture in the late 1990s was 135.4 t—only 0.84% of total use and 1.19% of veterinary use reported for 2006. The majority of aquaculture use of antibiotics in EU is for salmon; thus based on data presented in Table 9.4, aquaculture antibiotic use in the EU is probably a smaller percentage of total use than in the United States. It is not possible to estimate antibiotic use in aquaculture for other regions, but in some major aquaculture countries and especially in China, it probably would be comparatively greater than in the United States or the EU but still a small percentage of total use.

In spite of the relatively small quantities of antibiotics used in aquaculture, aquaculture tends to be concentrated in small areas. Places with a high density of aquaculture production facilities can discharge large quantities of antibiotics to the environment, for example, salmon farming area in Chile, and these areas could possibly be "hot spots" for the development of microbial resistance to antibiotics that could be spread much wider by horizontal transmission and cause resistance in microbes of human health concern (Smith et al. 2005).

Parasiticides

Internal and external parasites are common in cultured aquatic organisms, and various chemicals are used to combat them (Table 9.5). Parasiticide treatments are

Table 9.5 Parasiticides often used in aquaculture.

Class	Compound	Class	Compound
Avermectins	Emamectin benzoate	Organophosphate	Azamethiphos
	Ivermectin		Dichlorvos
Chitin synthesis	Teflubenzuron		Malathion
inhibitors	Diflubenzuron		Trichlorfon
General	Acetic acid	Pyrethroid	Cypermethrin
	Copper sulfate		Deltamethrin
	Formalin	Triarylmethane dye	Malachite green
	Hydrogen peroxide		
	Potassium permanganate		
	Sodium chloride		

applied in several ways. Animals may be dipped in baths of the parasiticide for a short period. Parasiticides may be applied directly to water in hatchery tanks, raceways, and other intensive culture units, and ponds. Curtains may be installed temporarily around cages to retain the chemical long enough for it to act on the parasites—similar to a bath treatment. Some parasiticides, for example, avermectins, are administered as feed additives. Regardless of the method of administration, the portion of the parasiticidal compound not retained in the animals enters the culture system and usually is discharged into the environment.

Formalin, salt, acetic acid, copper sulfate, hydrogen peroxide, potassium permanganate, and malachite green are used widely as parasiticides, and formalin is the most commonly employed. Other compounds listed in Table 9.5 are used most commonly in salmon culture and mainly for controlling sea lice.

Formalin is an aqueous solution consisting of 37–50% formaldehyde (CH_2O). Formaldehyde is toxic to fish, plants, and other animals with toxicity increasing with greater concentration and exposure time (Bills et al. 1977). In ponds formalin treatments of 15 mg/L or more may kill phytoplankton resulting in dissolved oxygen depletion (Allison 1962). Formalin is degraded by bacterial and chemical processes, and its half-life has been reported to range from 1 to 400 d (Masters 2004). Formalin is an irritant and has antigenic properties, and some studies suggest that it is a possible carcinogen but formalin residues have not been detected in aquaculture products (Wooster et al. 2005; Jung et al. 2001).

Acetic acid (CH_3COOH) is the principle component of vinegar. When applied to water it will dissolve and weakly dissociate, but at the concentrations used in aquaculture it would not depress the pH to a toxic level. Acetic acid will be decomposed quickly by bacteria, and large concentrations could cause dissolved oxygen depletion.

Malachite green is a triarylmethane dye with the formula $[C_6H_5C(C_6H_4N(CH_3)_2)_2]Cl$ made by condensation of benzaldehyde and dimethylaniline. It is highly effective against important protozoal and fungal infections and is used extensively in aquaculture (Alderman 1985). Malachite green is highly toxic to aquatic animals; LC50 values range from 2 to 7 mg/L after 3 hours to 0.04 to 2.0 mg/L after 96 hours (Srivastava et al. 2004). Aniline and chloroaniline compounds such as malachite green degrade microbially and by photolysis, and half-lives in water ranged from

27 to 173 hours (Hwang et al. 1987). Malachite green accumulates in tissues of exposed animals and can persist for long periods (Srivastava et al. 2004). This poses a food safety hazard to consumers because malachite green is a multi-organ toxin affecting kidneys, liver, spleen, heart, skin, eyes, lungs, and bones of rats and mice. In addition, malachite green can negatively affect immune and reproductive systems and it has genotoxic and carcinogenic potentials (Srivastava et al. 2004). Its use in aquaculture is viewed with alarm and it has been banned in several countries.

Environmental effects of parasiticides at salmon farms were discussed by Burridge et al. (2010). Although the compounds are toxic to aquatic life and persist in water and sediment for various lengths of time, the concentrations used do not appear to cause serious environmental problems. The main effect is toxicity to plankton, and the plankton community will rapidly recover when the concentrations of parasiticides decline.

There are few records of quantities of parasiticides used in aquaculture. Estimates of the amounts used for typical treatment concentrations could not be made because the volumes of water treated or the average frequency of applications could not be determined. Burridge et al. (2010) however, reported that the total use of parasiticides in salmon culture in Norway, Scotland, and Chile in 2007 was 600 kg, 195.6 kg, and 132 kg, respectively.

Parasiticides for aquaculture have applications in veterinary medicine, insect control, and sanitation. Thus aquaculture application of parasiticides—like for antibiotics—is a minor use.

Fungicides and disinfectant dips

Fungal infections are particularly troublesome in fish hatcheries and many compounds have been used to treat eggs and small fish (Marking et al. 1994). Some of these compounds also are used to disinfect small fish before stocking them in grow-out systems. Chemicals that have been used as antifungal agents and disinfectants include acetic acid, sodium chloride, formalin, glutaraldehyde, hydrogen peroxide, malachite green, povidone-iodine compounds, quaternary ammonium compounds such as benzalkonium chloride (ADBAC or BKC) and benzalkonium bromide (bromo-geramine), methylene blue (methylthioninium chloride), nystatin (an antifungal antibiotic produced by *Streptomyces noursei*), furazolidone, terramycin, streptomycin, erythromycin, chloramphenicol, and prefuran (Schnick 2001; Pathak et al. 2000; Yulin 2000; Lyle-Fritch et al. 2006; Marking et al. 1994).

The compounds are applied to relatively small vessels of water containing eggs or tanks containing small fish. Exposure usually is brief, but concentrations often are relatively high.

Disinfectants

Disease organisms can enter ponds via the water supply, and they can survive in puddles of water and moist soil between crops. Thus pond bottoms often are dried out between crops and bottoms also may be treated with lime to raise soil pH and kill unwanted organisms. Water for filling ponds may be disinfected in a reservoir before adding it to ponds or it may be disinfected in ponds before stocking culture animals.

Chlorine compounds

Chlorination is the most common means of destroying pathogenic microorganisms in municipal water supplies. Although chlorine gas (Cl_2) is commonly used in municipal water treatment this compound is extremely hazardous because it is a highly toxic gas that was used in warfare during World War I. It ordinarily is not and should not be used at aquaculture facilities.

Common household bleach or sodium hypochlorite (NaOCl) and calcium hypochlorite [$Ca(OCl)_2$] that are commonly used in swimming pools are used widely as disinfectants in aquaculture. These compounds are used in hatcheries at concentrations of 50 mg/L of active chlorine or more, and they may be used in ponds before stocking. The recommended treatment rate for disinfecting ponds is 10 mg/L chlorine or about 15 mg/L of calcium hypochlorite (Hill et al. 2013). Chlorine residuals are highly toxic to fish and other aquatic animals and applications of chlorine compounds to ponds containing fish or shrimp is extremely risky—even at concentrations of 1 or 2 mg/L (Potts and Boyd 1998).

Free chlorine residuals formed when hypochlorite is put into water quickly are expended in oxidizing organic matter and other reduced substances. Any remaining active chlorine is reduced to harmless chloride within 1 or 2 days through exposure to light.

The high toxicity of chlorine compounds represents a significant worker safety hazard. Workers should wear protective clothing and respirators. Hypochlorites should never be placed in an acidic solution because chlorine gas can be generated as illustrated in the following equation:

$$HOCl + H^+ = H_2O + Cl_2. \tag{9.11}$$

Insecticides

The organophosphorus insecticide Dichlorvos (dimethyl 2,2-dichlorovinylphosphate) or DDVP is often used as a disinfectant in shrimp farms. This compound is added to water in ponds before postlarvae are stocked at a rate of 2–3 mg/L. After about 1 week shrimp are stocked and no additional DDVP is applied to ponds. However, at some large farms water treated 7 days previously with DDVP is used for make-up water to replace evaporation and seepage or for minimum water exchange (1–2% pond volume per day). Treatment with DDVP in the reservoir is suspended at least 60 days before shrimp harvest.

Data provided the authors by a large shrimp farm revealed no detectable DDVP residue in pond water 5 days after treatment. Also no residue of DDVP could be detected in shrimp when ponds are harvested. Dichlorvos is more effective than lime as a disinfectant and cheaper and equally effective as chlorine.

Some farmers in Asia have used synthetic pyrethroid insecticides such as cypermethrin and deltamethrin to disinfect ponds. These compounds are highly toxic to shrimp and should not be used.

Miscellaneous compounds

During a recent trip to Thailand and Vietnam the senior author visited some shrimp supply stores and listed compounds sold as disinfectants. In addition to compounds mentioned above, the list included: BKC, chloramine, formalin, glutaraldehyde, iodophor (povidone-iodine and polyoxyethylene iodine complex), peroxymonopersulfate, potassium monopersulfate, hydrogen peroxide, trichlorocyanuric acid (TCCA), and sodium-N-chloro-para-toluenesulfonamide. Little information could be found on treatment rates, effectiveness, or possible negative impacts of these compounds.

Anesthetics

Anesthetics are used in research to immobilize fish for invasive, physiological investigations, and overdoses of these drugs may be used for euthanization. The main use of anesthetics in aquaculture is to sedate animals during grading, vaccination, transport, etc. Anesthetics are absorbed through the gills into the blood from which they reach the central nervous system. The stages of anesthesia include reduced motility and respiration (sedation), partial loss of equilibrium and reflex action (anesthesia), complete loss of equilibrium and reflex action (surgical anesthesia), and respiratory and cardiac failure (death) (Coyle et al. 2004). When removed from an anesthetic bath before irreversible respiratory and cardiac effects occur and placed in clean water, aquatic animals excrete the anesthetic drug and regain normal nervous system function.

Anesthetics approved for use in the United States are carbon dioxide, sodium bicarbonate, and tricaine methanesulfonate (MS-222 or TMS) (Schnick 2001). Several other compounds are in common use worldwide (Table 9.6). The wide range in dose (Table 9.6) for most anesthetics results because of species differences in susceptibility to specific compounds and depth of sedation or anesthesia necessary (Coyle et al. 2004).

Anesthetics are used at relatively high concentration in small volumes of water. Little is known about the rate of degradation of most of these substances or of their environmental toxicity. The cautious approach to disposal would be to release anesthetic baths into a wastewater treatment pond or into a municipal sewage system rather than discharge them directly into natural waters.

Most anesthetics have potential toxicity to humans, for example, even carbon dioxide can be toxic if used in a confined, poorly ventilated room. Those working with these materials should consult MSD sheets to ensure proper storage, handling, and personal safety.

Only one chemical anesthetic (MS-222) is approved by the FDA for use in the United States. A 21-day withdrawal period is required before fish treated with this

Table 9.6 Anesthetics used in aquaculture.

Common name	Chemical name	Dose range
MS-222	3-aminobenzoic acidethyl ester methanesulfonate	25–100 mg/L
Sodium bicarbonate	$NaHCO_3$ (treated with acetic acid to release carbon dioxide)	–
Carbon dioxide	CO_2	120–250 mg/L
Benzocaine	p-aminobenzoic acid ethyl ester	25–100 mg/L
Metomidate	1-(1-phenylethyl)-1 H-imidazole-5-carboxylic acid methyl ester	1–10 mg/L
Propoxate	propyl-DL-1-(phenylethyl) imidazole-5-carboxylate hydrochloride	0.5–10 mg/L
Quinaldine	2-methylquinoline sulfate	15–1000 mg/L
2-phenoxyethanol	1-hydroxy-2-phenoxyethane	200–300 μL/L
Clove oil	Active ingredients: eugenol (4-allyl-2-methoxyphenol) and isoeugenol (4-propenyl-2-methoxyphenol)	2–120 mg/L
Aqui-S	Mixture of isoeugenol and polysorbate 80 (a surfactant)	20–200 mg/L

Source: Coyle, S. D., R. M. Durborow, and J. H. Tidwell. 2004. Anesthetics in Aquaculture. Southern Regional Aquaculture Center Publication 3900, Mississippi State University, Stoneville, Mississippi.

compound can be harvested for sale as food fish (Coyle et al. 2004). Little is known about residues of anesthetics in fish and other aquaculture species and a similar withdrawal time probably should be used for anesthetics—other than bicarbonate and carbon dioxide gas—until data on residues are available.

Hormones

Hormones used to induce spawning in fish include human chorionic gonadotropin, follicle-stimulating hormone, pituitary hormone, piscine gonadotropins, luteinizing hormone (Stevens 1966; Chaudhuri 1976; Mylonas et al. 2010). Growth hormone also has been used to experimentally increase the growth rate of fish (Cavari et al. 1993; Agellon et al. 1988; Silverstein et al. 2000). Neither the common use of hormones in spawning nor the experiments on growth hormones have elicited much environmental or food safety concern.

The use of hormonal-induced sex reversal in fish has been much more controversial. Sex reversal can be used to produce same-sex fingerlings for stocking thereby avoiding reproduction in culture systems. In some cases hormonally sex-reversed fish may grow faster than normal fish. According to Pandian and Sheela (1995), sex reversal by hormonal treatment is possible in many species using one of a group of 31 steroids. The most widely known application of hormonal sex reversal in aquaculture is the use of 17-methyltestosterone with tilapia (Green and Teichert-Coddington 1994; Macintosh et al. 1985; Guerrero 1975). This compound usually is administered to tilapia fry in feed beginning soon after they are released from paternal care and continued for about 1 month. The 17-methyltestosterone concentration in feed typically is around 30–60 mg/kg diet.

The amount of 17-methyltestosterone used annually for sex reversing fish is not known, but maximum possible annual use of 100 kg was estimated (http://media.sustainablefish.org/MT_WP.pdf). This is a comparatively small amount when compared to hormone use in livestock production. The estimates of hormone excretion in livestock production were 33 t of estrogens and 7.1 t of androgens in the EU and 49 t of estrogens and 4.4 t of androgens in the United States (Lange et al. 2002). Nevertheless, there is much concern over the release of pharmaceutically active compounds into the environment via effluents, and efforts should be made to lessen the quantities released (Heberer 2002). Water treated with 17-methyltestosterone should be recycled until the sex-reversal task has been completed at a particular hatchery. Afterward the hatchery should hold the water in a waste treatment pond for 1 or 2 weeks before discharging to the environment. However, several studies have shown that 10–20% of pharmaceutical compounds such as 17-methyltestosterone may remain in wastewater that has passed through waste treatment facilities (Liu et al. 2012; Chang et al. 2011).

In the case of tilapia, fish are exposed to 17-methyltestosterone only during the early fry stage in amounts of about 0.02 mg/fish. The fry are then reared for at least 5 months with no further 17-methyltestosterone contact before they reach marketable size. The 17-methyltestosterone concentration in tilapia and trout declines rapidly after treatment and within 5 days the hormone concentration declined to 1% of the initial concentration (Johnstone et al. 1983) in selected organs. Several studies have not detected 17-methyltestosterone in the meat of trout and tilapia treated as fry after a few weeks (http://media.sustainablefish.org/MT_WP.pdf). It is reasonable to assume that 17-methyltestosterone used for sex reversal in fry will not contaminate marketable-sized fish.

A unique problem with 17-methyltestosterone is intentional consumption by farm workers because of its muscle-building properties and its perceived libido-enhancing characteristics. This hazard can be minimized by storage of 17-methyltestosterone in a locked cabinet and careful inventory control (http://media.sustainablefish.org/MT_WP.pdf).

Oxidants

Potassium permanganate

Potassium permanganate ($KMnO_4$) is used as an oxidant to detoxify rotenone and antimycin A used to eradicate wild fish in ponds (Boyd and Tucker 1998). It can also be used to oxidize hydrogen sulfide and ferrous iron in source water or in pond bottoms. It does not release dissolved oxygen into the water as some aquaculturists believe. Also the claim that potassium permanganate will oxidize organic matter, lessen oxygen demand, and improve dissolved oxygen concentration in pond water is not valid (Tucker and Boyd 1977). Nevertheless, potassium permanganate sometimes is used with the belief that it improves water quality.

There are relatively few uses of potassium permanganate in traditional agriculture. Possibly the largest amounts of potassium permanganate are used in municipal water treatment for removing reduced substances namely iron, manganese, and hydrogen sulfide from water supplies (Faust and Aly 1998). It has many niche uses as a reagent and disinfectant.

Potassium permanganate can be toxic to fish at concentrations of 4 or 5 mg/L or more (Tucker and Boyd 1977). However, potassium permanganate soon is transformed to manganese oxide and precipitates from water. Therefore there is little likelihood of toxic concentrations of this chemical in most aquaculture effluents.

Peroxides

Hydrogen peroxide (H_2O_2), sodium carbonate peroxyhydrate or SCP $(2NaCO_3 \cdot 3H_2O_2)$, and calcium peroxide (CaO_2) sometimes are used as emergency sources of dissolved oxygen during fish transport in rural areas in developing countries. The compounds decompose to release oxygen as follows:

$$2H_2O_2 = 2H_2O + O_2 \tag{9.12}$$
$$2CaO_2 + 2H_2O = 2Ca(OH)_2 + O_2 \tag{9.13}$$
$$2Na_2CO_3 \cdot 3H_2O_2 + 2CO_2 + 2H_2O = 4Na^+ + 4HCO_3^- + 6H_2O + 3O_2. \tag{9.14}$$

Peroxides also are used as general oxidants in the same manner as potassium permanganate. Because the compounds decompose quickly they seldom would be contained in aquaculture effluents.

Peroxides stronger than 10% strength are an explosion hazard. Concentrated solutions also are hazardous to humans, so protective clothing is a must when working with peroxides. Peroxides have many uses in industry and they are used as mouthwashes and disinfectants. Aquaculture is no doubt a very minor use of these compounds.

Some other oxidants as mentioned earlier are used in Asian shrimp ponds to kill vectors of diseases before stocking postlarvae.

Sodium sulfite

Sodium bisulfite $(NaHSO_3)$ and sodium metabisulfite $(Na_2S_2O_5)$ are used to prevent dark mottling in shrimp sold with heads on. The enzyme polyphenol oxidase that is responsible for black spotting or melanosis is still active in shrimp tissue after harvest—even in shrimp that have recently died. Placing shrimp on ice will not always stop black spotting in post-harvest shrimp, but exposure to sodium bisulfite will stop the process. It is a common procedure to dip shrimp for 2–5 minutes immediately after harvest in a tank on the pond bank that contains a 5% solution of either sodium bisulfite or sodium metabisulfite. Sodium bisulfite results from either compound because metabisulfite is transformed to bisulfite in water:

$$Na_2S_2O_5 + H_2O = 2NaHSO_3. \tag{9.15}$$

Waste sodium bisulfite solutions remaining after harvest of a shrimp pond represents an especially serious but seldom-recognized environmental hazard. Sodium bisulfite removes dissolved oxygen from water:

$$2NaHSO_3 + O_2 = 2NaHSO_4. \tag{9.16}$$

A 5% (50 000 mg/L) solution of sodium bisulfite can remove a large amount of dissolved oxygen quickly.

Acidity results when bisulfite ion disassociates:

$$NaHSO_4 = Na+ + HSO_4^-$$ (9.17)

$$HSO_4 \rightarrow H^+ + SO_4^{2-}$$ (9.18)

If enough sulfite enters water to completely neutralize the total alkalinity, pH can fall to a level toxic to aquatic organisms.

The easiest way to deactivate sodium sulfite solutions following their use as a shrimp dip is to hold them in a wastewater-treatment lagoon. If no treatment system is available, the solution can be placed in a tank, aerated to oxidize the remaining sulfite, and treated with calcium or sodium hydroxide to neutralize the resulting acidity. Each kilogram of sodium sulfite will require 0.36 kg calcium hydroxide or 0.38 kg sodium hydroxide for neutralization. Care should be taken not to add excess hydroxide because it could cause an unacceptably high pH. A neutralized sodium bisulfite solution can be safely discharged into natural water (Boyd and Gautier 2002).

Importing countries usually place a limit on the concentration of sodium sulfite to which shrimp are exposed but the use of this compound is usually not considered a food safety hazard.

Nitrate compounds

Sodium nitrate ($NaNO_3$), potassium nitrate (KNO_3), and calcium nitrate [$Ca(NO_3)_2$] are fertilizers but they also are oxidants. Sodium nitrate in particular has been promoted as a source of dissolved oxygen and a bottom soil oxidant. In reality it is not a source of dissolved oxygen for shrimp, but the oxygen in nitrate can be used by denitrifying bacteria. As long as nitrite is present, it will poise the redox potential in sediment above the threshold for hydrogen sulfide production (Boyd and Tucker 1998). This use of sodium nitrate is almost entirely restricted to shrimp culture in Mexico and Central and South America.

The hazards of nitrate compounds were discussed in the section on fertilizers but it should be repeated that these compounds should not be stored near petroleum products or near-open flames or sparks because of the danger of explosion.

Algicides

Copper compounds

Copper sulfate is widely used in swimming pools, recreational lakes and ponds, and municipal water supply lakes as an algicide. A large amount of copper sulfate also is used in aquaculture for off-flavor control (Boyd and Tucker 1998). Some species of blue-green algae can elaborate odorous compounds such as geosmin and methylisoborneol and excrete these substances into the water. Fish absorb odorous compounds resulting in an undesirable flavor in their flesh. The off-flavor problem has been particularly severe in channel catfish produced in ponds in the southern United States and it also has been reported in other species of pond fish and shrimp (Boyd and Tucker 1998).

The most common way of combating off-flavor in channel catfish culture is to treat ponds with copper sulfate 2 or 3 weeks before the intended harvest date to lessen the abundance of blue-green algae (Tucker and Hargreaves 2003; Tucker 2007).

Cupric ion (Cu^{2+}) is extremely toxic to fish and other aquatic organisms. The proportion of the total, soluble copper concentration consisting of Cu^{2+} decreases with increasing pH and total alkalinity concentration. The typical treatment rate for copper carbonate is 0.01 times the total alkalinity concentration (expressed in milligrams per liter). It is necessary to increase the copper sulfate dose as alkalinity increases, because the copper sulfate ion pair ($CuCO_3{}^0$) increases in proportion to cupric ion (Cu^{2+}) as pH and total alkalinity increases (Boyd and Tucker 1998). In the catfish farming region of the southeastern United States, pond waters typically have total alkalinities of 100–250 mg/L; hence, copper sulfate treatment rates usually are 1–2.5 mg/L. Ponds may be treated two or three times or more annually (Boyd et al. 2000), but these treatments may be made over a period of a few weeks before harvest.

The copper applied in copper sulfate disappears from the water quickly as a result of direct precipitation from the water as copper oxide, absorption by aquatic organisms, and adsorption by bottom soil (Han et al. 2001; McNevin and Boyd 2004). Studies have shown that copper concentrations decline to pre-treatment levels within 72 hours after copper sulfate application (Fig. 9.2). In addition, repeated applications of copper have not resulted in an upward trend in pre-treatment levels during a single growing season.

Soils in the catfish farming region of the United States have a high capacity to bind copper (Silapajarn and Boyd 2006), and accumulation of copper in

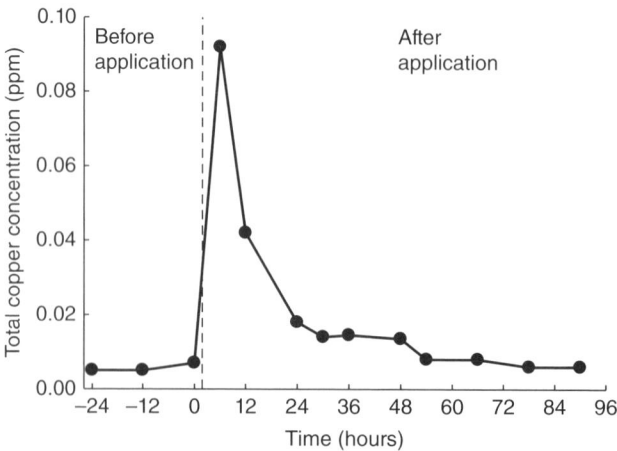

Figure 9.2 Copper concentrations before and after treatment of channel catfish ponds with copper sulfate. Modified from McNevin, A. A. and C. E. Boyd. Copper concentrations in channel catfish *Ictalurus punctatus* ponds treated with copper sulfate. *Journal of the World Aquaculture Society* 35:16–24. Copyright © 2004, John Wiley & Sons, Inc.

bottom soils does not appear to be increasing background copper concentrations in pond water. Copper concentration averaged over a 2-year period in a stream receiving the effluent from about 5000 ha of channel catfish ponds in Alabama did not differ from copper concentrations in control streams (Silapajarn and Boyd 2005).

Copper algicides made by combining elemental copper with triethanolamine or other chelating agents have been used to avoid high and potentially toxic concentrations of copper ion that could result when using copper sulfate in acidic, low-alkalinity water. The cost of chelated copper solutions is excessive for use in off-flavor control.

The annual world production of refined copper was about 16 Mt in 2011 (http://minerals.usgs.gov/minerals/pubs/commodity/copper/mcs-2012-coppe.pdf). Most is used for manufacturing wire, brass and other alloys. A small amount is used for manufacturing copper sulfate by treating copper metal or copper oxide with hot, concentrated sulfuric acid. Copper sulfate is used as a fungicide in agriculture and in swimming pools as an algicide. In addition, it has many niche uses as an analytical reagent in laboratories and as a reactant in industrial chemical processes. There are no estimates of the amount of copper sulfate and chelated copper used in aquaculture for algal control, but most is used in channel catfish farming. There were about 40 000 ha of ponds in 2010 (McCall 2010). The average depth of ponds is roughly 1 m; assuming three annual treatments with copper sulfate of 2 mg/L each, about 2400 t copper sulfate (612 t copper) might be used in catfish farming in the United States. Aquaculture certainly represents an insignificant percentage of the world copper market.

Copper concentrations must be kept at a low level in order to avoid toxicity to the culture species. The copper ion disappears from the water rapidly through precipitation and uptake by bottom soil. Thus there does not appear to be much likelihood that aquaculture effluents will contain toxic concentrations of copper. There is no evidence of elevated accumulation of copper in aquaculture species from production systems treated with copper sulfate.

Diuron

Diuron is a herbicide with the chemical name [3-(3;4-dichlorophenyl)-1,1-dimethylurea] that also has been used to control blue-green algae responsible for off-flavor in pond-reared channel catfish in the United States (Zimba et al. 2002). It is used at a very low dose—0.0125 mg/L active ingredient per application for a maximum of nine applications per year.

Studies of diuron-treated ponds revealed that diuron was effective in controlling blue-green algae and off-flavor. It caused changes in the composition of the plankton community in ponds but did not have a significant impact on water quality (Zimba et al. 2002). Catfish collected from diuron-treated ponds within 2–4 months after exposure did not contain diuron residues (Tucker et al. 2003). Although diuron can be used safely in culture ponds, discharge of waters containing diuron should be avoided because of its potent anti-algal activity.

Table 9.7 Aquatic herbicides used in the United States.

Common name	Chemical formula	Type
Copper sulfate	$CuSO_4 \cdot 5H_2O$	Contact
Chelated copper	2-aminoethanol copper and others	Contact
Diquat	6,7-dihydrodipyrido[1,2-a:2',1'-c]pyrazinediium dibromide	Contact
Endothall	(7-oxabicyclo[2.2.1]heptane-2,3-dicarboxylic acid	Contact
Fluridone	1-methyl-3-phenyl-5-3-(trifluoromethyl)phenyl|-41H|-pyridinone	Systemic
Glyphosate	N-(phosphonomethyl) glycine	Systemic
2,4-D	2,4-dichlorophenoxyacetic acid	Systemic

Sources: Masser, M. P., T. R. Murphy, and J. L. Shelton. 2001. Aquatic weed management, herbicides. Southern Regional Aquaculture Center Publication 361, Mississippi State University, Stoneville, Mississippi.
Avery, J. L. 2003. Aquatic weed management, herbicide safety, technology and application techniques. Southern Regional Aquaculture Center Publication 3601, Mississippi State University, Stoneville, Mississippi.

Aquatic herbicides

Herbicides are used to kill filamentous macroalgae and submergent, emergent, floating-leafed, and floating vascular plants that interfere with aquaculture management. The common herbicides used in the United States for aquatic weed control (Table 9.7) also are employed worldwide for this purpose. Other herbicides also are used; a list of aquatic herbicides applied in India (Ramaprabhu and Ramachandran 1984) included, in addition to those in Table 9.7, simazine (6-chloro-N,N'-diethyl-1,3,5-triazine-2,4-diamine) and paraquat (1,1'-dimethul-4,4'-bipyridinium dichloride).

Copper compounds used to control phytoplankton also can be effective against filamentous, macrophytic algae (Masser et al. 2001). Other herbicides are used primarily for controlling higher plants but simazine has been used for phytoplankton control (Tucker and Boyd 1978). Contact herbicides act quickly and kill all cells with which they come in contact. Systematic herbicides are absorbed into the plant and move to other tissues; they tend to act slowly. Contact herbicides are used mainly for annual herbaceous plants, while systematic herbicides are used mainly in perennial and woody plants (Avery 2003).

There are no published records of amounts of herbicide used in aquaculture. In the United States, aquatic herbicides are frequently used in sportfish ponds, but in aquaculture ponds, higher plants are usually eliminated by light limitation caused by dense plankton blooms (Boyd and Tucker 1998). Other than copper sulfate use to lessen the abundance of blue-green algae responsible for off-flavor, herbicide applications generally are limited to emergent vegetation around the edges of ponds (Boyd et al. 2000). The use of herbicides in aquaculture is miniscule in comparison to their use for weed control in agriculture, on highway and railroad right-of-ways, and in lawns, gardens, and golf courses.

Copper compounds can be highly toxic to fish and other aquatic animals but the other herbicides tend to have rather low toxicity to aquatic animals (Boyd and Tucker 1998). Possible negative impacts of copper compounds and precautions for their use have been discussed. The other herbicides have warnings on their labels that should be heeded. It is particularly important to prevent aerial drift of herbicides that could result in toxicity outside the treatment zone. Workers should wear protective clothing and avoid inhalation of fumes and contact with skin and eyes. Herbicides should be stored in their original containers in cool, well-ventilated buildings (Avery 2003).

Fish toxicants

Fish toxicants often are used to kill wild fish in ponds before stocking for the next crop. Sometimes the entire pond volume will be treated, while in other instances only puddles of water in the pond bottom are treated.

Rotenone

Rotenone is a complex organic compound ($C_{22}H_{22}O_{10}$) extracted from the roots of *Derris elliptica*, *Lonchocarpus* spp., and a few other leguminous plants (Shepard 1951). The primary use of rotenone is as an insecticide in agriculture and in vegetable and flower gardens. It is extremely toxic to fish and has been used by indigenous tribes to poison and capture fish for food, by aquaculturists to eliminate unwanted fish, and by biologists to sample fish populations (Boyd and Tucker 1998).

The concentration of active ingredient necessary to kill fish is about 0.05 mg/L but commercial rotenone preparations usually contain only 5% active ingredient. Treatment rates usually are 1–2 mg/L of commercial preparations (Boyd and Tucker 1998). Rotenone will detoxify within 1 or 2 weeks during warm weather, but a longer time is necessary in winter.

Saponin

Saponin is a complex glucoside found in the seeds and bark of a variety of plants. Saponin destroys red blood cells and is highly toxic to fish (Minsalan and Chiu 1986). It is more toxic to fish than to shrimp, and as a result it is often used to eliminate wild fish from shrimp ponds in Asia (Chen et al. 1996; Terazaki et al. 1980). Commercial saponin preparations are available (San Martín and Briones 1999), but teaseed cake and oil cake from mahua (*Bassia latifolia*) are more commonly used in aquaculture (Terazaki et al. 1980; Bhatia 1970). Treatment rates usually are about 15 mg/L, but Terazaki et al. (1980) reported that 1 mg/L crude saponin was sufficient to eliminate wild fish. Saponin degrades gradually in ponds, but the rate of degradation is not well understood.

Antimycin A

Antimycin A is a natural antibiotic produced by the bacterium *Streptomyces griseus*. It is a potent inhibitor of aerobic respiration but its toxicity to organisms varies greatly. It is more toxic to fish than other aquatic organisms and it is particularly toxic to scaly fish (Boyd and Tucker 1998). Antimycin A has sometimes been used to eliminate scaly fish from catfish ponds but it is not often used as a non-selective fish toxin because it is more expensive than rotenone.

Other fish toxicants

Other chemicals occasionally used to eradicate fish include anhydrous ammonia (Champ et al. 1973), calcium hydroxide–ammonium sulfate mixture (Subramanian 1983), sodium hypochlorite (bleach)–urea mixture (Ram et al. 1988), calcium hypochlorite, calcium hydroxide, potassium permanganate, and various synthetic insecticides (Boyd and Tucker 1998).

Precautions

Waters treated with fish toxicants should not be discharged until the compounds have lost their toxicity. When ammonia or ammonium sulfate is used as fish toxicants, several weeks may be necessary for unionized ammonia concentrations to fall to a safe level. Fish toxicants should not be applied when heavy rainfall is expected to avoid overflow. Potassium permanganate can be used to detoxify rotenone, antimycin A, and possibly saponin. Fish killed with fish toxicants should not be allowed for human consumption.

Coagulants

Coagulants are sometimes used to remove suspended soil particles from water to reduce turbidity (Boyd and Tucker 1998). The most effective way of avoiding turbidity is to control erosion of watersheds and farm infrastructure. In many cases aquaculture facilities take water from streams or estuaries and erosion control on catchments is not an option. In such situations water can be held in reservoirs for sedimentation before it is conveyed to production facilities. Nevertheless, soil particles may remain in suspension in ponds after sources of erosion have been controlled, and sedimentation in reservoirs may not affect adequate turbidity reduction. The tendency of soil particles to remain in suspension is greater in acidic, soft water than in harder freshwater and estuarine water.

Colloidal clay particles remain in suspension because they are negatively charged and repel each other. Coagulants with the opposite charge of the colloids are added to neutralize the charges on the colloid surfaces so that they will stick together—a process called coagulation. A second process called flocculation occurs when enough of the colloids stick together to form a visible floc that will settle from the water.

A time-honored method of removing turbidity from ponds is the application of organic matter (Boyd and Tucker 1998). Apparently, the decrease in pH caused by decomposition of the organic matter and the attachment of colloidal particles to decomposing particles of organic matter result in coagulation. Turbidity removal requires a large input of organic matter and one or more retreatments may be necessary. Decomposition of organic matter can result in dissolved oxygen depletion.

Routine liming increases calcium and magnesium concentrations in soft water and sometimes will initiate coagulation. The most effective coagulants are more soluble compounds, for example, calcium and magnesium sulfates and aluminum and ferric sulfates and chlorides. The two most commonly used coagulants are calcium sulfate (gypsum) and aluminum sulfate (alum). Usually treatment rates to clear pond water of turbidity are 100–300 mg/L of gypsum and 10–30 mg/L of alum (Boyd and Tucker 1998).

Gypsum has many uses such as a component of wallboard (sheetrock), plaster ingredient, fertilizer, and soil conditioner. Alum is used widely to remove turbidity from drinking water and it also has many niche applications. The amounts of gypsum and alum used in aquaculture are unknown but the total amounts used are thought to be extremely small in comparison to other applications. Experience suggests that coagulant use is rare in commercial aquaculture ponds but more common in sportfish ponds.

The increase in calcium ion concentration caused by gypsum increases total hardness but it is neutral in reaction and non-toxic to aquatic organisms. Alum is highly acidic (Boyd and Massaut 1999). Aluminum ion resulting from dissociation of alum hydrolyzes:

$$Al_2(SO_4)_3 \cdot 18H_2O \rightarrow 2Al^{3+} + 3SO_4^{2-} + 18H_2O \qquad (9.19)$$

$$2Al^{3+} + 6H_2O \rightarrow 2Al(OH)_3 + 6H^+. \qquad (9.20)$$

Each milligram per liter of aluminum sulfate can reduce alkalinity by about 0.5 mg/L. If the alkalinity is depleted, pH will fall to a level toxic to aquatic organisms. Aluminum sulfate applications should be limited to a concentration equal to 50% of the total alkalinity in order to avoid low pH. If this practice is followed environmental perturbations should not result from aluminum sulfate application. Also when using alum, workers should wear protective clothing because of the acidity of this coagulating agent.

Antifoulants

These substances are used in cage culture of fish and in cultivation of shellfish in enclosures. They are applied to the netting of enclosures in which the aquaculture organisms are confined. This practice is necessary because encrustation of the netting with fouling organisms reduces water exchange in the culture units and increases the weight that must be supported by flotation devices for cages (Paul and Davies 1986; Burridge et al. 2010). Netting material is impregnated with the antifouling chemical; and trays for shellfish culture constructed from or painted with non-fouling substances. The traditional antifouling substance was tri-n-butyltin (TBT). Because of the high environmental toxicity of TBT and its tendency to accumulate in the tissues of aquatic organisms (Davies et al. 1986; Short and Thrower 1986; Harino et al. 2000), it has been banned in most countries (Nehring 2001).

The main replacement for TBT is paint-containing copper (I) oxide (Cu_2O). There are no records of the total amount of copper-based antifoulants used worldwide in aquaculture, but Burridge et al. (2010) reports that up to 86.9 t of copper oxide were used for antifouling purposes in salmon aquaculture in Scotland in 2006. Farmed salmon production in Scotland is about 10% of world farmed salmon production (FAO 2012), and production in 2010 was about 7.5% greater than in 2006. Assuming a similar use of copper throughout salmon aquaculture, total copper use might be as much as 934 t copper oxide (747 t copper)—slightly more than the estimate for copper use in channel catfish farming. Of course aquaculture is not the only industry using antifouling paints. They are applied to hulls of boats and ships and many types of marine infrastructure.

Studies have shown that copper can accumulate in sediments near aquaculture facilities as a result of antifoulant use (Dean et al. 2007). There also is speculation about copper toxicity and copper accumulation in organisms. However, as discussed earlier, copper is rather insoluble in water and tightly bound in sediment. Copper-based antifoulants are more environmentally friendly than TBT, but efforts to find less toxic and bioaccumulative antifoulants should continue.

Osmoregulators and ionic balance adjustors

Aquaculture facilities sometimes have water supplies that are deficient in total ionic concentration or in the concentration of one or more major ions. Common salt (NaCl) is often used to increase total ionic concentration. Calcium sulfate ($CaSO_4 \cdot 2H_2O$) and potassium chloride (KCl) may be used when calcium or potassium ion concentrations are too low. Epsom salt ($MgSO_4 \cdot 7H_2O$) and potassium magnesium sulfate ($K_2SO_4 \cdot MgSO_4$) are common sources of magnesium (Boyd 2003).

Sodium chloride also is applied to pond waters to increase the chloride concentration and avoid nitrite toxicity (brown blood disease) in fish. Chloride interferes with nitrite uptake by fish and chloride:nitrite-nitrogen ratios of 20 or above essentially block nitrite uptake (Boyd and Tucker 1998). Channel catfish farmers in the United States usually maintain chloride concentrations of 50–100 mg/L in ponds by annual applications of common salt (Sipauaba-Tavares and Boyd 2003).

Phytoplankton can use bicarbonate as a carbon source when all free carbon dioxide has been removed. In waters with low calcium concentration but high alkalinity concentration, carbonate resulting from use of bicarbonate as a carbon source in photosynthesis accumulates because there is insufficient calcium to precipitate it as calcium carbonate. Hydrolysis of carbonate can cause pH to rise to 11 or more (Boyd and Tucker 1998). Gypsum is added to water to increase calcium concentration and limit pH rise; the usual treatment rate is twice the difference between total alkalinity and total hardness concentrations (Wu and Boyd 1990).

Increases in ionic concentration resulting from the use of mineral salts are small, and waters discharged from such facilities would not cause salinization. Moreover, the salts used are not sources of nitrogen and phosphorus that could cause eutrophication.

In Thailand and a few other countries, brine solutions from seawater evaporation ponds are transported inland and mixed with freshwater in ponds to provide water of 2–5 ppt salinity for marine shrimp culture, and in the United States and a few other countries, saline groundwater serves to supply inland ponds with water suitable for marine shrimp culture (Boyd et al. 2002). Salt is lost from such ponds by overflow, intentional discharge, and seepage and soil and water salinization can result (Braaten and Flaherty 2001; Boyd et al. 2006; Pine and Boyd 2011).

Huge amounts of sodium chloride, calcium sulfate, magnesium and potassium magnesium sulfates, and potassium chloride are used worldwide for a variety of purposes annually. Aquacultural use of these compounds is not known, but in comparison with other uses, it is no doubt quite insignificant.

Feed additives

Feeds contain nutrients from plant and animal meal ingredients. In addition, packs of several essential micronutrients, vitamins, and antioxidants typically are added to the feed. Studies have shown that 5–10% of organic carbon and 30–40% of nitrogen and phosphorus in fish feed are recovered in harvested fish. In shrimp, carbon and nitrogen recovery are similar, but only 5–10% of phosphorus is recovered. It is likely that the recovery of other nutrients and biochemical compounds at harvest is no greater than for carbon, nitrogen, and phosphorus.

The main water pollution concern about feed-based aquaculture is discharge of nitrogen and phosphorus that can cause eutrophication. Vitamins and antioxidants should not pose a water pollution problem but heavy metals such as zinc and copper are potentially toxic to aquatic life. A recent study of water of inland, low-salinity shrimp ponds did not reveal elevated concentration of zinc, copper, or other micronutrients (Prapaiwong and Boyd 2014). However, higher concentrations of zinc were measured in sediment near salmon cage culture operations than in control areas (Burridge et al. 1999; Chou et al. 2002). Whether or not elevation of sediment micronutrient metal concentrations near aquaculture operations constitutes an environmental hazard is unclear.

Bacterial products

These substances include living bacteria and enzyme preparations used with the belief that they will improve animal health and correct water and sediment quality imbalances. The effectiveness of these products—especially water and sediment quality aids—have not been demonstrated, but they are widely used in shrimp farming and other types of aquaculture (Boyd and Silapajarn 2006). A list of the kinds of bacteria claimed to be in packages of various bacteria products sold in Vietnam are listed in Table 9.8. The enzymes reported to be in the products were amylase, beta-glucosanase, cellulase, pectinase, and protease. These are probably the same products used worldwide in aquaculture because reference to product labels revealed that they had been imported from Vietnam. The bacteria and enzymes are common in nature and no environmental concerns have been expressed or would be expected related to their use.

Table 9.8 List of microbial organisms in probiotics sold for use in shrimp ponds.

Aspergillus oryzae	*Bacillus subtilis*
Bacillus amyloliquefaciens	*Lactobacillus acidophilus*
Bacillus cerevisicie	*Lactobacillus plantarum*
Bacillus circulans	*Nitrobacter winogradskyi*
Bacillus laterosporus	*Nitrosomonas eutropha*
Bacillus licheniformis	*Paracoccus panthropus*
Bacillus megaterium	*Rhodobacter* sp.
Bacillus polymyxa	*Rhodococcus* sp.
Bacillus pumilus	*Saccharomyces cerevisiae*

Source: Boyd (2012).

Fuels and lubricants

Fuels and lubricants needed for operating machinery often are stored in bulk at aquaculture facilities. Fires and explosions can result from careless handling of improper storage of petroleum products. Spills of fuels and lubricants can contaminate surrounding water and soil, or through runoff entering pond waters. Petroleum products can be directly toxic to fish, shrimp, and other animals. Moreover, the fish and other aquatic animals exposed to non-lethal concentrations of petroleum products may develop characteristic off-flavors variously described as "oily," "diesel fuel," "petroleum," or "kerosene," and be rejected from the market (Boyd and Tucker 1998).

Fuels should be stored above ground in storage tanks. These tanks should have vent pipes to remove vapor. The tanks should be clearly marked as containing flammable materials, and the nature of their contents identified. The tanks should be placed inside impervious containment basins of 1.5 times the tank volume. These basins will capture any fuel that is accidentally spilled during handling or if a tank ruptures. Rainwater must be drained from containment basins at regular intervals to prevent loss of containment volume. Lubricants should be stored in buildings with concrete floors and drains that empty into a containment area for capturing any spilled fluids. The buildings and areas where vehicles and other machines are repaired, serviced, or refueled also should have concrete floors with drains into containment areas. All used oils should be collected in suitable containers and shipped to a recycling center. Spilled fuel should be pumped from containment areas and stored in tanks until it can be properly disposed of.

Conclusions

There is considerable chemical use in aquaculture but this is not surprising because chemicals are used in traditional agriculture and most other human endeavors. Aquaculture is actually a rather minor contributor to the use of most chemicals. Nevertheless, aquaculture facilities tend to be concentrated in relatively small areas and their connections to natural waters are usually direct. Thus certain chemicals discharged from aquaculture have or potentially could cause serious environmental perturbations. The most serious issues appear to be release of nitrogen and phosphorus that can cause eutrophication in natural waters, and discharge of antibiotics that could lead to resistance to antibiotics in bacteria that could be transferred horizontally to bacteria of importance in human medicine.

A number of chemicals are toxic and their use in aquaculture or accidental spills into natural waters could cause mortality to aquatic species—especially plankton. Plankton communities, however, are resilient and will regrow once the concentration of the toxicant has diminished.

Most toxicity events are the result of improper application of treatment chemicals. Worker training in chemical use and strict adherence to warnings on chemical labels and MSD sheets can prevent toxicity incidents. Although the frequency and severity of accidents can be reduced through training and vigilance, accidents will eventually occur. As the old adage states—"to err is human." Also unpredictable natural events such as large amounts of rainfall, exceptionally strong wind storms, earthquakes, and tsunamis can undo the best attempts to lessen impacts of chemicals or other activities.

Sediment accumulation of nutrients, metals, antibiotics, and other chemicals have been demonstrated but much more research is needed to assess actual impacts of the accumulations.

Antibiotics or chemicals that are banned by governments for food safety reasons should not be used in aquaculture and producers should make efforts to lessen the use of permissible chemicals. In many cases the use of chemicals for disease control is not as effective as the application of better water quality management techniques to avoid stress that is frequently a precursor of disease outbreaks in culture species. Better biosecurity measures also lessen the frequency of disease in aquaculture facilities.

All chemical uses can lead to worker safety concerns. Fortunately worker hazards associated with aquaculture chemicals are known, and for most substances, these hazards and precautions for avoiding them are given on product labels and MSD sheets. The approach to avoiding worker hazards is to provide training on chemical safety and provide proper safety equipment. Accidents will occur and workers should be instructed in first aid or emergency response in the event of chemical exposure.

The eNGO perspective

The eNGO's perception of the chemicals used in aquaculture has evolved over the past two decades. Part of the evolution can be attributed to a greater understanding of the types of chemicals that are used and of the importance of these for the industry. Moreover, many of the chemicals used for aquaculture production are now considered benign or of little concern to the eNGOs. Substances such as liming materials and oxidizing agents are understood to have little if no environmental effect. However, the eNGOs remain cautious and reluctant to support the use of therapeutants such as antibiotics, parasiticides, piscicides, disinfectants, and herbicides because of impact that these chemicals can have on the environment outside of the farming activity. There is a general desire to limit the amount of organisms that must be destroyed to produce other organisms. Moreover, the human health issues related to some chemicals has become a mechanism for some eNGOs to lobby against certain aquaculture sectors.

Many mainstream eNGOs understand the need for antibiotics when disease outbreaks occur on farms. Typically, if an EU or US agency allows the chemicals to be used and there is a veterinary prescription, the eNGOs will tolerate antibiotic usage.

The early antifoulants used in cage culture such as tributyltin raised awareness that attention should be given to the specific compounds used as there could be toxic effects beyond the target organisms. In addition, the use of the parasiticide, emamectin benzoate, has raised concerns because there is the potential for environmental harm (http://www.farmedanddangerous.org/salmon-farming-problems/environmental-impacts/chemical-treatments-slice/).

There is also a concern of the bioaccumulation of emamectin benzoate in wild shrimp (http://www.dfo-mpo.gc.ca/csas-sccs/Publications/SAR-AS/2011/2011_082-eng.pdf).

There was an instance where an illegal parasiticide was used in the United States in an attempt to control sea lice at a salmon farm and its use resulted in the deaths of hundreds of lobsters (http://www.cbc.ca/news/technology/story/2013/04/26/nb-cooke-aquaculture-lobster-pesticide.html).

Whether the chemicals used to kill pests are legal or illegal, their use in culture systems that are open to the environment (cages) will always cause concern for eNGOs and they will likely campaign against their use.

The use of 17-methyltestosterone has been discussed, but it is useful to understand the debates that have occurred between the tilapia industry and the eNGOs. Through the Tilapia Aquaculture Dialogue (discussed in Chapter 14), producers and eNGOs actively engaged in dialogue on the use of this hormone. Although there was a general disapproval of the use of a human hormone in food fish culture, eNGOs recognized that the minor use of this hormone during the first few weeks of fry rearing resulted in an improved FCR and a reduction in the amount of escapes from tilapia farms. It appeared to be tolerable for the industry to use this hormone to address other impacts that were identified as higher priorities by the eNGOs. Some niche retailers have rejected the use of 17-methyltestosterone in products they sell simply because their markets generally require that no chemicals be used. The eNGOs would likely prefer this situation, but there is a greater understanding of the justification for the use of 17-methyltestosterone.

Although the aquaculture industry has made significant improvements and reductions in the chemicals used in culture activities, there still is use of banned products by some producers in certain regions of the world. It is difficult for some industry advocates to acknowledge this fact, but as long as there are producers that use illegal substances, eNGOs will still identify these issues as challenges for the aquaculture industry as a whole. Cao (2007) attributed the use of banned substances in China to efforts to resolve the problems that have risen from poor environmental waste management. There is also an aspect of scale that likely increases the probability of using banned and illegal chemicals at aquaculture operations. Small-scale producers in less-developed countries generally have less accountability than do the processing plants and exporters because of a non-integrated supply chain where farmed products can be aggregated at one or more points for transport to processing plants. Thus the product is rendered untraceable and attribution to a specific farm is nearly impossible.

The most notable illegal therapeutants used in aquaculture feeds are ciprofloxacin, chloramphenicol, enrofloxacin and various fluoroquinolones. There is little evidence that these chemicals are maliciously used; however, there remains systemic problems in certain regions of the world, particularly in Asia which, not coincidentally, has the highest proportion of small-scale producers. There have been efforts by various institutions and governments to educate the small-scale producers on the appropriate use of therapeutants.

There is still use of illegal anti-microbial agents; the most common substances identified are malachite green and gentian violet. Although the use of illegal or banned chemicals is more commonly identified with risks to human health, their persistence in the environment raises concerns of bioaccumulation and the health of other organisms.

The use of agricultural pesticides has also been observed in some countries. Their use is primarily to prepare shrimp ponds for stocking of postlarvae. They have been used to eradicate unwanted fish and also to kill snails which may be vectors of disease. In addition, several farmers have expressed to the junior author that if these pesticides are applied, the natural worms and other food items for shrimp will be released from the soil into the water column. This myth demonstrates that there still exist areas and communities of farmers that require a greater understanding of which chemicals are safe and approved for use in aquaculture and for what purpose these chemicals should be used.

No industry has an untarnished record. The rise of aquaculture followed a time of increased international trade. There were instances when inappropriate chemicals were used and lessons have been learned about the problems resulting from residues in imported products. Nevertheless, eNGOs feel that there is blind support from industry advocates who declare that banned and illegal chemicals are no longer used in aquaculture. Further, eNGOs do not shy away from the larger issue of traceability and accountability of small-scale producers. Some in the aquaculture industry are reticent to admit that more of the chemical contamination issues reside in the small-holder sector of aquaculture. This is likely a result of strong advocacy for small-scale aquaculture producers by some organizations that contend the large-scale industry is marginalizing the small-holder communities.

References

Abdolazizi, S., E. Ghaderi, N. Naghdi, and B. B. Kamangar. 2011. Effects of clove oil as an anesthetic on some hematological parameters of *Carassius auratus*. *Journal of Aquaculture Research and Development* 2:108.

Agellon, L. B., C. J. Emery, J. M. Jones, S. L. Davies, A. D. Dingle, and T. T. Chen. 1988. Promotion of rapid growth of rainbow trout (*Salmo gairdneri*) by a recombinant fish growth hormone. *Canadian Journal of Fisheries and Aquatic Sciences* 45:146–151.

Alderman, D. J. 1985. Malachite green: a review. *Journal of Fish Diseases* 8:289–298.

Allison, R. 1962. The effects of formalin and other parasiticides upon oxygen concentrations in ponds. *Proceedings Southeastern Association of Game and Fish Commissioners* 16:446–449.

Arthur, J. R., C. R. Lavilla-Pitogo, and R. P. Subasinghe (editors). 2000. Use of chemicals in aquaculture in Asia. *Proceedings of the Meeting on the Use of Chemicals in Aquaculture in Asia.* Iloilo, Philippines: Southeast Asian Fisheries Development Center.

Avery, J. L. 2003. Aquatic weed management, herbicide safety, technology and application techniques. Southern Regional Aquaculture Center Publication 3601, Mississippi State University, Stoneville, Mississippi.

Benbrook, C. M. 2002. *Antibiotic drug use in U.S. aquaculture.* Sandpoint, Idaho: The Northwest Science and Environmental Policy Center.

Bhatia, H. L. 1970. Use of Mahua oil cake in fisheries management. *Indian Farming* 20:39–40.

Bills, T. D., L. L. Marking, and J. H. Chandler, Jr. 1977. Formalin: its toxicity to non-target aquatic organisms, persistence, and counteraction. U.S. Fish and Wildlife Service, investigations in fish Control, Washington, DC.

Boyd, C. A., C. E. Boyd, A. A. McNevin, and D. B. Rouse. 2006. Salt discharge from an inland farm for marine shrimp in Alabama. *Journal of the World Aquaculture Society* 37:345–355.

Boyd, C. E. 1990. *Water Quality in Ponds for Aquaculture.* Auburn University: Alabama Agricultural Experiment Station.

Boyd, C. E. 2003. Mineral salts correct imbalances in culture water. *Global Aquaculture Advocate* 6(4):56–57.

Boyd, C. E. 2012. Assessment of relationships between environmental factors and the new shrimp disease in Vietnam. Report to FAO Fisheries and Aquaculture, Rome, Italy.

Boyd, C. E. and D. Gautier. 2002. Sodium bisulfite treatments improve shrimp appearance but require proper disposal. *Global Aquaculture Advocate* 5(4):70–71.

Boyd, C. E. and L. Li. 2012. Environmental issues in pond fertilization. In C. C. Mischke, editor, *Aquaculture Pond Fertilization*, pp. 65–72. Ames: Wiley-Blackwell.

Boyd, C. E. and L. Massaut. 1999. Risks associated with use of chemicals in pond aquaculture. *Aquacultural Engineering* 20:113–132.

Boyd, C. E. and O. Silapajarn. 2006. Influence of microorganisms on water and sediment quality in aquaculture ponds. In R. C. Ray, editor, *Microbial Biotechnology in Agriculture and Aquaculture*, pp. 261–285. Enfield: Science Publishers.

Boyd, C. E. and C. S. Tucker. 1998. *Pond Aquaculture Water Quality Management.* Boston: Kluwer Academic Publishers.

Boyd, C. E., T. Thunjai, and M. Boonyaratpalin. 2002. Dissolved salts in waters for inland, low-salinity shrimp culture. *Global Aquaculture Advocate* 5(3):40–45.

Boyd, C. E., J. Queiroz, J. Lee, M. Rowan, G. N. Whitis, and A. Gross. 2000. Environmental assessment of channel catfish, *Ictalurus punctatus*, farming in Alabama. *Journal of the World Aquaculture Society* 31:511–544.

Braaten, R. O. and M. Flaherty. 2001. Salt balances of inland shrimp ponds in Thailand: implications for land and water salinization. *Environmental Conservation* 28:357–367.

Bumb, B. L. and C. A. Baanante. 1996. *The role of fertilizer in sustaining food security and protecting the environment to 2020.* Washington: International Food Policy Research Institute.

Burridge, L., J. S. Weis, F. Cabello, J. Pizarro, and K. Bostick. 2010. Chemical use in salmon aquaculture: a review of current practices and possible environmental effects. *Aquaculture* 306:7–23.

Burridge, L. E., K. Doe, K. Haya, P. M. Jackman, G. Lindsay, and V. Zitko. 1999. Chemical analysis and toxicity tests on sediments under salmon net pens in the Bay of Fundy. *Canadian Technical Report of Fisheries and Aquatic Sciences*, Publication 2297, Department of Fisheries and Oceans, St. Andrews, Canada.

Cabello, F. C. 2006. Heavy use of prophylactic antibiotics in aquaculture: a growing problem for human and animal health and for the environment. *Environmental Microbiology* 8:1137–1144.

Cao, L. 2007. Environmental impact of aquaculture and countermeasures to aquaculture pollution in China. *Environmental Science and Pollution Research International* 14:452–462.

Cavari, B., B. Funkenstein, T. T. Chen, L. I. Gonzalez-Villasenor, and M. Schartl. 1993. Effect of growth hormone on the growth rate of the gilthead seabream (*Sparus aurata*), and use of different constructs for the production of transgenic fish. *Aquaculture* 111:189–197.

Champ, M. A., J. T. Lock, C. D. Bjork, W. G. Klussmann, and J. D. McCullough, Jr. 1973. Effects of anhydrous ammonia on a central Texas pond, a review of previous research with ammonia in fisheries management. *Transactions of the American Fisheries Society* 102:73–82.

Chang, H., Y. Wan, S. Wu, Z. Fan, and J. Hu. 2011. Occurrence of androgens and progestogens in wastewater treatment plants and receiving river waters: Comparison to estrogens. *Water Research* 45:732–740.

Chaudhuri, H. 1976. Use of hormones in induced spawning of carps. *Journal of the Fisheries Research Board of Canada* 33:940–947.

Chen, J.-C., K.-W. Chen, and J.-M. Chen. 1996. Effects of saponin on survival, growth, molting and feeding of *Penaeus japonicas* juveniles. *Aquaculture* 144:165–175.

Chou, C. L, K. Haya, L. A. Paon, L. Burridge, and J. D. Moffatt. 2002. Aquaculture-related trace metals in sediments and lobsters and relevance to environmental monitoring program ratings for near-field effects. *Marine Pollution Bulletin* 44:1259–1268.

Christensen, A. M., F. Ingerslev, and A. Baun. 2006. Ecotoxicity of mixtures of antibiotics used in aquacultures. *Environmental Toxicology and Chemistry* 25:2208–2215.

Costello, M. J., A. Grant, I. M. Davies, S. Cecchini, S. Papoutsoglou, D. Quigley, and M. Saroglia. 2001. The control of chemicals used in aquaculture in Europe. *Journal of Applied Ichthyology* 17:173–180.

Coyle, S. D., R. M. Durborow, and J. H. Tidwell. 2004. *Anesthetics in Aquaculture*. Southern Regional Aquaculture Center Publication 3900, Mississippi State University, Stoneville, Mississippi.

Darwish, T., C. Khater, I. Jomaa, R. Stehouwer, A. Shaban, and M. Hamze. 2011. Environmental impact of quarries on natural resources in Lebanon. *Land Degradation and Development* 22:345–358.

Davies, I. M., J. C. McKie, and J. D. Paul. 1986. Accumulation of tin and tributyltin from anti-fouling paint by cultivated scallops (*Pecten maximus*) and Pacific Oysters (*Crassostrea gigas*). *Aquaculture* 55:103–114.

Dean, R. J., T. M. Shimmield, and K. D. Black. 2007. Copper, zinc and cadmium in marine cage fish farm sediments: an extensive survey. *Environmental Pollution* 145:84–95.

Dery, P. and B. Anderson. 2007. Peak phosphorus. Energy Bulletin, August 2007 (http://www.energybulletin.net/ node/33164).

Edwards, P. 1980. A review of recycling organic wastes into fish, with emphasis on the tropics. *Aquaculture* 21:261–279.

Edwards, P. 1988. Tilapia raised on septage as high protein animal feed, pp. 7–13. In R. S. V. Pullin, T. Bhukaswan, K. Tonguthai, and J. L. Maclean, editors, 2nd ed. International Symposium on Tilapia in Aquaculture. ICLARM Conference Proceedings 15, Department of Fisheries, Bangkok, Thailand and International Center for Living Aquatic Resources Management, Manila, Philippines.

Ericksen, G. E. 1983. The Chilean nitrate deposits. *American Scientist* 71:366–374.

FAO (Food and Agriculture Organization of the United Nations). 1997. *Aquaculture Development. FAO Technical Guidelines for Responsible Fisheries 5*. Rome: FAO.

FAO (Food and Agriculture Organization of the United Nations). 2012. *The State of World Fisheries and Aquaculture*. Rome: FAO.

Faust, S. D. and O. M. Aly. 1998. *Chemistry of Water Treatment*, 2nd ed. Chelsea: Ann Arbor Press, Inc.

Fixen, P. E. and A. M. Johnston. 2012. World fertilizer nutrient reserves: a view to the future. *Journal of the Science of Food and Agriculture* 92:1001–1005.

Goncalves-Ferreira, C. S., B. A. Nunes, J. M. de Melo Henriques-Almeida, and L. Guilhermino. 2007. Acute toxicity of oxytetracycline and florfenicol to the microalgae *Tetraselmis chuii* and to the crustacean *Artemia parthenogenetica*. *Ecotoxicology and Environmental Safety* 67:452–458.

Gräslund, S., K. Holmström, and A. Wahlström. 2003. A field survey of chemicals and biological products used in shrimp farming. *Marine Pollution Bulletin* 46:81–90.

Green, B. W. and D. R. Teichert-Coddington. 1994. Growth control and androgen-treated Nile tilapia, *Oreochromis niloticus* (L.), during treatment, nursery and grow-out phases in tropical fish ponds. *Aquaculture Research* 25:613–621.

Guerrero, R. D. 1975. Use of androgens for the production of all-male *Tilapia aurea* (Steindachner). *Transactions of the American Fisheries Society* 104:342–348.

Han, F. X., J. A. Hargreaves, W. L. Kingery, D. B. Huggett, and D. K. Schlenk. 2001. Accumulation, distribution, and toxicity of copper in sediments of catfish ponds receiving periodic copper sulfate applications. *Journal of Environmental Quality* 30:912–919.

Harino, H., M. Fukushima, and S. Kawai. 2000. Accumulation of butyltin and phenyltin compounds in various fish species. *Archives of Environmental Contamination and Toxicology* 39:13–19.

Heberer, T. 2002. Occurrence, fate, and removal of pharmaceutical residues in the aquatic environment: a review of recent research data. *Toxicology Letters* 131:5–17.

Hektoen, H., J. A. Berge, V. Hormazabal, and M. Yndestad. 1995. Persistence of antibacterial agents in marine sediments. *Aquaculture* 133:175–184.

Hill, B. J., F. Berthe, D. V. Lightner, and R. E. Sais. 2013. Methods for disinfection of aquaculture establishments. In *Manual of Diagnostic Tests for Aquatic Animals*, pp. 28–39. Paris: Office International des Epizooties.

Holten-Lützhøft, H.-C., B. Halling-Sørensen, and S. E. Jørgensen. 1999. Algal toxicity of antibacterial agents applied in Danish fish farming. *Archives of Environmental Contamination and Toxicology* 36:1–6.

Hwang, H.-M., R. E. Hodson, and R. F. Lee. 1987. Degradation of aniline and chloroanilines by sunlight and microbes in estuarine water. *Water Research* 21:309–316.

IFDC (International Fertilizer Development Center). 2008. *Global shortage of sulfuric acid contributes to rising fertilizer costs – China is world's largest sulfur importer. Focus on Fertilizers and Food Security*. Muscle Shoals, Alabama: IFDC.

Johnstone, R., D. J. Macintosh, and R. S. Wright. 1983. Elimination of orally administered 17 α-methyltestosterone by *Oreochromis mossambicus* (tilapia) and *Salmo gairdneri* (rainbow trout) juveniles. *Aquaculture* 35:249–259.

Jung, S. H., J. W. Kim, I. G. Jeon, and Y. H. Lee. 2001. Formaldehyde residues in formalin-treated olive flounder (*Paralichthys olivaceus*), black rockfish (*Sebastes schlegeli*), and sea-water. *Aquaculture* 194:253–262.

Kemper, N. 2008. Veterinary antibiotics in the aquatic and terrestrial environment. *Ecological Indicators* 8:1–13.

Lange, I. G., A. Daxenberger, B. Schiffer, H. Witters, D. Ibarreta, and H. H. D. Meyer. 2002. Sex hormones originating from different livestock production systems: fate and potential disrupting activity in the environment. *Analytical Chimica Acta* 473:27–37.

Langer, W. H. and B. F. Arbogast. 2003. Environmental impacts of mining natural aggregate. In A. G. Fabbri, G. Gaal, and R. B. McCammon, editors, *Deposit and Geoenvironmental Models for Resource Exploitation and Environmental Security*, pp. 151–169. Boston: Kluwer Academic Publishers.

Liu, J., L. You, M. Amini, M. Obersteiner, M. Herrero, A. J. B. Zehnder and H. Yang. 2010. A high-resolution assessment on global nitrogen flow in cropland. *Proceedings of the National Academy of Science* 107:8035–8040.

Liu, S., G Ying, J. Zhao, L. Zhou, B. Yang, Z. Chen, and H. Lai. 2012. Occurrence and fate of androgens, estrogens, glucocorticoids and progestogens in two different types of municipal wastewater treatment plants. *Journal of Environmental Monitoring* 14:482–491.

Lyle-Fritch, L. P., E. Romero-Beltrán, and F. Páez-Osuna. 2006. A survey on use of the chemical and biological products for shrimp farming in Sinaloa (NW Mexico). *Aquacultural Engineering* 35:135–146.

Macintosh, D. J., T. J. Varghese, and G. P. S. Rao. 1985. Hormonal sex reversal of wildspawned tilapia in India. *Journal of Fish Biology* 26:87–94.

Marking, L. L, J. J. Rach, and T. M. Schreier. 1994. Evaluation of antifungal agents for fish culture. *Progressive Fish-Culturist* 56:225–231.

Masser, M. P., T. R. Murphy, and J. L. Shelton. 2001. Aquatic weed management, herbicides. Southern Regional Aquaculture Center Publication 361, Mississippi State University, Stoneville, Mississippi.

Masters, A. L. 2004. A review of methods for detoxification and neutralization of formalin in water. *North American Journal of Aquaculture* 66:325–333.

McCall, M. 2010. Falling acreage expected to push live price higher. *The Catfish Journal* 25:1–3.

McNevin, A. A. and C. E. Boyd. 2004. Copper concentrations in channel catfish *Ictalurus punctatus* ponds treated with copper sulfate. *Journal of the World Aquaculture Society* 35:16–24.

Minsalan, C. L. O. and Y. N. Chiu. 1986. Effects of teaseed cake on selective elimination of finfish in shrimp ponds. In J. L. Maclean, L. B. Dizon, and L. V. Hosillos, editors, *The First Asian Fisheries Forum*, pp. 79–82. Manila: Asian Fisheries Society.

Miranda, C. D. and R. Zemelman. 2002. Bacterial resistance to oxytetracycline in Chilean salmon farming. *Aquaculture* 212:31–47.

Mylonas, C. C., A. Fostier, and S. Zanuy. 2010. Broodstock management and hormonal manipulations of fish reproduction. *General and Comparative Endocrinology* 165:516–534.

Nehring, S. 2001. After the TBT era: alternative anti-fouling paints and their ecological risks. *Senckenbergiana Maritima* 31:341–351.

Nygaard, K., B. T. Lunestad, H. Hektoen, J. A. Berge, and V. Hormazabal. 1992. Resistance to oxytetracycline, oxolinic acid and furazolidone in bacteria from marine sediments. *Aquaculture* 104:31–36.

Pandian, T. J. and S. G. Sheela. 1995. Hormonal induction of sex reversal in fish. *Aquaculture* 138:1–22.

Parise, M. and V. Pascali. 2003. Surface and subsurface environmental degradation in the karst of Apulia (southern Italy). *Environmental Geology* 44:247–256.

Pathak, S. C., S. K. Ghosh, and K. Palanisamy. 2000. The use of chemicals in aquaculture in India. In J. R. Arthur, C. R. Lavilla-Pitogo, and R. P. Subasinghe, editors, *Use of Chemicals in Aquaculture in Asia: Proceedings of the Meeting on the Use of Chemicals in Aquaculture in Asia*, May 20–22, 1996, pp. 87–112. Iloilo, Philippines: Southeast Asian Fisheries Development Center.

Paul, J. D. and I. M. Davies. 1986. Effects of copper- and tin-based anti-fouling compounds on the growth of scallops (*Pecten maximus*) and oysters (*Crassostrea gigas*). *Aquaculture* 54:191–203.

Pine, H. J. and C. E. Boyd. 2011. Stream salinization by inland brackish-water aquaculture. *North American Journal of Aquaculture* 73:107–113.

Potts, A. C. and C. E. Boyd. 1998. Chlorination of channel catfish ponds. *Journal of the World Aquaculture Society* 29:432–440.

Prapaiwong, N. and C. E. Boyd. 2014. Trace elements in waters of inland, low-salinity shrimp ponds in Alabama. *Aquaculture Research* 45: 327–333

Primavera, J. H., C. R. Lavilla-Pitogo, J. M. Ladja, and M. R. Dela Peña. 1993. A survey of chemical and biological products used in intensive prawn farms in the Philippines. *Marine Pollution Bulletin* 26:35–40.

Ram, K. J., G. R. M. Rao, S. Ayyapan, C. S. Purushothaman, P. K. Saha, K. C. Pani, and H. K. Muduli. 1988. A combination of commercial bleaching powder and urea as a potential piscicide. *Aquaculture* 72:287–293.

Ramaprabhu, T. and V. Ramachandran. 1984. Developments in aquatic weed control research in India relating to fisheries. *Journal of Aquatic Plant Management* 22:97–100.

Rhodes, G., G. Huys, J. Swings, P. McGann, M. Hiney, P. Smith, and R. W. Pickup. 2000. Distribution of oxytetracycline resistance plasmids between aeromonads in hospital and aquaculture environments: implications of Tn1721 in dissemination of the tetracycline resistance determinant tet A. *Applied Environmental Microbiology* 66:3883–3890.

San Martín, R. and R. Briones. 1999. Industrial uses and sustainable supply of *Quillaja saponaria* (Rosaceae) saponins. *Economic Botany* 53:302–311.

Sarter, S., H. N. K. Nguyen, L. T. Hung, J. Lazard, and D. Montet. 2007. Antibiotic resistance in Gram-negative bacteria isolated from farmed catfish. *Food Control* 18:1391–1396.

Schmidt, A. S., M. S. Bruun, I. Dalsgaard, K. Pedersen, and J. L. Larsen. 2000. Occurrence of antimicrobial resistance in fish-pathogenic and environmental bacteria associated with four Danish rainbow trout farms. *Applied and Environmental Microbiology* 66:4908–4915.

Schnick, R. A. 1988. The impetus to register new therapeutants for aquaculture. *Progressive Fish-Culturist* 50:190–196.

Schnick, R. A. 2001. Aquaculture chemicals. In *Kirk-Othmer Encyclopedia of Chemical Technology*, 4th ed., Vol. 3, pp. 209–225. New York: John Wiley & Sons.

Shepard, H. H. 1951. *The Chemistry and Action of Insecticides*. New York: McGraw-Hill Book Company.

Short, J. W. and F. P. Thrower. 1986. Accumulation of butyltins in muscle tissue of chinook salmon reared in sea pens treated with tri-n-butyltin. *Marine Pollution Bulletin* 17:542–545.

Silapajarn, O. and C. E. Boyd. 2005. Effects of channel catfish farming on water quality and flow in an Alabama stream. *Reviews in Fisheries Science* 13:109–140.

Silapajarn, O. and C. E. Boyd. 2006. Copper adsorption capacity of pond-bottom soils. *Journal of Applied Aquaculture* 18:85–92.

Silverstein, J. T., W. R. Wolters, M. Shimizu, and W. W. Dickhoff. 2000. Bovine growth hormone treatment of channel catfish: strain and temperature effects on growth, plasma IGF-I levels, feed intake and efficiency and body composition. *Aquaculture* 190:77–88.

Sipauaba-Tavares, L. H. and C. E. Boyd. 2003. Possible effects of sodium chloride on quality of effluents from Alabama channel catfish ponds. *Journal of the World Aquaculture Society* 34:217–222.

Smil, V. 2000. Phosphorus in the environment: natural flows and human interferences. *Annual Review of Energy and the Environment* 25:53–88.

Smith, D. L., J. Dushoff, and J. G. Morris, Jr. 2005. Agricultural antibiotics and human health. *PLOS Medicine* 2:3232.

Spanggaard, B., F. Jorgensen, L. Gram, and H. H. Huss. 1993. Antibiotic resistance in bacteria isolated from three freshwater fish farms and an unpolluted stream in Denmark. *Aquaculture* 115:195–207.

Srivastava, S., R. Sinha, and D. Roy. 2004. Toxicological effects of malachite green. *Aquatic Toxicology* 66:319–329.

Stevens, R. E. 1966. Hormone-induced spawning of striped bass for reservoir stocking. *The Progressive Fish-Culturist* 28:19–28.

Subramanian, S. 1983. Eradication of fishes by application of ammonia. *Aquaculture* 35:273–275.

Supriyadi, H. and A. Rukyani. 2000. The use of chemicals in aquaculture in Indonesia. In J. R. Arthur, C. R. Lavilla-Pitogo, and R. P. Subasinghe, editors, *Use of Chemicals in Aquaculture in Asia*, pp. 113–118. Iloilo, Philippines: Aquaculture Department, Southeast Asian Fisheries Development Center.

Tendencia, E. A. and L. D. de la Peña. 2001. Antibiotic resistance of bacteria from shrimp ponds. *Aquaculture* 195:193–204.

Terazaki, M., P. Tharnbuppa, and Y. Nakayama. 1980. Eradication of predatory fishes in shrimp farms by utilization of Thai tea seed. *Aquaculture* 19:235–242.

Tucker, C. S. 2007. Managing preharvest off-flavors in pond-raised catfish. *Global Aquaculture Advocate* 10(5):56–57.

Tucker, C. S. and C. E. Boyd. 1977. Relationships between potassium permanganate treatment and water quality. *Transactions of the American Fisheries Society* 106:481–488.

Tucker, C. S. and C. E. Boyd. 1978. Effects of simazine treatment on channel catfish and bluegill production in ponds. *Aquaculture* 15:345–352.

Tucker, C. S. and J. A. Hargreaves. 2003. Copper sulfate to manage cyanobacterial off-flavors in pond-raised channel catfish. In A. M. Rimando and K. K. Schrader, editors, *Off-Flavors in Aquaculture*, pp. 133–146. Washington: American Chemical Society.

Tucker, C. S., S. K. Kingsbury, and R. L. Ingram. 2003. Tissue residues of diuron in channel catfish *Ictalurus punctatus* exposed to the algicide in consecutive years. *Journal of the World Aquaculture Society* 34:203–209.

Vaccari, D. A. 2009. Phosphorus: a looming crisis. *Scientific American*, 300:54–59.

van Beynen, P. and K. Townsend. 2005. A disturbance index for karst environments. *Environmental Management* 36:101–116.

Verdegem, M. C. J. and R. H. Bosma. 2009. Water withdrawal for brackish and inland aquaculture, and options to produce more fish in ponds with present water use. *Water Policy* 11:52–68.

Webber, J. T., E. D. Mintz, R. Cañizares, A. Semiglia, I. Gomez, R. Sempértegui, A. Dávila, K. D. Greene, N. D. Puhr, D. N. Cameron, F. C. Tenover, T. J. Barrett, N.

H. Bean, C. Ivey, R. V. Tauxe, and P. A. Blake. 1994. Epidemic cholera in Ecuador: multidrug-resistance and transmission by water and seafood. *Epidemiology and Infection* 112:1–11.

West, T. O. and A. C. McBride. 2005. The contribution of agricultural lime to carbon dioxide emissions in the United States: dissolution, transport, and net emissions. *Agriculture, Ecosystems and Environment* 108:145–154.

Wooster, G. A., C. M. Martinez, and P. R. Bowser. 2005. Human health risks associated with formalin treatments used in aquaculture: initial study. *North American Journal of Aquaculture* 67:111–113.

Wu, R. and C. E. Boyd. 1990. Evaluation of calcium sulfate for use in aquaculture ponds. *Progressive Fish-Culturist* 52:26–31.

Yulin, J. 2000. The use of chemicals in aquaculture in the People's Republic of China. In J. R. Arthur, C. R. Lavilla-Pitogo, and R. P. Subasinghe, editors, *Use of Chemicals in Aquaculture in Asia: Proceedings of the Meeting on the Use of Chemicals in Aquaculture in Asia,* May 20–22, 1996, pp. 141–153. Iloilo, Philippines: Southeast Asian Fisheries Development Center.

Zimba, P. V., C. S. Tucker, C. C. Mischke, and C. C. Grimm. 2002. Short-term effect of diuron on catfish pond ecology. *North American Journal of Aquaculture* 64:16–23.

Zhang, X-X, T. Zhang, and H. H. P. Fang. 2009. Antibiotic resistance genes in water environment. *Applied Microbiology and Biotechnology* 82:397–414.

Chapter 10

Water pollution

Aquaculture facilities generate wastes of which all or a portion are discharged into the environment. The most significant pollutants from aquaculture production units are: (1) nutrients added in fertilizers or feeds and not recovered in harvested biomass; (2) organic matter resulting from uneaten feed and excrement of culture animals or produced by phytoplankton photosynthesis in response to nutrient inputs; (3) suspended solids consisting of plankton and soil particles suspended by erosion of pond infrastructure, and resuspension of sediment during harvest. Three additional but less common types of pollution are dissolved salts in seepage, overflow, and intentional discharge from saline-water culture facilities in freshwater areas; toxic substances such as copper sulfate, chlorine compounds, and pesticides used to control undesirable organisms in culture systems; and residues of pharmaceutical products for disease control. Pollutants found in aquaculture effluents are in general the same types found in effluents and runoff from agriculture and municipal wastewater treatment plants.

The presence of potential pollutants does not necessarily mean that an effluent will cause environmental degradation in receiving water. Water quality deterioration and associated biological perturbations occur only if the input of pollutants exceeds the capacity of the receiving water to render them harmless through dilution and assimilation. Moreover, there often are multiple sources of pollution in water bodies receiving aquaculture effluents. These two factors make it difficult to assess the effect of aquaculture facilities on receiving water bodies.

This chapter provides information about aquaculture and water pollution, and discusses methods for reducing the volume and improving the quality of aquaculture effluents. Application of government-imposed aquaculture effluent regulations as well as the voluntary adoption by producers of best management practices (BMP) or participation in "eco-label" certification programs will be considered in other chapters.

Aquaculture, Resource Use, and the Environment, First Edition. Claude E. Boyd and Aaron A. McNevin.
© 2015 John Wiley & Sons, Inc. Published 2015 by John Wiley & Sons, Inc.

Pollution potential of production systems

The main types of aquaculture production systems were described in Chapter 1, but in summary there are three basic types: (1) ponds to which fertilizers are applied, and ponds, flow-through units, cages, and net pens in which feeds are used to increase production; (2) rearing of bivalve molluscs, seaweed, and a few other species by "planting" their propagules on artificial structures or by confining organisms in containers at specific sites in open water where the culture organisms rely upon natural sources of nutrients for growth; (3) highly intensive, water-recirculating systems that often are located inside greenhouses.

Many ponds discharge only in response to large rainfall events or when drained for harvest; others that receive stream or spring flow or in which water exchange is applied discharge more frequently or continuously. The level of production in ponds varies greatly with species and culture method, but there is a world-wide tendency toward greater production intensity. Annual production of channel catfish (*Ictalurus punctatus*) in ponds in the southern United States that averaged less than 2000 kg/ha in the early 1970s increased to 3713 kg/ha in 2000 and to 5544 kg/ha in 2010 (http://www.aces.edu/dept/fisheries/aquaculture/catfish-database/catfish-2013.php). Production of the whiteleg shrimp (*Litopenaeus vannamei*) was traditionally less than 1000 kg/ha per crop, but with feeding and aeration that is commonly used today, yields of 5000 kg/ha and above are common (Chamberlain 2010). The tendency toward greater production levels has increased concern over water pollution.

In feed-based aquaculture about 80–90% of feed applied typically is eaten, and of this, 80–90% is absorbed across the intestine. Much of the absorbed material will become metabolic wastes; biomass of the culture species will contain only 10–20% of the dry matter added in feed (Boyd and Tucker 1998). In a culture unit where feed conversion ratio (FCR) = 1.7, 1700 kg feed (90% dry matter) yields 1000 kg live fish (25% dry matter), the distribution of the added dry matter might be as follows: uneaten feed, 230 kg; feces, 195 kg; metabolic wastes of culture species, 855 kg; biomass, 250 kg. Metabolic wastes are mainly carbon dioxide, water, ammonia, and phosphate, and decomposition of uneaten feed and feces results in these same metabolites.

Although the greatest concern over water pollution is directed at feed-based culture systems (Boyd et al. 2005), ponds treated with inorganic or organic fertilizers also may have effluents with high concentrations of nutrients (Boyd and Tucker 1998). Water drained from extensive fish ponds for harvest also may have elevated concentrations of suspended solids (Banas et al. 2008).

In raceways, tanks, and other flow-through units, water may be passed through at rates of two to three times the volumes of culture units per hour essentially flushing out all wastes. An alternative in flow-through culture is to use less flushing to exchange water between production units and a treatment pond in which water purification occurs. Of course when heavy rains occur, treatment ponds will overflow.

In cage and net pen culture dissolved and suspended solids resulting from feed input are flushed from culture units into surrounding areas by water currents. Settleable solids accumulate on the bottom near cages or inside pens. Because cage and net pen farms usually are located in rivers, reservoirs, lakes, or the sea, the entire waste load typically enters public waters. Sometimes cages or net pens are installed in ponds on private property. Ponds serve to treat the waste load but effluents from ponds will enter public water bodies.

Systems in which propagules are planted in open water do not result in increased nutrient inputs. They are not considered sources of water pollution, but pollution of water bodies into which these systems are superimposed can negatively impact the production and quality of the culture species (Shumway et al. 2003; Boyd et al. 2005).

In spite of relying on internal, mechanical, and biological water treatment, water recirculation systems release effluents of considerable pollutional strength when filters are cleaned or dissolved solids concentration diluted with source water (Timmons et al. 2002). These systems may someday be commonplace, but they presently represent a minute fraction of global aquaculture production and will not be considered here.

Methods for enhancing production capacity

In traditional pond aquaculture only a small proportion of the pond area and volume is necessary for the culture species—most of the area and volume function to assimilate waste. Production capacity is restricted by the amount of feed that may be applied without causing water quality to deteriorate below acceptable limits for culture species (Boyd and Tucker 1998; Boyd et al. 2007). Production capacity is increased by liming acidic ponds to enhance microbial decomposition, applying mechanical aeration to supply supplemental dissolved oxygen for the culture species and to oxidize waste, and water exchange to flush wastes from ponds. Of course water exchange has the environmentally undesirable effect of exporting some of the waste load to the outside environment.

Ponds may be lined with plastic to prevent aerator-induced erosion allowing more aeration and greater production. Production may be further intensified by dredging organic sediment from bottoms during the crop to reduce oxygen demand within ponds. The senior author has experience with a shrimp (*L. vannamei*) farm in Indonesia where yields typically reach 12–15 t/ha per crop. An Asian catfish (*Pangasius hypophthalmus*) farm in Vietnam produced 10 000 t/year in 12 ha of ponds in 2010 (Boyd et al. 2011). In both cases high rates of water exchange are used and sediment is pumped from bottoms of ponds and discharged to the environment. Moreover, it is common practice in shrimp farming in Asia to wash bottoms of empty ponds between crops with water from high-pressure nozzles to remove sediment and lessen oxygen demand during the next crop (Yuvanatemiya et al. 2011). This sediment is released into adjacent water bodies.

System waste loads

In ponds treated with organic manures, nitrogen and phosphorus from decomposing manure stimulate phytoplankton productivity increasing the base of the natural food web leading to greater fish or crustacean production. Chemical fertilizers such as urea and triple superphosphate dissolve in pond water increasing nitrogen and phosphorus concentrations to stimulate phytoplankton growth.

There is variation in the proportions of nitrogen and phosphorus applied to ponds in organic manures and chemical fertilizers recovered in biomass at harvest. However, calculations based on data from several studies described by Boyd and Tucker (1998) suggested that differences between nitrogen and phosphorus inputs and outputs in biomass at harvest (the system loads) typically ranged between 25–75 kg nitrogen/t production and 10–30 kg phosphorus/t production.

Much aquaculture production is derived from natural food-based systems in India, China, Bangladesh, and other developing countries. The use of feed is increasing but it likely will be several decades before the majority of production is from feed-based culture. Meanwhile, it does not seem wise to ignore the water pollution potential of natural food-based culture systems.

System loads of carbon, nitrogen, and phosphorus for several common aquaculture species in feed-based culture are presented (Table 10.1). Loads for organic carbon tended to decline as the FCR decreased and ranged from 408 to 951 kg/t. The greatest system loads for nitrogen were 90.4 kg/t for black tiger prawn (*Penaeus monodon*) and 88.4 kg/t for *I. punctatus*. Nitrogen loads were similar (45.4–59.4 kg/t) for the other species. Fish species had system phosphorus loads of 10–13 kg/t. Bone is made mainly of calcium phosphate; thus shrimp that have no bone had greater system loads of phosphorus than found for fish. It is interesting to note that system loads of nitrogen and phosphorus for ponds treated with chemical and organic fertilizers were similar to those for ponds with feeding.

Nitrogen and phosphorus added to ponds in fertilizers and feeds stimulate phytoplankton growth resulting in large amounts of particulate organic matter. Moreover, particulate organic matter results from uneaten feed and feces. It is not feasible to compute system loads for organic matter because net phytoplankton productivity varies greatly and seldom is measured. Nevertheless, the discharge of organic suspended solids from culture systems tends to increase with the intensity of culture.

In ponds organic suspended solids consist mostly of plankton and concentrations of organic matter in water seldom exceed 50 mg/L. The load of organic matter discharged to the environment when a typical pond of 1.5 m average depth is completely drained usually would not exceed 750 kg/ha.

In raceways, cages, and net pens, organic suspended solids consist mainly of small particles of feed and feces. As mentioned above, the amount of solids generated often would be around 425 kg/t production.

Ponds act as sedimentation basins, but when they are drained, large amounts of sediment—mostly mineral matter—may be resuspended and discharged. Schwartz and Boyd (1994) reported that an average of 9362 kg/ha of particulate solids (2302 kg/t production) were released when channel catfish ponds in the United States were completely drained. Banas et al. (2008) reported an average discharge of 1231 kg/ha of particulate solids from extensively managed fish ponds in France, and Szabo (1994) estimated that 1875 kg/ha of particulate solids were released during draining of an extensive carp pond in Hungary. Production was not given for the extensive

Table 10.1 System loads of organic carbon, nitrogen, and phosphorus for six aquaculture species.

Species	Feed (%)			Typical FCR	Whole body composition (%)			System load (kg/t fish)			System load (% of feed input)		
	C	N	P		C	N	P	C	N	P	C	N	P
Channel catfish *Ictalurus punctatus*	48	5.1	0.9	2.2	10.5	2.38	0.68	951	88.4	13.0	90	79	66
Blue tilapia *Oreochromis aureus*	48	4.8	1.0	1.7	10.9	2.22	0.70	707	59.4	10.0	87	73	59
Atlantic salmon *Salmo salar*	48	7.0	1.3	1.1	12.0	2.96	0.40	408	47.4	10.3	77	62	72
Rainbow trout *Oncorhynchus mykiss*	48	6.4	1.3	1.1	12.0	2.50	0.35	408	45.4	10.8	77	65	75
Whiteleg shrimp *Litopenaeus vannamei*	48	5.6	1.2	1.5	11.7	2.86	0.32	603	55.4	14.8	84	66	82
Black tiger prawn *Penaeus monodon*	48	6.7	1.5	1.8	12.4	3.02	0.25	740	90.4	24.5	86	75	91

Source: Modified from Boyd and Queiroz (2001a) and Boyd et al. (2007).

ponds, but it was probably less than 1 t/ha, and discharge of particulate solids per ton of production likely was similar to that reported for channel catfish ponds.

Sodium chloride is used in some types of aquaculture to counteract nitrite toxicity in culture animals. Salt inputs are low (50–100 mg/L) and the increase in salinity is not environmentally harmful (Sipauaba-Tavares and Boyd 2003). Ponds do not retain a significant amount of dissolved solids introduced in source water. Thus most of the salt contained in water of low-salinity, inland aquaculture ponds is discharged through seepage, overflow, and draining for harvest (Braaten and Flaherty 2001; Boyd et al. 2006a).

Other substances such as algicides and other pesticides, antibiotics and other therapeutic agents, and oxidizing agents also tend to be assimilated to a large extent by natural processes in ponds (Boyd and Massaut 1999). However, such chemicals are of greater concern in raceway and cage or net pen culture. Toxicity, however, is usually not a concern with aquaculture effluents because chemicals used in aquaculture facilities are obviously not applied at great enough concentrations to cause toxicity in the culture animals. Thus effluents from such facilities would not be expected to have high concentrations of potential toxins.

Assimilation and removal of wastes

The part of the system nutrient load assimilated inside a culture system or removed from it by other means lessens the environmental waste load of an aquaculture facility. When water is retained in ponds for a considerable period of time natural processes will assimilate wastes (Boyd and Tucker 1998). Nutrients released when organic matter is decomposed stimulate additional organic matter synthesis by phytoplankton, but phytoplankton die and are decomposed quickly by microorganisms (Boyd 1985; Boyd and Tucker 1998). Organic matter also settles to pond bottoms to become bottom soil organic matter (Steeby et al. 2004; Boyd et al. 2010). Ammonia nitrogen is oxidized to nitrate by nitrifying bacteria, and nitrate nitrogen that enters anaerobic zones in pond bottoms is denitrified to nitrogen gas that diffuses into the atmosphere (Hargreaves 1998). Ammonia also diffuses from pond water to the atmosphere (Weiler 1979; Gross et al. 2000). Phosphorus contained in particulate solids settles to the bottom and inorganic phosphorus is strongly adsorbed by pond soil. For example, about 60% of phosphorus applied to experimental ponds at the E. W. Shell Fisheries Center of Auburn University near Auburn, Alabama over a 22-year period could be accounted for in bottom soils (Masuda and Boyd 1994), and 68% of phosphorus applied in feed to three shrimp (*P. monodon*) ponds in Madagascar during a single crop was bound in bottom soil (Boyd et al. 2006b). Percentages of elements applied to ponds in feed that are subsequently discharged in effluents (environmental waste loads) ranged from 10 to 20% for carbon and from 10 to 40% for nitrogen and phosphorus (Avnimelech and Lacher 1979; Boyd 1985; Pengseng 2007; Schwartz and Boyd 1994; Gross et al. 1998; Gross et al. 2000).

MacMillan et al. (2003) attributed a 40% reduction in the amount of phosphorus discharged from trout raceways in Idaho to improved feeding practices and exclusion of fish from ends of raceways to provide a quiescent zone in which particulate solids

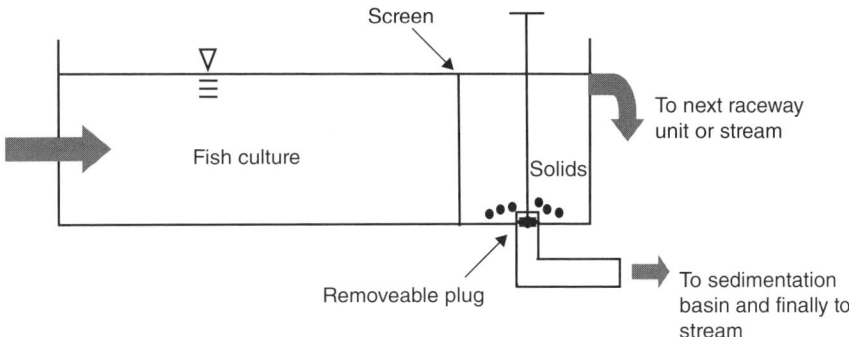

Figure 10.1 Illustration of the use of the end of a raceway for solids removal. Drawing not to scale.

settle and are removed (Fig. 10.1). Soderberg (2007) found optimum conditions for solids collection in quiescent zones to be 2.4–3.4 cm/sec velocity and (1) a fish density of 160 kg/m^3 at an overflow rate of 0.5 to 1.0 cm/sec or (2) a fish density of 80 kg/m^3 at an overflow rate of 0.5 cm/s. Under these conditions up to 20% of the total solids load was collected in the quiescent zone. Solids from raceways can be further concentrated by sedimentation in an off-line basin, removed, and applied to land as a soil amendment (Fornshell and Hinshaw 2008). Particulate solids also can be removed via center drains in tanks and other flow-through systems. However, particulate solids that do not settle and dissolved solids are flushed from flow-through units into the receiving water.

Nitrogen cannot be removed as effectively as phosphorus from flow-through units because only about 9% of nitrogenous waste of salmonids—the typical species group produced in flow-through culture—is particulate (Foy and Rosell 1991). Screens are sometimes used to remove nutrients in particulate matter from discharge of flow-through systems in Europe. Bergheim and Brinker (2003) found that particulate nitrogen and phosphorus removal by screening reduced nitrogen and phosphorus concentrations in effluent by 3% and 21%, respectively.

In cage culture it is imperative to have a good rate of water flow through densely stocked cages to avoid low dissolved oxygen and high ammonia concentrations. Particulate solids either quickly fall through the cage netting or are flushed out by the current. Various methods of collecting solid wastes from cages have been investigated but all have proved impractical and they are excessively expensive to operate (Belle and Nash 2008).

Effluent quality

Overflow and water released during water exchange usually is of relatively good quality, but with higher concentrations of several key variables than typically found in receiving waters (Boyd 1978; Boyd and Gross 1999; Silapajarn and Boyd 2005; Soongsawang and Boyd 2012). Pond overflow usually is within the following concentration ranges: pH, 7–9; dissolved oxygen, >3 mg/L; total suspended solids, 30–120 mg/L; total ammonia nitrogen, 0.3–5 mg/L; total nitrogen, 2–10 mg/L; total phosphorus 0.1–0.5 mg/L; 5-d biochemical oxygen demand, 5–20 mg/L. Nevertheless, other than for total suspended solids, pond effluent is relatively mild in comparison to domestic wastewater. Tchobanoglous et al. (2003) reported the average composition of medium-strong domestic wastewater as follows: pH of 6.5 to 8, <1 mg/L dissolved oxygen, 210 mg/L total suspended solids, 25 mg/L total ammonia nitrogen, 40 mg/L total nitrogen, 7 mg/L total phosphorus, and 190 mg/L 5-d biochemical oxygen demand.

Effluents at harvest are similar to overflow in composition until near completion of draining. During final stages of draining, crowding of frightened culture animals and operations to capture the animals resuspend sediment. Although pH and dissolved oxygen concentration remained about the same, other variables mentioned above increased in concentration in the final 25% of draining effluents (Boyd 1978; Schwartz and Boyd 1994; Teichert-Coddington et al. 1999; Soongsawang and Boyd 2012). An example of the changes in concentrations of water quality variables during harvest of inland, low-salinity shrimp ponds (production averaged 3574 kg/ha) is shown in Fig. 10.2. Nevertheless, with the possible exception of total suspended solids concentration, pond draining effluents are not highly polluted relative to domestic or other common wastewater sources.

Reducing concentrations and loads of pollutants

Pollutants will not cause environmental degradation unless they are at high enough concentration to cause adverse impacts in the mixing zone around the effluent outfall or if pollutant loads exceed the assimilative capacity of the receiving waters. Efforts to reduce the pollution potential of aquaculture facilities should focus on three main issues: (1) effective use of fertilizers and feeds to reduce the amount of waste; (2) implementation of procedures for removing or treating wastes on the farm to lessen the amount discharged to the environment; (3) minimizing overflow and water exchange.

An obvious way to lessen nutrient input is to restrict fertilizer and feed use. Although this approach may be necessary to avoid pollution of a given body of water, methods that lessen the pollution load per unit of production without limiting the amount of production are more useful.

More pollution is generated by some culture species than by others, and some culture methods have a greater pollution potential than others. Thus environmental advocacy groups tend to promote culture methods that cause less pollution, for example, ponds versus cages, and herbivorous fish versus omnivorous or carnivorous ones. However, methods for reducing pollution in general will be considered rather than encouraging use of particular species or culture methods over others.

Nutrient input management

Fertilizers should only be used as needed to promote phytoplankton growth. When ponds have a heavy phytoplankton bloom, fertilizer application should be delayed. There is a mistaken idea that pond fertilizers should contain about seven times as

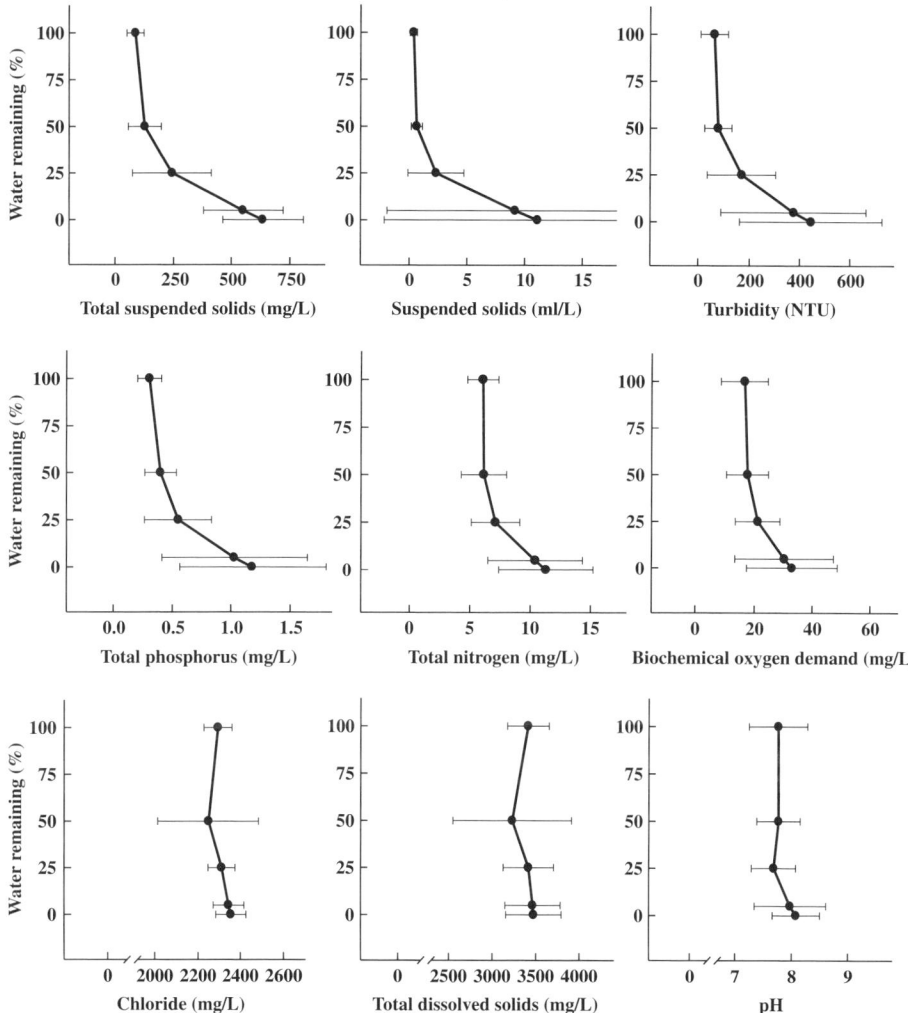

Figure 10.2 Mean concentrations and standard deviations for water quality variables measured at different stages of pond water level drawdown for harvest at an inland, low-salinity shrimp farm in Alabama. *Source:* Prapaiwong, "Water quality in inland ponds for low-salinity culture of Pacific white shrimp *Litopenaeus vannamei*" Auburn University, 2011.

much nitrogen as phosphorus (Boyd et al. 2008). This leads to excessive fertilization with nitrogen and increases nitrogen output in effluents (Boyd and Li 2012). In most cases a fertilizer nitrogen:phosphorus ratio of around 5:1 and even lower in older ponds is adequate (Boyd 1976).

Feed quality and feeding practices are key issues in pollution reduction because feed is the primary source of nutrients and organic matter in many aquaculture effluents. Several factors influence efficiency of feed use as follows: the contribution of natural food organisms to production; the proportion of the feed that is consumed by culture animals; proportions of the nutrients in consumed food absorbed across the intestine; proportions of absorbed nutrients contained in biomass at harvest.

The amount of feed eaten by the culture species declines if fish are stressed by disease or adverse environmental conditions. For example, fish eat and grow less as dissolved oxygen concentration declines (Collins 1984; Torrans 2005, 2008).

The proportion of feed absorbed across the intestine is a function of feed digestibility, and the efficiency of conversion of absorbed nutrients to biomass depends upon how well the feed composition matches nutritional requirements. Moreover, the percentage of feed intake used for growth decreases when animals are stressed. This is illustrated by data from Wang et al. (1997) for effects of salinity on common carp (*Cyprinus carpio*); grown at 0.5 g/L salinity, carp converted 33.4% of their food energy intake to growth as compared to 10.4% at 8.5 g/L salinity.

Assuming that environmental conditions are adequate and culture organisms are healthy, high FCR usually is caused by poor quality feed and overfeeding. Although feed is expensive, many producers overfeed because they feel that production will suffer unless the culture organisms have all the feed that they can eat.

Lowering FCR will reduce feed cost and diminish the quantity of feeding wastes that enter the culture system per unit of production. Each 0.1-unit reduction in FCR lowers feed input by 100 kg/t of production. Assuming that composition of culture animals is unaffected by FCR, the corresponding reduction in outputs in waste for carbon, nitrogen, and phosphorus in kilograms per tonne production for each 0.1 unit improvement in FCR would be equal to the percentage of each element in the feed (Table 10.2). An improvement in FCR from 2.2 to 1.8 would lessen total system waste load by 20.2%, 23.1%, and 27.7% for organic carbon, nitrogen, and phosphorus, respectively.

System loads of nitrogen and phosphorus also can be lessened by using feed with adequate, but not excessive, nitrogen and phosphorus. For example, reducing the crude protein content of feed by 1% (0.16% reduction in nitrogen) at an FCR of 1.8 would lessen system nitrogen load by 2.88 kg/t (Table 10.3). The corresponding reductions in system phosphorus load caused by a 0.1% reduction in feed phosphorus concentration would be 1.8 kg/t (Table 10.3). Most feeds have similar organic carbon concentrations and it likely is not feasible to change carbon concentration appreciably.

Table 10.2 Effects of feed conversion ratio (FCR) on system loads of organic carbon, nitrogen, and phosphorus in channel catfish ponds.

FCR	Carbon (kg/t)	Nitrogen (kg/t)	Phosphorus (kg/t)
2.20	951	88.4	13.0
2.10	903	83.3	12.1
2.00	855	78.2	11.2
1.90	797	73.1	10.3
1.80	759	68	9.4
1.70	711	62.9	8.5
1.60	663	81.6	7.6
1.50	615	76.5	6.7

Feed composition: 48% C; 5.1% N; 0.9% P.
Fish composition: 10.5% C; 2.38% N; 0.68% P.

Table 10.3 Effect of reducing concentration of crude protein (%N × 6.25) and phosphorus in feed at a feed conversion ratio of 1.8 on system waste loads of nitrogen and phosphorus.

Feed crude protein (%)	Feed nitrogen (%)	System N load (kg/t)	Feed phosphorus (%)	System P load (kg/t)
32	5.12	68.36	1.2	14.8
31	4.96	65.48	1.1	13.0
30	4.80	62.60	1.0	11.2
29	4.64	59.72	0.9	9.4
28	4.48	56.84	0.8	7.6

Fish composition: 38% N; 0.68% P.

Effluent management in ponds

Erosion control should begin at the farm design and construction stages. Side slopes of embankments and canals and bottom slopes of canals should be in accordance with soil properties. Guidelines for design of embankment slopes and sides of canals can be found in Yoo and Boyd (1994). It is particularly important to compact soil well at its optimum moisture content. The standard Proctor test may be used to ascertain the optimum moisture content for compaction but the usual optimum moisture values are 6–10% for sand, 8–12% for mixtures of sand and silt, 11–15% for silt, and 13–21% for clay (Boyd 2008). Embankments should be planted with grass, and highly vulnerable areas should be reinforced with rip-rap, gabion, geofabric, or other material.

Erosion on watersheds can be minimized by grading of steep, erosion-prone areas and installation of vegetative cover. Turbid runoff from specific areas may be diverted by terraces or ditches. Where farms have no control over watersheds the only option often is to use a sedimentation basin to remove solids from incoming water before transferring it to ponds.

Overflow following rainfall events is a major cause of discharge. The ratio of pond watershed area to pond storage capacity regulates the amount of inflow from runoff and thereby determines overflow volume and hydraulic retention time (HRT) in watershed ponds (Yoo and Boyd 1994). A long HRT favors natural assimilation of wastes within ponds to lessen the amount of waste discharged (Boyd and Tucker 1998). Excess runoff often may be diverted from a pond by ditches, terraces, or a combination thereof (Boyd et al. 2003).

Embankment ponds receive little runoff because watershed area consists only on tops and above water slopes on insides of levees. In embankment ponds the practice of maintaining water level 15–20 cm below the tops of overflow structures (Fig. 10.3) will prevent most rainwater entering ponds from overflowing (Boyd 1982; Cathcart et al. 1999). This method of water-level control also is effective during drier parts of the year in avoiding overflow from watershed ponds. Obviously reduction in water exchange also lessens the amount of effluent from ponds and lengthens HRT.

In pond aquaculture systems procedures that encourage the assimilation of wastes during the crop should be applied. Use of adequate mechanical aeration is particularly important in oxidizing organic matter and ammonia. Based on the feed

Figure 10.3 Illustration of method of retaining storage volume in ponds to avoid overflow after rainfall events.

oxygen demand and the aerator efficiency equation (Boyd 2009), the amount of aeration required to oxidize feeding waste and maintain dissolved oxygen concentration above 3 mg/L in a system without waste removal is about 1 hp aeration for each 10 kg/ha per day of feed input.

Aerators should be installed at least 1 m beyond the inside toes of embankments and they should not impinge water currents against embankment. Aerators also should be in water of at least 1 m depth. Areas that are susceptible to aerator erosion should be reinforced with stone or geofabric.

Channel catfish and some other aquaculture species can be harvested by seining and without draining ponds (Boyd et al. 2000), but eventually ponds must be drained completely to repair earthwork. If ponds are drained from the bottom, the velocity and turbulence of water entering the discharge structure will resuspend sediment resulting in increased suspended solids until the area around the structure is swept clean (Hargreaves et al. 2005). Whether ponds are drained from the surface or the bottom, effluents will become more elevated in suspended solids during the last 20–25% of drawdown (Prapaiwong and Boyd 2012). It is sometimes possible to close the drain during the final stages of harvest and remove animals by seining or dipping. The water can then be held for 1 or 2 days for sedimentation of solids and then released slowly to avoid resuspension of solids (Seok et al. 1995). Of course final draining effluent could be passed through a sedimentation basin to remove suspended solids.

Water from a pond may be pumped to an adjacent pond during harvest operations and then returned after harvest (Boyd 2003; Boyd et al. 2006a). A reservoir also may be used to facilitate harvest without discharge of effluents to natural waters (Fig. 10.4).

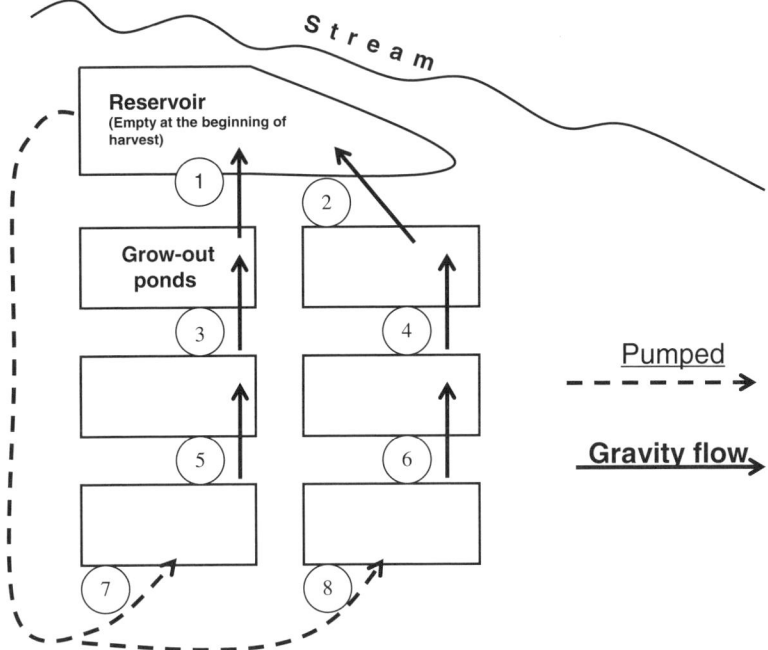

Figure 10.4 Illustration of use of reservoir to allow water from pond draining to be reused.

Resuspension of sediment from empty ponds by rainfall can be a source of suspended solids in farm effluents. Pond drains should be closed after harvest to avoid discharge after rainfall events while ponds are empty (Boyd et al. 2003). Effluents from washing pond bottoms with high-pressure streams of water is particularly concentrated in suspended solids and should be held in a sedimentation basin before release to the environment (Yuvanatemiya et al. 2011).

Sedimentation basins

Sedimentation basins detain water to provide time for suspended solids to settle from effluents before final discharge to natural waters. The HRT needed for removal of particles of a specific size by sedimentation can be estimated using the Stoke's law equation (Boyd 1995). The settling velocity of particles depends upon several factors but particle diameter and particle density are the most important. Settling velocities of fine sand, silt, and clay size particles are 2.3×10^{-3}, 9×10^{-5}, and 9×10^{-7} m/sec, respectively. The critical settling velocity is the minimum settling velocity necessary for a particle to settle before flowing out of a sedimentation basin (Fig. 10.5).

The critical settling velocity is related to settling pond inflow rate and areas:

$$V_{cs} = \frac{Q}{A} \tag{10.1}$$

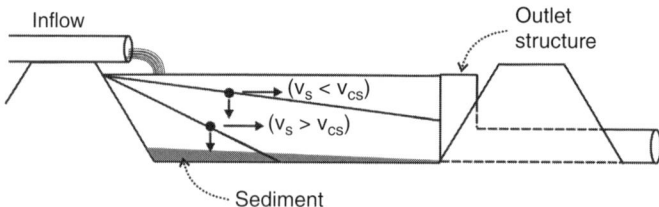

Figure 10.5 Illustration of sedimentation basin (not to scale). V_s, terminal settling velocity; V_{cs}, critical settling velocity.

where V_{cs} is the critical settling velocity (m/sec), Q is the inflow (m³/sec), and A is the surface area (m²). To illustrate the use of Equation 10.1 if it is desired to remove particles with critical settling velocities greater than 9×10^{-5} m/sec from a discharge of 0.11 m³/sec, a 1-m deep settling basin would need an area of 1222 m² to provide sufficient HRT for sedimentation. However, settling basins should be larger than the minimum size by at least 50% to allow sediment storage and maintain sufficient HRT.

Sand and coarse silt particles can be removed effectively by sedimentation but an exceedingly long HRT is necessary to remove clay particles. For example, it would require a 122 222 m² settling basin to remove clay particles ($V_{cs} = 9 \times 10^{-7}$ m/sec) from a discharge of 0.11 m³/sec. Organic particles and particularly plankton have a low particle density relative to mineral particles. Thus organic matter usually cannot be removed effectively from pond effluents by sedimentation (Boyd and Queiroz 2001b; Ozbay and Boyd 2003).

There are guidelines for effective operation of settling basins. Effluent should enter at the surface on one side and exit at the surface on the opposite side (Fig. 10.5). A baffle can be installed in a basin to prevent the effluent from passing directly from the inflow point to the outflow point. Mechanical aerators or other devices that mix water should not be placed in settling basins, for the turbulence they create will inhibit sedimentation. If aeration of final discharge is required it should be done in a separate basin.

Settling basins will fill with solids over time and sediment removal is occasionally necessary to maintain HRT. Sediment removal will resuspend solids and it is desirable to construct dual settling basins so that one may be operated while the other is being cleaned. Sediment removed from settling basins or production ponds can cause environmental degradation. Sediment piles create an ecological nuisance by disrupting or destroying natural vegetation, and rainfall erosion of sediment piles can lead to turbid runoff. Sediment from ponds with saline water has a salt burden, and salt leached by rainfall can lead to salination of soils, surface water, and groundwater (Boyd et al. 1994).

Sediment removed from settling basins, ponds, or canals should be confined in a bunded area with enough storage volume to avoid overflow after rains. After solids resuspended by rainfall have resettled, standing water can be left to evaporate or slowly drained away with care to avoid resuspension of solids.

Sediment usually consists of sand and silt particles and it is not of good quality for repairing erosion damage to farm earthwork or for new construction. However, it can be used for earth fill, and non-saline sediment may be spread over agricultural land and incorporated into the soil.

Preventing salination

Inland ponds in freshwater areas that are filled with saline water or coastal ponds that are constructed above a freshwater aquifer are likely to cause salination. Such ponds ideally should be lined with plastic membranes but at least they should not be constructed on highly permeable soil (Boyd et al. 2006a). Because ponds will likely seep regardless of precautions taken to prevent seepage loss, they should be constructed as far as possible from freshwater streams (Pine and Boyd 2011). A lined ditch with bottom elevation deeper than pond bottoms can capture lateral seepage from ponds and avoid salination of nearby surface soils. Water reuse should be practiced for inland, saline water aquaculture (Fig. 10.4). When such farms must discharge water, the water should be discharged slowly to avoid large spikes in salinity of the receiving water body.

Best management practices

A BMP is a practice that is considered to be the best available method for preventing or lessening a specific, negative environmental impact (Hairston et al. 1995). Normally a single BMP will not be sufficient to reduce negative impacts to an acceptable level; rather a suite of BMPs will be necessary. Suggestions for reducing the volume and improving the quality of effluents such as have been discussed in this chapter can be formalized and presented as BMPs (Boyd 2003) as discussed in Chapter 13.

Perspectives on pollution by aquaculture

An estimate of discharge of nitrogen and phosphorus from aquaculture systems was made using 2009 global aquaculture statistics and the following assumptions: 30% of freshwater production was feed-based; marine and brackish water fish were reared in cages; marine shrimp were reared in ponds; average environmental loads of nitrogen and phosphorus were 40% of average system loads for ponds and equal to system loads for cage and raceway culture (Table 10.4). The estimated global discharges of

Table 10.4 Average environmental loads of nitrogen and phosphorus.

| | (kg nutrient/t production) | | |
| | Feed-based | Natural-food-based | |
Variable	pond culture	pond culture	Cage culture
Nitrogen	25.76	20	64.4
Phosphorus	5.6	8	13.9

nitrogen and phosphorus to the environment by aquaculture were 1 038 209 t and 308 966 t, respectively.

Globally, an estimated 50 Mt of anthropogenic nitrogen enters natural water bodies annually via effluents and runoff (Schlesinger 2009), while the estimate for anthropogenic phosphorus was 10.5 Mt (Liu et al. 2008). Aquaculture inputs are only 2% and 2.9% of total estimated anthropogenic inputs to natural waters of nitrogen and phosphorus, respectively. Much more aquaculture is conducted in Asian countries—especially in China—than in other countries and the contribution of aquaculture to total anthropogenic discharges of nitrogen and phosphorus to natural waters is possibly greater in Asia than elsewhere. Cao et al. (2007) determined that marine aquaculture effluents contributed 2.8% as much nitrogen, 5.3% as much phosphorus, and 1.8% as much organic matter to the Yellow Sea and Bohai Sea in China as did land-based sources (agriculture, industry, and sewage). Considering that agriculture is the leading source of anthropogenic nitrogen and phosphorus, and aquaculture represents about 8 or 9% of world animal protein production, aquaculture does not seem to discharge disproportionately large amounts of nitrogen and phosphorus. Areas outside Asia with large amounts of aquaculture could have percentages close to the global averages. For example, Páez-Osuna et al. (1998) concluded that shrimp farming in coastal states of Mexico contributed 1.5% and 0.9% of total anthropogenic discharges of nitrogen and phosphorus, respectively.

Most studies of aquaculture and water pollution have focused on concentrations and loads of potential pollutants in effluents and did not confirm whether or not negative environmental impacts occurred. However, there is increasing information on the effects of aquaculture effluents on water quality and aquatic plant and animal communities at specific locations. Some studies have focused mainly on water quality (Soongsawang and Boyd 2012; Silapajarn and Boyd 2005; Biao et al. 2004; Boaventura et al. 1997; Nordvarg and Johansson 2002), others have given primary attention to effects on plant and animal communities (Selong and Helfrich 1998; Guo and Li 2003; Stephens and Farris 2004; Kutti et al. 2007; Loch et al. 1996), and a few have evaluated effects on sediment quality (Carroll et al. 2003; Wu et al. 1994). Depending upon the variable considered, the observed effects have ranged from insignificant to severe. Nevertheless, it is clear that aquaculture can cause significant water pollution, and educational and regulatory steps should be taken to reduce these negative impacts to the lowest that are technologically and economically feasible.

Fisheries processing effluents

About 60–70% of global fisheries production is processed instead of reaching the consumer as a completely intact organism (Islam et al. 2004). Processing results in waste, a portion of which is recovered and rendered into a useful material such as animal feed, biofuel, dietetic products, natural pigments, cosmetics, enzymes, and fertilizer (Arvanitoyannis and Kassaveti 2008). But processing plants also discharge wastewater. The composition of shrimp and fish processing effluents are variable; Carawan et al. (1986) reported the following ranges: biochemical oxygen demand, 200–1,000 mg/L; total suspended solids, 100–800 mg/L; fats, oils, and grease,

40–300 mg/L. Sirianuntapiboon and Nimnu (1999) reported that a seafood process-ing plant in Thailand had effluent concentrations of total nitrogen and total phospho-rus of 346 mg/L and 31 mg/L, respectively. Concentration ranges for total nitrogen and total phosphorus in other seafood processing plant effluents have been reported, respectively, as 95–124 mg/L and 58–63 mg/L (Sohsalam et al. 2008); 126–703 mg/L and 32–42 mg/L (Prasertsan et al. 1994); 138–162 mg/L and 15–32 mg/L (Sridang et al. 2008). The average is about 242 mg/L of total nitrogen and 39 mg/L of total phosphorus. The wastes from processing plants are more concentrated and of greater pollution potential than those from culture systems.

There is no reason to suspect that aquaculture animals and wild-caught animals would differ with respect to processing waste and effluents. Moreover, many process-ing plants often accept both aquaculture and wild-caught animals. The contribution of aquaculture to the waste load in processing effluents should roughly be in pro-portion to the world production of the two types of products. According to Islam et al. (2004) during the period 1992–2001 the average annual effluent waste loads of seafood processing globally were biochemical oxygen demand, 159 120 Mt; total suspended solids, 794 090 Mt; fats, oil, and grease, 859 690 Mt. The wastewater volume was estimated at 1 010 000 000 m^3/year. Based on average total nitrogen and total phosphorus concentrations in seafood processing effluents estimated above, loads of these two variables might have been as much as 244 420 Mt and 39 390 Mt, respectively. Possibly 50% of the nitrogen and phosphorus loads could be attributed to aquaculture. Thus processing of aquaculture products may discharge about 10% as much nitrogen and phosphorus as enters natural waters from production facilities.

Conclusions

Effluents from aquaculture production facilities are more concentrated than most natural waters into which they are discharged. Although aquaculture discharges are relatively mild compared to industrial and municipal wastewater, they still have potential to cause eutrophication, turbidity, and sediment accumulation in receiv-ing water bodies. The pollution potential of aquaculture can be greatly reduced by adopting practices to reduce nutrient inputs to culture systems, enhancing the capac-ity for waste assimilation in ponds, and treating the most concentrated effluents in settling basins. Aquaculture effluents possibly could contribute as much as 2 and 3%, respectively, of total anthropogenic nitrogen and phosphorus entering natural waters. Processing of aquacultural products might contribute about 10% as much nitrogen and phosphorus to anthropogenic inputs as does aquaculture production.

The eNGO perspective

Water quality issues related to different types of aquaculture facilities have been discussed in this chapter. Many of these facilities are relatively benign to the envi-ronment, and industry advocates claim that eNGOs should do more to recognize more benign aquaculture species and culture systems. There is, however, a fundamental misunderstanding of the eNGOs by those who believe the environmentalists will channel their science, marketing, and campaign efforts toward heralding certain types of aquaculture. Plainly stated this is not the role of the eNGOs. The discussion that follows focuses on the development of a novel way of thinking about effluents and the environment and how this came about within the eNGO community.

The issue of water pollution from aquaculture facilities has gone through several levels of iterations by the eNGOs. As discussed previously many of the eNGOs feel that the overall goal of the aquaculture industry should be to transition toward closed systems. This view is held primarily because of the intense campaigns against salmon aquaculture, and the subsequent view that cage culture has greater impacts on water quality than does other types of aquaculture systems. Of course there are eNGOs who understand that in some instances and for some species recirculating aquaculture is not currently feasible.

There has been a greater awareness in the eNGO community of the complexities and challenges in determining if an aquaculture facility is "polluting" a natural water body. Nevertheless the eNGOs do not feel they are responsible for finding solutions to impacts generated by aquaculture activities. Discharge of nutrients and other substances from aquaculture facilities used to be thought of from the "point-source" perspective in the eNGO community. It was also surmised that cage culture was the ultimate water polluter. As the aquaculture industry and the eNGOs began to collaborate further it was realized that aquaculture is more akin to agriculture and they adopted a "non-point source" perspective.

Because water availability for culture is the main aspect of farm siting, large water sources where aquaculture is suitable tend to have numerous producers in the vicinity. Each of these producers contributes to the effects of the cumulative waste release from farms. Thus eNGOs have found that attributing water quality impacts to a specific farm is nearly impossible unless the farm is the only producer in the watershed which is unlikely.

With little means to attribute certain water quality impacts to specific farms, there has been a push within the aquaculture industry and some eNGOs to establish limits for allowable concentrations and loads discharged from facilities. Undoubtedly this is the most effective way to quantify the amount of wastes being discharged into the environment; however, these limits do not depict the condition of the receiving waters nor the other farms or unrelated industries and households discharging into the same receiving water body. The limits are an excellent way for producers to determine their level of efficiency with regards to feed and other nutrient impacts, but they fail to promote stewardship of the surrounding water resources. There was a time when discharge limits were seen as the only strategy by some of the eNGOs to protect the environment. Limits were quantitative and succinct; one could compare one farm to another farm as to the amount of nitrogen or phosphorus that was being discharged per unit of production. Of course there were some eNGOs that could not put these discharge limits into a relative context and thought that pushing for stricter limits than those proposed was the best option. This push for more stringent discharge limits basically brought the eNGOs to the stance of no discharge, hence the desire for recirculating systems.

There came to be recognition in the eNGO community that the environment rather than the farm was where limits should be set—meaning the receiving waters should have the restrictions rather than limiting the farm's effluent. This recognition appears to be quite an obvious oversight, but it was always perceived to be too difficult to institutionalize. It should be understood that many aquaculture producers conduct monitoring in their respective receiving waters, but adhering to some limit in an environment where they had limited control was not common. Probably the first to embrace this idea were a few eNGOs and tilapia producers in the Tilapia Aquaculture Dialogue. Many criticisms of this approach were leveled by industry and academia because producers could not control what others were discharging in the watershed. This small group in the Tilapia Aquaculture Dialogue thought it was a win-win for both the environmentalists and the industry.

There are certain regions and water bodies that are simply unfit for aquaculture because of the level of pollution. Producers would not choose these sites to grow fish or other aquatic organisms. Further better producers desire information about their surrounding environment as it may affect their productivity. From an eNGO perspective this was the ultimate macro-environmental success, as some responsibility for being "stewards" of the receiving water body was being placed on producers. As was discovered many producers were already attempting, for many reasons including profitability, to ensure that their surrounding environment was not being degraded.

The recognition of the benefits to an approach that was receiving water driven by the eNGOs has been largely accepted within the community as it appears to be the most logical way to protect natural water resources. Shifting emphasis from effluent quality to effects of effluents on receiving water bodies allows evaluation of cage culture and other production systems on an equal basis. This was an important innovation in the approach of the eNGOs to the pollution issue, because it allows fairer and equitable comparison of pollution effects of different culture methods. If cage culture operations could show that the environment was not being degraded or changing, they would be considered as acceptable as pond culture from a water quality perspective. Further, pond aquaculture operations were now forced to identify the impacts in their receiving waters, which in some cases can be challenging. For example, many fish ponds in China are connected by a myriad of government-constructed canals, and to truly address the impacts on receiving waters these canals had to be followed to rivers and even to the ocean to begin to monitor their impacts. Where once the ability to quantify and limit effluents in ponds was the primary means of determining impact on the environment, now it is more broadly accepted that regardless of the system, quantity, and concentration of discharge producers are going to have to show that the receiving water body environment will need to be at the least maintained but more likely improved.

References

Arvanitoyannis, I. S. and A. Kassaveti. 2008. Fish industry waste: Treatments, environmental impacts, current and potential uses. *International Journal of Food Science and Technology* 43:726–745.

Avnimelech, Y. and M. Lacher. 1979. A tentative nutrient budget for intensive fish ponds. *Bamidgeh* 31:3–8.

Banas, D., G. Masson, L. Leglize, P. Usseglio-Polatera, and C. E. Boyd. 2008. Assessment of sediment concentration and nutrient loads in effluents drained from extensively-managed fishponds in France. *Environmental Pollution* 152:679–685.

Belle, S. M. and C. E. Nash. 2008. Better management practices for net-pen aquaculture. In C. S. Tucker and J. A. Hargreaves, editors, *Environmental Best Management Practices for Aquaculture*, pp. 261–330. Ames: Blackwell Publishing.

Bergheim, A. and A. Brinker. 2003. Effluent treatment for flow through systems and European Environmental Regulations. *Aquacultural Engineering* 27:61–77.

Biao, X., D. Zhuhong, and W. Xiaorong. 2004. Impact of the intensive shrimp farming on the water quality of the adjacent coastal creeks from Eastern China. *Marine Pollution Bulletin* 48:543–553.

Boaventura, R., A. M. Pedro, J. Coimbra, and E. Lencastre. 1997. Trout farm effluents: Characterization and impact on the receiving streams. *Environmental Pollution* 95:379–387.

Boyd, C. A., C. E. Boyd, A. A. McNevin, and D. B. Rouse. 2006a. Salt discharge from an inland farm for marine shrimp in Alabama. *Journal of the World Aquaculture Society* 37:345–355.

Boyd, C. A., P. Pengseng, and C. E. Boyd. 2008. New nitrogen fertilization recommendations for bluegill ponds in the southeastern United States. *North American Journal of Aquaculture* 70:308–313.

Boyd, C. E. 1976. Nitrogen fertilizer effects on production of tilapia in ponds fertilized with phosphorus and potassium. *Aquaculture* 7:385–390.

Boyd, C. E. 1978. Effluents from catfish ponds during fish harvest. *Journal of Environmental Quality* 7:59–63.

Boyd, C. E. 1982. Hydrology of small experimental fish ponds at Auburn, Alabama. *Transactions of the American Fisheries Society* 111:638–644.

Boyd, C. E. 1985. Chemical budgets for channel catfish ponds. *Transactions of the American Fisheries Society* 114:291–298.

Boyd, C. E. 1995. *Bottom Soils, Sediment, and Pond Aquaculture*. New York: Chapman and Hall.

Boyd, C. E. 2003. Guidelines for aquaculture effluent management at the farm-level. *Aquaculture* 226:101–112.

Boyd, C. E. 2008. Better management practices in marine shrimp aquaculture. In C. S. Tucker and J. A. Hargreaves, editors, *Environmental Best Management Practices for Aquaculture*, pp. 227–259. Ames: Blackwell Publishing.

Boyd, C. E. 2009. Estimating mechanical aeration requirement in shrimp ponds from oxygen demand of feed. In C. L. Browdy and D. E. Jory, editors, *The Rising Tide, Proceedings of the Special Session on Suitable Shrimp Farming*, pp. 230–234. Baton Rouge: World Aquaculture Society.

Boyd, C. E. and A. Gross. 1999. Biochemical oxygen demand in channel catfish *Ictalurus punctatus* pond waters. *Journal of the World Aquaculture Society* 30:349–356.

Boyd, C. E. and L. Li. 2012. Environmental issues in pond fertilization. In C. C. Mischke, editor, *Aquaculture Pond Fertilization: Impacts of Nutrient Input on Production*, pp. 65–72. New York: Wiley Blackwell.

Boyd, C. E. and L. Massaut. 1999. Risks associated with use of chemicals in pond aquaculture. *Aquacultural Engineering* 20:113–132.

Boyd, C. E. and J. Queiroz. 2001a. Nitrogen and phosphorus loads by system, USEPA should consider system variables in setting new effluent rules. *Global Aquaculture Advocate* 4(6):84–86.

Boyd, C. E. and J. Queiroz. 2001b. Feasibility of retention structures, settling basins, and best management practices in effluent regulation for Alabama channel catfish farming. *Reviews in Fisheries Science* 9:43–67.

Boyd, C. E. and C. S. Tucker. 1998. *Pond Aquaculture Water Quality Management*. Boston: Kluwer Academic Publishers.

Boyd, C. E., J. Firth, and F. Rajts. 2011. Effluent and sludge management at a BAP-certified *Pangasius* farm. *Global Aquaculture Advocate* 14:40–42.

Boyd, C. E., P. Munsiri, and B. F. Hajek. 1994. Composition of sediment from intensive shrimp ponds in Thailand. *World Aquaculture* 25:53–55.

Boyd, C. E., K. Corpron, E. Bernard, and P. Penseng. 2006b. Estimates of bottom soil and effluent load of phosphorus at a semi-intensive marine shrimp farm. *Journal of the World Aquaculture Society* 37:41–47.

Boyd, C. E., A. A. McNevin, J. Clay, H. M. Johnson. 2005. Certification issues for some common aquaculture species. *Reviews in Fisheries Science* 13:231–279.

Boyd, C. E., C. W. Wood, P. Chaney, and J. F. Queiroz. 2010. Role of aquaculture pond sediments in sequestration of annual global carbon emissions. *Environmental Pollution* 158:2537–2540.

Boyd, C. E., C. Tucker, A. McNevin, K. Bostick, and J. Clay. 2007. Indicators of resource use efficiency and environmental performance in fish and crustacean aquaculture. *Reviews in Fisheries Science* 15:327–360.

Boyd, C. E., J. Queiroz, J. Lee, M. Rowan, G. N. Whitis, and A. Gross. 2000. Environmental assessment of channel catfish, *Ictalurus punctatus*, farming in Alabama. *Journal of the World Aquaculture Society* 31:511–544.

Boyd, C. E., J. F. Queiroz, G. N. Whitis, R. Hulcher, P. Oakes, J. Carlisle, D. Odom, Jr., M. M. Nelson, and W. G. Hemstreet. 2003. Best management practices for channel catfish farming in Alabama. Special Report 1, Alabama Catfish Producers, Montgomery, Alabama.

Braaten, R. O. and M. Flaherty. 2001. Salt balances of inland shrimp ponds in Thailand: implications for land and water salinization. *Environmental Conservation* 28:357–367.

Cao, L., W. Wang, Y. Yang, C. Yang, Z. Yuan, S. Xiong, and J. Diana. 2007. Environmental impact of aquaculture and countermeasures to aquaculture pollution in China. *Environmental Science and Pollution Research* 14:452–462.

Carawan, R. E., D. P. Green, F. B. Thomas, and S. D. Thomas. 1986. *Reduction in Waste Load from a Seafood Processing Plant*. Raleigh: North Carolina Agricultural Extension Service, North Carolina State University.

Carroll, M. L., S. Cochrane, R. Fieler, R. Velvin, and P. White. 2003. Organic enrichment of sediments from salmon farming in Norway: environmental factors, management practices, and monitoring techniques. *Aquaculture* 226:165–180.

Cathcart, T. P., J. W. Pote, and D. W. Rutherford. 1999. Reduction in effluent discharge and groundwater use in catfish ponds. *Aquacultural Engineering* 20:163–174.

Chamberlain, G. 2010. History of shrimp farming. In V. Alday-Sanz, editor, *The Shrimp Book*, pp. 1–34. Nottingham: Nottingham University Press.

Collins, G. 1984. Fish growth and lethality versus dissolved oxygen. In *Proceeding Specialty Conference on Environmental Engineering*, pp. 750–755. Los Angeles: ASCE.

Fornshell, G. and J. M. Hinshaw. 2008. Better management practices for flow-through aqua-culture systems. In C. S. Tucker and J. A. Hargreaves, editors, *Environmental Best Man-agement Practices for Aquaculture*, pp. 331–388. Ames: Wiley Blackwell.

Foy, R. H. and R. Rosell. 1991. Fractionation of phosphorus and nitrogen loadings from a Northern Ireland fish farm. *Aquaculture* 96:31–42.

Gross, A., C. E. Boyd, and R. T. Lovell. 1998. Phosphorus budgets for channel catfish ponds receiving diets with different phosphorus concentrations. *Journal of the World Aquaculture Society* 29:31–39.

Gross, A., C. E. Boyd, and C. W. Wood. 2000. Nitrogen transformations and balance in channel catfish ponds. *Aquacultural Engineering* 24:1–14.

Guo, L. and Z. Li. 2003. Effects of nitrogen and phosphorus from fish cage-culture on the communities of a shallow lake in middle Yangtze River basin of China. *Aquaculture* 226:201–212.

Hairston, J. E., S. Kown, J. Meetze, E. L. Norton, P. L. Dakes, V. Payne, and K. M. Rogers. 1995. Protecting water quality on Alabama farms. Alabama Soil and Water Conservation Committee, Montgomery, Alabama.

Hargreaves, J. A. 1998. Nitrogen biochemistry of aquaculture ponds. *Aquaculture* 166:181–212.

Hargreaves, J. A., C. S. Tucker, E. Thornton, and S. K. Kingsbury. 2005. Characteris-tics and sedimentation of initial effluent discharged from excavated channel catfish pond. *Aquacultural Engineering* 33:96–109.

Islam, M. S., S. Khan, and M. Tanaka. 2004. Waste loading in shrimp and fish processing effluents: Potential source of hazards to the coastal and nearshore environments. *Marine Pollution Bulletin* 49:103–110.

Kutti, T., P. K. Hansen, A. Ervik, T. Hoisaeter, and P. Johannessen. 2007. Effects of organic effluents from a salmon farm on a fjord system. II. Temporal and spatial patterns in infauna community composition. *Aquaculture* 262:355–366.

Liu, Y., G. Villalba, R. U. Ayres, and H. Schroder. 2008. Global phosphorus flows and envi-ronmental impacts from a consumption perspective. *Journal of Industrial Ecology* 12:229–247.

Loch, D. D., J. L. West, and D. G. Perimutter. 1996. The effect of trout farm effluent on the taxa richness of benthic macroinvertebrates. *Aquaculture* 147:37–55.

MacMillan, J. R., T. Huddleston, M. Woolley, and K. Fothergill. 2003. Best management practice development to minimize environmental impact from large flow-through trout farms. *Aquaculture* 226:91–99

Masuda, K. and C. E. Boyd. 1994. Phosphorus fractions in soil and water of aquaculture ponds built on clayey, Ultisols at Auburn, Alabama. *Journal of the World Aquaculture Society* 25:379–395.

Nordvarg, L. and T. Johansson. 2002. The effects of fish farm effluents on the water quality in the Aland archipelago, Baltic Sea. *Aquacultural Engineering* 25:253–279.

Ozbay, G. and C. E. Boyd. 2003. Particle size fractions in pond effluents. *World Aquaculture* 34:56–59.

Páez-Osuna, F., S. R. Guerrero-Galván, and A. C. Ruiz-Fernández. 1998. The environmen-tal impact of shrimp aquaculture and the coastal pollution in Mexico. *Marine Pollution Bulletin* 36:65–75.

Pengseng, P. 2007. Resource use and waste production at a semi-intensive black tiger prawn *Penaeus monodon* farm. Ph.D. diss., Auburn University, Alabama.

Pine, H. J. and C. E. Boyd. 2011. Stream salinization by inland brackish-water aquaculture. *North American Journal of Aquaculture* 73:107–113.

Prapaiwong, N. 2011. Water quality in inland ponds for low-salinity culture of Pacific white shrimp *Litopenaeus vannamei*. Ph.D. diss., Auburn University, Alabama.

Prapaiwong, N. and C. E. Boyd. 2012. Effluent volume and pollutant loads at an inland, low-salinity, shrimp farm in Alabama. *Aquacultural Engineering* 48:1–5.

Prasertsan, P., S. Jung, and K. A. Buckle. 1994. Anaerobic filter treatment of fishery wastewater. *World Journal of Microbiology and Biotechnology* 10:11–13.

Schlesinger, W. H. 2009. On the fate of anthropogenic nitrogen. *Proceedings of the National Academy of Sciences* 106:203–208.

Schwartz, M. and C. E. Boyd. 1994. Effluent quality during harvest of channel catfish from watershed ponds. *Progressive Fish-Culturist* 56:25–32.

Selong, J. H. and L. A. Helfrich. 1998. Impacts of trout culture effluent on water quality and biotic communities in Virginia headwater streams. *Progressive Fish-Culturist* 60:247–262.

Seok, K., S. Leonard, C. E. Boyd, and M. Schwartz. 1995. Water quality in annually drained and undrained channel catfish ponds over a three-year period. *Progressive Fish-Culturist* 57:52–58.

Shumway, S. E., C. Davis, R. Downey, R. Karney, J. Kraeuter, J. Parsons, R. Rheault, and G. Wikfors. 2003. Shellfish aquaculture—in praise of sustainable economies and environments. *World Aquaculture* 34:8–10.

Silapajarn, O. and C. E. Boyd. 2005. Effects of channel catfish farming on water quality and flow in an Alabama stream. *Reviews in Fisheries Science* 13:109–140.

Sipauaba-Tavares, L. H. and C. E. Boyd. 2003. Possible effects of sodium chloride on quality of effluents from Alabama channel catfish ponds. *Journal of the World Aquaculture Society* 34:217–222.

Sirianuntapiboon, S. and N. Nimnu. 1999. Management of water consumption and wastewater of seafood processing industries in Thailand. *Journal of Science and Technology* 6:158–167.

Soderberg, R. W. 2007. Efficiency of trout raceway quiescent zones in controlling suspended solids. *North American Journal of Aquaculture* 69:275–280.

Sohsalam, P., A. J. Englande, and S. Sirianuntapiboon. 2008. Seafood wastewater treatment in constructed wetland: Tropical case. *Bioresource Technology* 99:1218–1224.

Soongsawang, S. and C. E. Boyd. 2012. Effects of effluents from a fisheries research station on stream water quality. *North American Journal of Aquaculture* 74:73–79.

Sridang, P. C., A. Pottier, C. Wisniewski, and A. Grasmick. 2008. Performance and microbial surveying in submerged membrane bioreactor for seafood processing wastewater treatment. *Journal of Membrane Science* 317:43–49.

Steeby, J. A., J. A. Hargreaves, C. S. Tucker, and S. Kingsbury. 2004. Accumulation, organic carbon and dry matter concentration of sediment in commercial channel catfish ponds. *Aquacultural Engineering* 30:115–126.

Stephens, W. W. and J. L. Farris. 2004. Instream community assessment of aquaculture effluents. *Aquaculture* 231:149–162.

Szabo, P. 1994. Quality of effluent from earthen fish ponds in Hungary. *Journal of Applied Ichthyology* 10:326–334.

Tchobanoglous, G., F. L. Burton, and H. D. Stensel. 2003. *Wastewater Engineering*. New York: McGraw Hill.

Teichert-Coddington, D. R., D. B. Rouse, A. Potts, and C. E. Boyd. 1999. Treatment of harvest discharge from intensive shrimp ponds by settling. *Aquacultural Engineering* 19:147–161.

Timmons, M. B, J. E. Ebeling, F. W. Wheaton, S. T. Summerfelt, and B. J. Vinci. 2002. *Recirculation Aquaculture Systems*, 2nd ed. Ithaca: Cayuga Aqua Ventures.

Torrans, E. L. 2005. Effect of oxygen management on culture performance of channel catfish in earthen ponds. *North American Journal of Aquaculture* 67:275–288.

Torrans, E. L. 2008. Production responses of channel catfish to minimum daily dissolved oxygen concentration in earthen ponds. *North American Journal of Aquaculture* 70:371–381.

Wang, J. Q, H. Lui, H. Po, and L. Fan. 1997. Influence of salinity on food consumption, growth, and energy conversion efficiency of common carp (*Cyprinus carpio*) fingerlings. *Aquaculture* 148:115–124.

Weiler, R. R. 1979. Rate loss of ammonia from water to the atmosphere. *Journal Fisheries Research Board of Canada* 36:685–689.

Wu, R. S. S., K. S. Lam, D. W. MacKay, T. C. Lau, and V. Yam. 1994. Impact of marine fish farming on water quality and bottom sediment: A case study in the sub-tropical environment. *Marine Environmental Research* 38:115–145.

Yoo, K. H. and C. E. Boyd. 1994. *Hydrology and Water Supply for Aquaculture*. New York: Chapman and Hall.

Yuvanatemiya, V., C. E. Boyd, and P. Thavipoke. 2011. Pond bottom management at commercial shrimp farms in Chantaburi Province, Thailand. *Journal of the World Aquaculture Society* 42:618–632.

Chapter 11

Biodiversity

Biodiversity is a relatively new term, but a much older concept. The concept originally was referred to as biological diversity, but in the mid-1980s, the expression was shortened to biodiversity (http://plato.stanford.edu/entries/biodiversity/). The original concept of biological diversity or species diversity was related to species richness; a habitat with a large number of species was said to have greater species diversity than a habitat with a fewer number of species. The relative abundance of species proved to be a better estimate of species diversity than the simple enumeration of the total number of species present. Thus species diversity can be used to explain how the individual organisms in a habitat are distributed among the species present. An index of species diversity usually is computed for a particular group of organisms. For example, Margalef (1958) proposed the following equation for estimating plankton diversity:

$$\text{Diversity} = \frac{\text{Number of species} - 1}{\ln(\text{total number of plankton organisms})}. \tag{11.1}$$

Many other equations for estimating species diversity have been developed because of differences in the types of communities within and between ecosystems and in the interests of investigators.

Odum (1971) discussed species diversity and related it to the structure and function of ecosystems. Ecosystems of high species diversity were viewed by Odum as being more stable than those of lower species diversity. Removing a species from a community of high species diversity would not likely have a large influence on the function of the ecosystem because another species capable of conducting the same function usually would be present. As species diversity decreases the likelihood of a negative influence on the ecosystem function following the removal of a single species would increase. It has been observed in many studies that nutrient

Aquaculture, Resource Use, and the Environment, First Edition. Claude E. Boyd and Aaron A. McNevin.
© 2015 John Wiley & Sons, Inc. Published 2015 by John Wiley & Sons, Inc.

enrichment of ecosystems leads to a low species diversity, but to a high population density of the species present. Disturbed environments also may have low diversity because many species cannot tolerate the conditions created by the disturbance.

The concept of diversity has been expanded to include the diversity of ecosystems, communities, and habitats. Ecosystems have unique geological, edaphic, hydrologic, and climatic regimes that affect the types and abundance of species living in them. For example, an aquatic ecosystem is much different from a terrestrial ecosystem, but a nutrient-poor aquatic ecosystem also differs greatly from a nutrient-rich one. Moreover, the species of organisms in an ecosystem all have genetic diversity, and the interaction of ecosystem diversity and genetic diversity result in gradual changes in the characteristics of organisms over time. The new concept of biodiversity is so broad that it is almost impossible to find a suitable, single index of biodiversity. As a result it is still popular to count and enumerate species as a way of assessing changes in biodiversity.

It is generally accepted that tropical ecosystems are more diverse than ecosystems in temperate and cold climates. Moreover, invertebrate, plankton, and bacterial communities tend to be more diverse than higher plant and vertebrate communities.

The major factors that work to lessen biodiversity are as follows: deforestation and other land use changes; overexploitation of species by hunting or fishing; introduction of exotic species that may be highly aggressive, competitive, or predaceous and for which there are no natural controls; introduction of new diseases with which one or more organisms have no resistance; pollution with organic matter, nutrients, or toxins; modification of gene pools by introduction of organisms that are genetically different from local organisms. Aquaculture can be a source of all of the major factors contributing to lower biodiversity. This chapter will provide a discussion of the influence of aquaculture on biodiversity, and make some comparisons of these effects to those of agriculture, capture fisheries, and a few other human activities.

Major concerns about biodiversity

The eNGO statement about aquaculture sustainability to the United Nations Commission on Sustainable Development (UNCSD) in May 1996 (http://darwin. bio.uci.edu/~sustain/_shrimpecos/declare2.html) used the term biodiversity only once and specifically in relation to the collection of larval organisms for use in stocking ponds and other culture facilities. The Choluteca Declaration of October 1996 by eNGOs about unsustainable shrimp farming used the term biodiversity twice and the expression diversity of species once in referring to a lack of planning for coastal zone development, capture of wild shrimp postlarvae in ponds, and the use of harmful substances (http://darwin.bio.uci.edu/~sustain/_shrimpecos/declare1.html). The meaning that has evolved for the concept of biodiversity, however, is broad and covers almost all aspects of the environment and its interactions with life forms and their genetic codes (http://plato.stanford.edu/entries/biodiversity/). Therefore biodiversity was alluded to in 10 of the 16 statements to the UNCSD and in 11 of the 18 demands listed in the Choluteca Declaration.

The most thorough lists of the major negative impacts on biodiversity likely to result from aquaculture are those of Beardmore et al. (1997) and Diana (2009). Beardmore and his collaborators gave the following major impacts:

- Habitat destruction for constructing ponds,
- Water pollution by farm effluents,
- Effects of antibiotics and other chemicals on local microfauna and macrofauna,
- Intensive collection of wild seeds,
- Competition with endemic fauna by exotic species that escape from culture units,
- Introduction of pathogens and parasites to natural fisheries stocks via farm stocks,
- Genetic effects of escaped farm animals on the local fauna.

Diana's list was similar, but it is more refined because greater clarity with respect to effects had been achieved during the decade since Beardmore et al. made their list. Diana's list follows:

- Hazards of escaped aquatic crops as invasive species,
- Eutrophication and associated changes in flora and fauna in waters receiving effluents from aquaculture facilities,
- Conversion of mangroves and wetlands to ponds,
- Excessive freshwater use,
- Exploitation of wild fish stocks for fish meal to use in feeds or live food for culture species,
- Transfer of disease and parasites from farmed animals to wild stock,
- Genetic alteration of wild stock resulting from escaped farm stock,
- Destruction of predators that feed on farm stock,
- Effects of antibiotics and hormones used in aquaculture facilities on the flora and fauna of water bodies receiving farm effluents.

Aquaculture effluents and biodiversity

The main biodiversity issue with effluents relates to eutrophication; Chapter 10 provides information on the concentrations of water quality variables in aquaculture effluents. That chapter also discusses the quantities of pollutants released by aquaculture and the potential of effluents to cause eutrophication, turbidity, sedimentation, toxicity, etc. in receiving water bodies. There are few studies documenting actual instances of eutrophication resulting from aquaculture effluents. Of course several studies have reported increases in eutrophication in coastal waters in aquaculture areas (Páez-Osuna et al. 1998, 2003; Eng et al. 1989) and near cages in lakes (Guo and Li 2003), but aquaculture was only one of several sources of effluents impacting the water bodies. This is the typical situation and studies have not clearly revealed the relative contribution of the different sources of pollutants on eutrophication of receiving waters.

One study (Thomas et al. 2010) stated that eutrophication of the New Caledonia Lagoon was caused by wastewater discharges from shrimp farms. The study showed a shift in picoplankton with the disappearance of some species and an increase in other species. Although this study actually does not clearly show that aquaculture was the only source of the nutrients causing eutrophication, other sources were not considered important.

A study in Brazil compared two reservoirs—one impacted by cage aquaculture and one without cage aquaculture (Pistori et al. 2010). The chlorophyll *a* concentration was higher in the reservoir with cage aquaculture than in the other reservoir. Also growth trials with the floating macrophyte *Salvinia molesta* revealed that the plant grew much faster in the reservoir with cages than in the other. Certainly this study revealed that eutrophication in the reservoir with cages resulted from aquaculture, and eutrophication typically leads to a decline in biodiversity.

A notable case of the impacts of aquaculture on water quality in a natural water body is cage culture of *Oreochromis niloticus*, *Chanos chanos*, and *Caranx ignobilis* in Lake Taal, Philippines. In 1993 cage production in the lake was approximately 4000 t but it increased to nearly 70 000 t by 2008 (http://pemsea.org/eascongress/special-events/seminar-on-ecosystem-approach-to-conservation-and-fisheries-management/presentation_ea_macas.pdf). The increase in production was linked to numerous fish kills including a massive kill in 2006 in which 80% of cage operations were affected (Vista et al. 2006). Biodiversity in the lake also diminished dramatically. In 1927 the fish inventory of Lake Taal recorded 76 migratory and many endemic species. In 1975 the inventory decreased to 15 migratory and only 4 endemic species. In 2003 the catch of *Sardinella tawilis*, the most popular endemic fish species from Lake Taal, reportedly dropped by 80% (http://www.philstar.com/nation/773730/kalikasan-writ-sought-taal-lake).

There is considerably more information of the effects of aquaculture on benthic diversity in receiving water bodies. Many studies have been conducted on cage culture, and changes in sediment chemical and physical composition resulting in alterations of benthic communities near cages typically have been observed (Islam 2005; Kutti et al. 2007). For example, the study by Kutti et al. (2007) showed that impacts on benthos were restricted to a distance of about 250 m beyond the cage farm and the impacts were greatest when the farm was at maximum production. The zone near the farm was dominated by seven benthic species, but at distances of 550–900 m from the farm, there were 20–40 species of benthic organisms. Of course the total biomass of benthos was much greater near the farm.

Studies of streams receiving effluents from trout farms also have revealed changes in benthic communities (Brown and King 1995; Loch et al. 1996; Amirkolaie 2008; Živić et al. 2009; Selong and Helfrich 1998). The results of the study by Selong and Helfrich (1998) are typical of the findings of other studies. Periphyton increased for about 400 m downstream, and macroinvertebrate richness and abundance of sensitive taxa (mayflies, stoneflies, and caddisflies) were also reduced over this distance. The decrease in sensitive species was countered by an increase in pollution-tolerant isopods and gastropods.

A study of in-stream communities in streams receiving effluents from channel catfish farms in Arkansas in the United States (Stephens and Farris 2004) revealed little difference in benthic communities upstream and downstream of outfalls. A similar study of streams receiving aquaculture farm effluents in Ghana also showed no or relatively small differences in fish species and benthic communities upstream and downstream of farm outfalls (Ansah et al. 2012). The failure to find much influence of farm discharges in these two studies as compared to streams receiving trout farm effluent may be related to a greater sensitivity to pollution by organisms in cold water habitats than in warm water habitats. However, a study of cage culture in subtropical Hong Kong (Gao et al. 2005) showed that the diversity of benthic macrofauna was significantly less and benthic community structure was different when sampling sites near cages were compared to the control sampling locations.

A study conducted in Brazil (Sousa et al. 2006) considered the impact of shrimp farm effluent on bacterial communities in mangrove habitats. This investigation found no difference in abundance of species richness of microbial communities associated with shrimp farm effluent.

In summary, aquaculture effluents have the potential to contribute to eutrophication and the loss of biodiversity. But the contribution of aquaculture effluents relative to agricultural, municipal, and industrial wastes has not been assessed. There are documented cases of decreases in benthic biodiversity resulting solely from aquaculture effluents but these effects have been localized to areas near farms.

Many methods are used to eradicate disease vectors, wild fish, and predaceous insects from ponds, and herbicides, algicides, and other chemicals are used in ponds for pest and disease control as discussed in Chapter 9. There is no evidence that residues of these treatments are at high-enough concentration in farm effluents to have toxic effects on the flora and fauna of receiving water bodies. Antibiotic residues may be contained in aquaculture effluents but there is not enough information available to assess the actual influences of these residues on biodiversity. However, the substances certainly have the potential to reduce biodiversity and to lead to the development of antibiotic resistance in microorganisms.

Effects of land and water use on biodiversity

The main issues related to land use by aquaculture are conversion of agricultural land to ponds, construction of ponds in mangrove forests and other wetlands, and ecological disturbances in terrestrial habitats resulting from disposal of pond sediment of solid wastes from farms (Chapter 5). The amount of land used by aquaculture is insignificant compared to land use in agriculture, and the loss of mangroves and other wetlands by conversion to ponds also is small compared to conversion for other uses. There are no data on the extent of improper disposal of sediments and solid wastes from aquaculture facilities, but the impacts are near aquaculture farms and the affected areas are relatively small.

The influence on biodiversity of land use changes by aquaculture mainly is the change of land surface covered by vegetation to a water surface. The area converted to a pond still functions as an aquatic ecosystem, but with characteristics much different than most natural water bodies. In short, ponds are eutrophic and of comparatively low biodiversity when compared to most natural water bodies. The area affected by aquaculture, however, is relatively small and the impacts on biodiversity are relatively small on a global basis. On a local basis the development of aquaculture projects usually causes a decline in biodiversity in the area where they are sited.

Withdrawal of freshwater for use in aquaculture (Chapter 6) may sometimes conflict with other water uses but aquaculture seldom is responsible for reducing the amount of freshwater available for natural ecological communities. Ponds capture runoff but they overflow; thus their main influence is to reduce the peak of downstream hydrographs rather than reducing the overall volume of downstream flow (Schoof and Gander 1982; Boyd et al. 2009). There is a possibility that diversion of water from some streams for trout farming may cause a serious reduction in downstream flow where water exiting the farms is released into a different drainage (Boyd et al. 2005). Of course water pollution and salination resulting from aquaculture (see Chapter 10) could degrade the quality of water for other uses.

Escapes of farm stock

At almost any aquaculture facility some proportion of the farm stock will eventually escape into natural waters—escapes can result from human error, infrastructure failure, overflow after unpredictably large storms or floods, live escapes from predators, tsunamis and hurricane-driven storm surges, earthquakes, etc. despite the efforts of aquaculturists to avoid escapes.

The magnitude of losses of salmon by escapes from cages was discussed in a review by Gross (1998). In an instance in Norway 700 000 salmon escaped from cages into the Atlantic Ocean during a single storm. In another example, 101 000 and 360 000 salmon escaped from cages into the Puget Sound, Washington State (USA). The assessment by the salmon industry in British Columbia, Canada that about 2% of fish stocked in cages escape each year gives some perspective on the quantity of fish that may be lost annually from cage culture globally.

A report of huge losses of shrimp from ponds was provided by managers of farms located near Choluteca, Honduras. During Hurricane Mitch in 1998 heavy rainfall lead to flood levels that were 0.5 to 1.0 m above the tops on pond embankments and hundreds of tons of shrimp escaped. Shrimp and fish also were lost from coastal ponds impacted by the huge tsunami that struck Indonesia, Thailand, and other Asian countries in December 2004. Of course normal losses of fish and shrimp from ponds usually are linked to small numbers of animals that escape during harvest. Instances of fish and other species escaping from ponds during floods also have been related to the senior author by farmers in several countries.

Animals that escape from aquaculture facilities may be exotic species not endemic to the area or they may have a different genetic makeup than endemic shrimp resulting from selective breeding for stock improvement or genetic manipulation by gene transfer.

Escapes of farm animals, aquatic plants, and animals from aquaria and zoological parks, and ornamental plants from their intended place of use are commonplace. The fact that aquaculture species escape should not be surprising but there is no way to totally prevent these escapes.

Exotic species

Introduction of animals and plants into areas beyond their natural range also is nothing new; the first recorded intentional introductions were the movement of trees, vines, figs, and roses from central Asia Minor into Mesopotamia about 2500 BC by the Sumerians (Ryerson 1933). Such introductions have continued until today and include almost all taxonomic categories. Many species originally found on one continent are now on all continents.

The more adaptable an organism the more likely it is to become established when translocated to a place outside its natural range. Some introductions were nonintentional resulting from hitchhikers in the world's human and cargo transportation system. Many introductions were intentional and made or sanctioned by governments, while others were made by individuals. The introductions were made for many reasons but most were made to provide new agricultural crops and livestock, novel or desirable ornamental plants and animals, predator or pest control, or game animals. The literature on introductions of non-native species is huge because many introductions have been made and many positive and negative impacts have resulted. In the United States alone there are currently about 50 000 non-native species (Pimentel et al. 2000). These introductions make up about 98% of the plants and animals used in US agriculture. Introduced species also are important as house and landscape plants, in forestry, for pets, and to serve as sportfish and game animals. Other species have become integrated into natural ecosystem structure and function quite harmoniously (Richardson et al. 2000; Sagoff 2005). However, other species have become major pests resulting in serious losses to agriculture, homeowners, businesses, and municipalities as well as causing ecological perturbations. The economic damage caused by introduced species is huge and difficult to assess because it includes damages that result in immediate economic losses and ecological damage that is more insidious and difficult to quantify. Pimentel et al. (2000) placed the total damage caused by nonindigenous species in the United States at $137 billion annually. The situation with non-native species is similar in most countries, and reviews of the problems with invasive species abound, for example, Yan et al. (2001); Weber and Gut (2004); Lonsdale (1994); and Weber (1997).

Introduction of exotic species also is a common practice in aquaculture. Nearly 200 aquatic species are farmed (Avault 1996) but about 30 to 40 species make up most of the world's aquaculture production. The most important species have been introduced throughout the regions of the world suitable for their culture. According to Bartley and Casal (1998) aquaculture development has been responsible for most records of fish introductions and their establishment. Naylor et al. (2001) stated that most aquatic species cultured in the United States are not native to their farm sites. They cite examples of seaweed species introduced into Hawaii, carp from Asia that have become common in the Mississippi River Basin, salmon farm escapes that have become common along the coastline from Maine in the United States to New Brunswick in Canada, and molluscan shellfish from Asia endemic in many coastal areas of the United States.

There recently has been a massive conversion of the shrimp farming industry in Asia from the native *Penaeus monodon* to *Litopenaeus vannamei* introduced from South and Central America (Senanan et al. 2009). To illustrate Thailand formerly produced *P. monodon* almost exclusively, but presently nearly all of the production has shifted to *P. vannamei*. Tilapia have been intentionally dispersed throughout the subtropical and tropical world—mainly as food fish—but also for aquatic weed control, baitfish, and aquarium fish (Canonico et al. 2005).

The common opinion that aquaculture has played the dominant role in introducing exotic, aquatic species is a fallacy. The aquarium trade has been a major source of introductions (Tlusty 2002). Courtenay et al. (1974) reported that 38 species and several hybrids of exotic fish had been found in fresh and brackish water in Florida, and 20 species and five hybrids were reproducing. The aquarium trade was the major source of these exotics.

Sportfishing also is a major source of fish introductions in the United States. Crossman and Cudmore (1999) found that 57% of 214 intentional fish species introductions in the United States were for sportfish or forage fish introduced as food for sportfish. Also they reported that 61% of 901 introductions of fish species in a US state where the species had not previously occurred were for sportfish or their forage species. The reasons for the other introductions to states were bait species 7%, aquarium trade 10%, aquaculture 5%, and miscellaneous 14%. According to Rahel (2004) federal agencies authorized most of the US introductions in the past, but many introductions are now by unauthorized action of the public—mostly for sportfishing and the aquarium trade.

Introduction of non-native species is common in many countries as a few examples will show. The inland waters of Australia have been colonized by 20 introduced species from other continents (Arthington 1991). Over 160 invasive species of aquatic organisms have been introduced into Ontario (Canada). Some of these introductions have resulted from fish stocking programs, aquaculture, bait for sportfishing, and aquarium and ornamental pond use. Others have been caused by recreational boating, canals, and other water diversions, and commercial shipping (mainly in ballast water of ships) (Kerr et al. 2005). Kumar (2000) determined that 6 species of sportfish, 14 species of food fish for aquaculture, 2 species of fish that eat mosquito larvae, and 228 species of ornamental fish had been introduced into India. In Norway 11 of 43 self-sustaining fish species in freshwaters were introduced—most were originally introduced for sportfishing (Hesthagen and Sandlund 2007). Other countries in Europe also have had many species introductions (Strayer 2010). In the state of Minas Gerais, Brazil, fish have been introduced by reservoir stocking programs, sportfishing, control of disease vectors, aquarium trade, and aquaculture (Alves et al. 2007). About 40% of the fish found in the waters of Minas Gerais state are non-native species.

Marine organisms—especially organisms with planktonic and sessile stages—have been spread throughout the world by international shipping (Minchin et al. 2005). Aquatic organisms may be transported as fouling or boring organisms on the hulls of ships. Ship hulls must be submersed enough to provide stability during changes in course or in heavy seas. Ballast water to adjust buoyance and depth of hull submergence is pumped into and out of tanks in the hull. Ballast water contains planktonic organisms, small fish, and other animals that are often taken on in one continent and discharged in another. Although not mentioned as frequently bacteria and viruses are particularly susceptible to spread by international shipping (Drake et al. 2002, 2007). A good example of the spread of aquatic species by international shipping is afforded by European shore crabs *Carcinus maenas* and *Carcinus aestuarii*. These organisms spread during the period 1817–1993 from their native range along the western coasts of Europe and northern Africa to Australia, South Africa, Madagascar, south Asia, Japan, Brazil, Hawaii, and both the Atlantic and Pacific coasts of North America (Carlton and Cohen 2003).

The effects of introduction of exotic aquatic organisms—including aquaculture species—are difficult to assess. The most likely effect of molluscs that feed on plankton is disruption of the food web by reducing primary productivity, although there are examples of the shellfish industry introducing novel diseases with culture stock (Andrews 1996; Ewart and Ford 1993). Species of freshwater crustaceans may consume macrophytes and benthic animals causing disruption of food webs, while exotic fish often cause large reductions in the abundance of their prey organisms including native fish, also negatively affecting food webs. All types of organisms can bring new diseases that may infect native organisms (Strayer 2010). Exotic aquatic plants may produce dense infestations that can affect water movement, lessen primary productivity by native species, and change the structural aspects of aquatic habitats.

There are many case studies on impacts of introductions of aquatic species. Carp introductions in the United States have been associated with increased turbidity of water, loss of macrophytes, and a decline in the abundance of sportfish species (Cahn 1929). Grass carp can greatly reduce the amount and change the species composition of macrophytes in water bodies (Bain 1993; Pipalova 2006). Introduction of the Nile perch (*Lates nilotica*) into Lake Victoria in Africa resulted in the elimination of many species of cichlids (Lowe-McConnell 1993).

The most famous cases of exotic fish being introduced involve the tilapias. They have been introduced throughout the world, and where climate is warm enough to allow them to survive year around, they have become endemic in freshwater and brackish water environments. Because of their tolerance to a wide range of environmental factors, high fecundity, aggressiveness, and omnivorous nature, tilapia are highly competitive and cause a decline in the number of other fish species and their abundance (Canonico et al. 2005). There are many examples of introduced tilapia becoming the dominant species in aquatic ecosystems. In spite of the widespread introductions and spread of tilapia, there are still areas in most countries that tilapia have not colonized.

Genetically modified organisms

Selective breeding of plant and animal species has been used for many years to improve stock, and hybrids have been produced that have superior characteristics. In recent years genetically modified organisms (GMOs) have been produced by incorporating a gene from one species into another species to express a desired trait. This procedure often is referred to as gene splicing, and the resulting organisms are

Table 11.1 Purposes for production of interspecific hybrids of fish.

Increase growth rate	Obtain sexual dimorphism
Transfer desirable traits between species	Increase harvestability
Combine desirable traits of two species	Increase environmental tolerances
Obtain sterile fish or mono-sex offspring	Increase overall hardiness in culture species

Source: Bartley et al. (2001).

often said to be genetically modified, genetically engineered, or transgenic. The first food produced by this technique was the Flavr Savr Tomato that remains firm when it ripens (http://ohioline.osu.edu/hyg-fact/5000/5058.html). There has been much research on GMOs in agriculture, and particularly with respect to corn, soybean, cotton, alfalfa, sugar beet, and several other major crops. The improvements in traits include insect and virus resistance, herbicide resistance, and improved nutrition that result in greater yields (Sachs 2012).

According to Sachs (2012) the first GMO organisms for agriculture were produced in 1983. Field testing began in 1987 and commercialization was initiated in 1996. About 102 million ha of transgenic crops were planted in 22 countries by 10.3 million farmers in 2006 (http://www.ornl.gov/sci/techresources/Human_Genome/elsi/gmfood.shtml), and according to Roberts (2007) 55% of soybeans worldwide and 80% in the United States were transgenic. By 2010 there were 148 million ha of transgenic crops in 29 countries (Sachs 2012). Transgenic grains are used in human food and animal feed. Thus aquaculture feed may contain plant meals from transgenic crops.

Inter-specific hybrids of fish have been used for various purposes (Table 11.1). According to Bartley et al. (2001) some hybrids are an important contribution to aquaculture production, and hybrid striped bass (USA), hybrid clarid catfish (Thailand), hybrid characids (Venezuela), and hybrid tilapia (Israel) were cited as examples. Hybrid ictalurid catfish also are widely used in the US catfish farming industry (Dunham et al. 2008). Hybrids are genetically modified in a sense because genes originating from different species are combined in a single organism. However, by international definition GMO organisms are the result of recombinant DNA technology, and hybrids are not GMOs by this definition (Bartley et al. 2001).

Gene transfer technology also has been applied to aquaculture animals. Gene transfer was first accomplished with goldfish (Zhu et al. 1985). Within 20 years gene transfer had been accomplished for more than 30 species of fish including carps, tilapia, catfish, and salmonids (Dunham 1999; Devlin et al. 2006). Research efforts have produced both transgenic bivalve molluscs (Kapuscinski 2005) and penaeid shrimps (Zheng et al. 2010). No references could be found to the extent of production of transgenic species in aquaculture.

Farm animals that escape into natural water bodies typically are genetically different from wild animals of the same species because of genetic improvement through selective breeding, hybridization, and possibly even genetic modification. Breeding of farm animals with natural populations can affect the gene pools of natural populations. This has been demonstrated most clearly with salmon for which in many areas the farmed population outnumbers the wild population. Studies reviewed by Gross

(1998) revealed that: escaped farm fish contribute 44% of total salmon genes in the Vosso River and 21% in the Imsa River of Norway; farmed salmon have shifted the gene pool of a wild Atlantic salmon population in Ireland; about 55% of salmon redds in the Magaguadavic River, New Brunswick, Canada were made by escapees. Atlantic salmon used on farms in the Pacific coast of the United States and Canada have escaped and their establishment is probable.

Predator control

Predator control and its equivalent, the control of herbivores that feed on crop plants, are common in all kinds of agriculture. Wild animals such as wolves, foxes, coyotes, wild dogs, and other predators cause substantial losses of livestock worldwide (Eldridge et al. 2002; Dorrance and Roy 1976; Greentree et al. 2000; Norbury and McGlinchy 1996). Various methods—shooting, trapping, poisoning, guard dogs, etc.—are used in attempts to control the predators. Large grazing animals such as deer can cause damage to crops in many parts of the world (Hygnstrom and Craven 1988). Public hunting and controlled shooting is a recognized control method but other techniques such as live removal, scare devices, repellents and fencing are encouraged. In rural areas of Africa, elephants cause considerable damage to crops (O'Connell-Rodwell et al. 2000) and predators, especially lions, destroy much livestock (Butler 2000; Ogada et al. 2003). There is much conflict between farmers and wildlife in rural Africa that results in the destruction of wild animal species whose populations already are in global decline. Thus much effort is being devoted to nonlethal control methods.

The control of small animals such as rodents and insect pests commonly rely upon the use of poisons and especially chemical pesticides that can affect many species other than the target species. Biological control is the preferred method of eradication, but in some cases, the biological control organism may adversely affect other species (Strong and Pemberton 2000). Integrated pest management that combines biological, chemical, and mechanical based on knowledge of the pest and its environment often may be superior to biological control alone.

Aquaculture has wildlife conflicts similar to those presented above for agriculture (Beveridge 2001). Predators that feed on aquatic animals often are attracted to aquaculture facilities because of the abundance of prey. Birds are the most common predators in most types of aquaculture. According to a summary by Pillay (2004) a single pelican may eat 1–3 t of fish annually and a heron may consume 100 kg fish/year. A pair of nesting cormorants will catch nearly 500 kg of fish in 1 year (du Plessis 1957). The most destructive predators in aquaculture worldwide were considered by Pillay (2004) to be cormorants, fish eagles, herons, and kingfishers. Cormorants, anhingas, pelicans, wood storks, herons, egrets, ibis, and waterfowl such as scarps, mergansers, mallards, gulls, terns, ospreys, great horned owls, bald eagles, kingfishers, and common grackle are listed as avian predators of aquaculture species in the United States (Barras 2007). However, cormorants, pelicans, and various wading birds are considered the most serious avian predators in the United States (Littauer et al. 1997).

Birds also can transmit fish parasites and virus diseases of shrimp. Thus bird control has become an issue in biosecurity at some aquaculture facilities and especially at marine shrimp farms. Some species of mammals such as freshwater otters (Kloskowski 2005), sea lions, sea otters, and seals (Nash et al. 2000) prey on fish and other aquaculture species. Even bullfrogs and alligators may be serious predators at aquaculture facilities (Avault 1996). Molluscs also have predators such as oyster drills, starfish, and crabs in addition to fish and ducks (Avault 1996). Insects and small, wild fish are predacious on fry and larvae in ponds.

Many birds, mammals, and alligators are killed—legally or illegally (see Chapter 12)—by irritated farmers. It is common practice to use various chemicals to control insects and wild fish in ponds before stocking.

There are few records of the quantities of birds and mammals killed at aquaculture facilities. Some countries have regulations on lethal control of predators—especially birds—causing farmers to use nonlethal control except where permits are issued to kill some animals when losses to predators are excessively great (see Chapter 12). Nevertheless, in many countries aquaculturists destroy predators indiscriminately on a regular basis. Lacking data on the extent of lethal predator control it is impossible to assess the overall influence of this practice on biodiversity. Nevertheless in countries that impose regulations, predator control is not thought to negatively influence biodiversity.

Impingement of aquatic animals

Facilities such as power plants, desalination plants, and other industries that take in large amounts of water to use in cooling or processing have screens to avoid the entrance of animals along with the water (Gray et al. 1986; Chow et al. 1981). The intake rate is high resulting in strong currents of inflowing water and fish and other animals are impinged on screens and die. This is a major issue and much effort has been devoted to modifying water intake practices to lessen impingement, but impingements cannot be totally avoided (Thompson 2000; Mayhew et al. 2000).

Large pumps often are used to obtain water for irrigation in agriculture and for supplying aquaculture facilities (Gray et al. 1986; Yoo and Boyd 1994). The intakes of pumps used in agriculture and aquaculture may or may not be screened. Thus animals may be either impinged on screens or entrained in pumps. The extent of aquatic animal mortality resulting from water intake practices in agriculture and aquaculture has not been studied, but the need for screens has been pointed out (Boyd 1999). Aquaculture is probably a rather minor contributor to aquatic animal deaths by impingement and entrainment because many industries and municipalities use large pumps to obtain water from natural water bodies.

Movement of aquaculture species

According to Reantaso (2002) movement of aquaculture species dates back to the Greeks and Romans who moved oysters, carp, and perch throughout Europe. The practice has continued and became increasingly common as the aquaculture industry grew. There is movement of eggs, broodstock, and larval and postlarval stages of various species that typically do not represent new introductions. For example, shrimp hatcheries may sell shrimp broodstock or shrimp postlarvae to hatcheries or farms in other countries and sometimes on other continents. This practice is not limited to shrimp; it also is common in fish and bivalve aquaculture.

There are international and national regulations for movements of aquaculture species (see Chapter 12), but within countries, there often are no restrictions in moving animals from one state or province to another. To illustrate in the US catfish fingerlings produced in hatcheries in Mississippi frequently are delivered to ponds in other states without need for governmental permission.

Capture of wild larvae and broodstock

There are many types of coastal aquaculture that have collected wild juveniles from the sea for seed stock. The most important are penaeid shrimp, milkfish, eels, yellowtail, southern bluefin tuna, oysters, and mussels (Hair et al. 2002). The use of wild-caught larvae was essential before hatchery techniques for producing larvae were developed, and even after that, wild-caught broodstock were necessary to provide eggs. As the culture techniques for species improve there is less need for wild-caught animals. Shrimp aquaculture is a good example: originally wild postlarvae were captured and stocked in ponds; next hatcheries were developed and wild broodstock were obtained from the sea; finally the cycle was closed allowing farm-raised broodstock to be used to produce hatchery postlarvae for stocking in grow-out ponds. A study by Sonnenholzner et al. (2002) stated that in the 1970s and 1980s the shrimp farmers in Ecuador used wild-caught postlarvae almost entirely. Records obtained from farmers revealed that between 1995 and 2000 the use of wild postlarvae declined from 61 to 6% of pond stockings. Today less than 1% of ponds in Ecuador are stocked with wild-caught postlarvae. A similar trend in the use of wild-caught broodstock has also occurred. Thus the use of wild-caught shrimp in aquaculture is no longer a mainstream practice. Other species have or will in the future follow the same trend as shrimp aquaculture.

The production of hatchery seed stock is highly desirable in aquaculture because it allows a timely and dependable supply and permits the application of selective breeding for stock improvement. It also avoids the environmental damage done to natural ecosystems by fishing for wild broodstock and postlarvae. Capture of wild seed stock reduces the abundance of the target species, and there is a catch of non-target species that usually is destroyed (Hair et al. 2002). Certainly one practice to lessen the effects of aquaculture on biodiversity is to use seed stock from hatcheries that rely on farm-reared broodstock.

Disease transmission between farmed and wild animals

The topic of animal health is beyond the scope of this book, but an increase in the disease of wild plants and animals as a result of aquaculture would negatively impact the biodiversity of natural waters. It is well known that confining and concentrating culture species provides an increased opportunity for diseases and parasites—especially where environmental conditions are stressful. Moreover, diseases may be introduced into aquaculture facilities via the stock, and the stock are traded internationally. There are many cases of aquaculture diseases spreading throughout a region within a period of a few months or spreading among countries and continents within a few years.

It also is known that diseases and parasites can spread from wild species to cultured species and vice versa (Bondad-Reantaso et al. 2005). Pillay (2004) gave several examples of diseases that spread from aquaculture animals to wild animals. These include two diseases caused by protozoan parasites of oysters, parasitic copepods of shellfish, and oyster drills, *Gyrodactylus salaris* or the salmon fluke, and the fungal agent *Aphanomyces astaci* or crayfish plaque that decimated native crayfish in English rivers. The koi herpes virus has caused mass mortalities in both farmed and wild koi and common carp (Garver et al. 2010). Viral diseases of shrimp also can spread to wild shrimp populations (Walker and Winton 2010).

There are few studies of the significance of aquaculture in spreading diseases to wild animals. Bergh (2007) explains that wild fish are no healthier than aquacultured fish—wild fish and other aquatic organisms have diseases naturally. Nevertheless, Bergh states that aquaculture may cause the spread of pathogens into new geographical areas or increase the number of hosts of parasites. A study by Ford and Meyers (2008) claimed that transmission of diseases from farmed salmon have in many cases reduced the survival of wild salmon by 50% or more. Arthur and Subasinghe (2002) provided the following list of possible negative impacts of aquatic animal diseases—which could be transmitted from aquaculture stock—on wild populations and biodiversity:

- Changes in predator and prey populations and relationships,
- Changes in pathogen host abundance,
- Reduction in intra-specific genetic variation,
- Local extirpation of components of aquatic communities,
- Extinctions of species.

The control of diseases is important for profitability in aquaculture. Thus there is much effort to enhance biosecurity, develop specific pathogen-free stock, avoid stressful culture conditions that increase disease susceptibility in culture species, and to develop vaccines and medicines. Certainly the spread of diseases from aquaculture to wild stock is an important issue, but it is not possible to make an estimate of the magnitude of the effect on biodiversity.

Beneficial effects of aquaculture on biodiversity

Capture fisheries certainly are not considered to be environmentally friendly but the Marine Stewardship Council has rated some fisheries as sustainable. Most of the world's fisheries, however, are subjected to heavy fishery pressure and many are overfished. Many types of capture fisheries concentrated on larger fish—usually carnivorous ones—representing the higher trophic levels of the food web. As the catch of larger fish declines in response to fishing pressure, fish of lower trophic levels make up a larger proportion of the total landings—a phenomenon described by Pauly et al. (1998) as "fishing down marine food webs".

Fishing gear is not highly selective; other species often of no commercial value also are captured along with the target species. These unintended captures known as by-catch often are discarded back to the sea and most of the organisms die. The global capture fisheries production 94.3 Mt annually for the period 1988–1991 had an associated annual by-catch of 27 Mt or 28.6% (Pauly and Christensen 1995). In addition to capturing unintended fish and invertebrate marine animals, some fishing gear such as driftnets, trawls, and long lines result in the deaths of whales, seals, turtles, and millions of birds (http://www.eionet.europa.eu/gemet/concept?cp=2834). Trawling and dredging for fish and other aquatic species is particularly harmful to the sea floor. It disrupts the surface destroying benthic communities and altering the shape of the surface. In a review of the effects of trawling on the sea floor Puig et al. (2012) concluded the extent of trawling was so great globally that many parts of the ocean bottom were subjected to effects similar to those caused by plowing of agricultural land. The effects of trawling heal over time but there is repeated trawling in areas where this procedure is used for fishing.

Other negative effects of fishing include adding artificial structures to attract marine organisms, use of explosives to capture fish, and dumping of debris and loss of fishing gear that can entangle or otherwise harm aquatic animals. In addition, ozone-depleting refrigerants and carbon dioxide are released from fishing vessels (http://www.fao.org/docrep/006/Y4773E/y4773e05.htm).

Besides impacts discussed above—several of which are inherent to capture fisheries—it must not be forgotten that many fisheries are simply being overfished (Pauly et al. 2002). Overfishing is exacerbated by excessive fishing capacity, subsidies, and insufficient regulations. Fisheries products from aquaculture supplement the world supply of fisheries products and should lessen fishing pressure on the oceans and inland waters. The negative impacts of aquaculture apparently could be reduced to acceptable levels by use of better practices without restricting production. However, there is little possibility for increasing the capture fishery, and to make is sustainable would no doubt require a significant reduction in the present levels of global, capture fisheries production.

Aquaculture presently is highly dependent on fish meal and fish oil for use in feeds (see Chapter 8). Thus aquaculture increases the demand for small pelagic fish for making fish meal and places an increased demand on this fishery. Thus efforts by the aquaculture industry to limit the use of marine fish meal and oil from pelagic fish by finding suitable substitutes are critical to conservation of pelagic fisheries to allow aquaculture to continue to lessen fishing pressure on natural stocks and supplement capture fisheries to meet world demand for fisheries products for human consumption.

Conclusions

Several aspects of aquaculture can have negative impacts on biodiversity of both terrestrial and aquatic ecosystems. These impacts include: eutrophication; land use changes and destruction of wetland habitats; introduction of obnoxious, nonendemic species; excessive use of marine resources for fish meal and oil to include in feeds; destruction of fish-eating birds and other predators; spread of diseases of aquatic animals; capture of wild aquatic animals for broodstock and seedstock; impingement of aquatic animals at water intakes. In fact, essentially all of the negative impacts of aquaculture can be summed up as perturbations that potentially could lead to reduced biodiversity and cause undesirable changes in ecosystems.

With the exception of the use of small marine fish to make fish meal and fish oil in feeds, most of the perturbations caused by aquaculture are rather minor and only affect biodiversity in localized areas. These perturbations should be avoided or minimized through better management, but aquaculture has not had a major influence on biodiversity at country or global levels as has agriculture, the capture fishery, industries, and cities. The problem with biodiversity stems from the huge global population and increasing rates of consumption, and the use of resources and associated production activities necessary to sustain it.

Fisheries production is an important sector of world food production, and aquaculture is essential to supplement the capture fishery in meeting the world demand for fisheries products. Aquaculture appears to have less overall negative impact on biodiversity than does fishing of wild species, many of which are overexploited. Aquaculture reduces the fishing pressure on some species of marine and inland aquaculture species and this is beneficial to the biodiversity of marine ecosystems from which these species are captured. But aquaculture must reduce its dependence on fish meal and fish oil from small pelagic fish in order to reduce the exploitation of the sea and lessen negative impacts on marine biodiversity.

The eNGO perspective

The conservation of natural species and the preservation of biodiversity is principle in the mission statements of nearly all eNGOs. The most notable example of success for the eNGO community related to this principle has been the establishment of protected areas (PAs). Through the collaboration of eNGOs, governments, the private sector and multilateral institutions, the number of PAs has exceeded 100 000 sites, covering 12.5% of the earth's land surface and 0.5% of the ocean surface (Chape et al. 2003). However, Moral and Sale (2011) noted that globally the use of PAs is not going to be sufficient, by itself, to offset the ongoing loss of biodiversity. They offer the following reasons for this insufficiency:

- Large gaps in the coverage of critical ecological processes related to individual home ranges of species and their propagule dispersal;
- The overall failure of such areas to protect against the broad range of threats affecting ecosystems;
- Budget constraints of management authorities;
- Conflicts with human development;
- Lack of forethought of the integration of these areas with a growing human population and the accompanying anthropogenic stressors.

Although PAs represent some of the eNGOs' greatest successes on a global scale, there is recognition that securing these preserves will only reduce environmental degradation in a specific area and consequently not address the systemic issues associated with natural resource extraction and utilization. Aside from PAs there are limited tools that eNGOs can use to foster greater biodiversity preservation. The primary alternative approaches to biodiversity preservation are encouraging policies that protect biodiversity, integration of environmental restrictions into multilateral development efforts, and market-based incentive programs (standards and certification). Advocacy, partnerships, and joint projects with actors in the private, governmental, and multilateral sectors are the primary tools used to achieve biodiversity preservation. Aquaculture is touched by all three of these strategies.

As stated in this chapter, it is often very challenging if not impossible to attribute certain biodiversity impacts to aquaculture operations in habitats with multiple resource users. The eNGOs expend significant effort attempting to make these attributions. However, the technical capacity of the aquaculture industry to provide feasible alternative explanations for certain environmental conditions is often superior to that of the eNGOs. For example, there are few in the eNGO sector that could engage in a technical discussion on the appropriate levels of nitrogen in farm wastewater as it relates to the environmental condition of a specific receiving water body. This is not to say that there are no forward-thinking leaders in the eNGO community who have or are acquiring a technical capacity in specific industries to maneuver through complex issues, but it is more common for eNGOs to use consultants to strengthen their technical capacity. Even if technical sophistication was present, there are often data deficiencies in the methodology to describe the biodiversity of areas surrounding aquaculture facilities.

To illustrate the difficulty mentioned above imagine a situation in which an eNGO is attempting to partner with an international development agency to foster the inclusion of a biodiversity preservation component as a condition of project implementation. This is not an uncommon scenario. From an aquaculture development perspective the likely questions posed to the eNGO would relate to the limits on land conversion, water use, wastewater discharge, densities, escapes, etc. As has been made clear in previous chapters there are few circumstances where there are "hard and fast numbers" to answer these questions—complicated further by shifting seasonal and developmental conditions that maintain a dynamic local environment. Answering these questions would require a professional with a significant amount of experience in the sector. The eNGO could invest in obtaining the technical capacity needed to address the situation, but likely the less sophisticated and easiest remedy for the eNGO is to request some form of an EIA for the farm site that by default will address all of these questions.

The context for EIAs has been provided in Chapter 4. What is critical to note about this discussion is that an EIA is no "silver bullet" for attempting to maintain biodiversity. An EIA is simply a document—paper and ink. It may or may not have bearing on the potential biodiversity impacts of construction or operations of a particular aquaculture facility. The authors have both encountered numerous examples in aquaculture where there was a wide range of quality and applicability of EIAs. Some in the eNGO community are aware of the potential quality concerns and the

Table 11.2 A suggested checklist for farmers and guideline for auditors on a complete Biodiversity Environmental Impact (B-EIA) process and report completed by the Shrimp Aquaculture Dialogue.

Item	Biodiversity impact	
	Yes	No
1. Quality of the B-EIA process (e.g., was it participatory and transparent?).		X
(a) B-EIA carried out by a valid expert in accordance with the above table.		X
(b) The B-EIA was publicly (locally) communicated with sufficient time for interested parties to participate and/or get informed.		X
(c) Stakeholders are listed and impact descriptions are documented and in preparation of the final B-EIA report, meetings with the listed stakeholders (or by stakeholders chosen representatives) have taken place.		X
(d) These meetings have been recorded and the minutes are attached to the final report; names and contact details of participating stakeholders included.		X
(e) Evidence is provided that draft and final B-EIA reports have been submitted to local government representatives and, if requested by stakeholders, a legally registered civil organization chosen by these stakeholders.		X
(f) Evidence is provided that the final B-EIA reports have been submitted and reviewed by a specialist with appropriate expertise on biodiversity issues.		X
(g) B-EIA completed according to guidance on B-EIA and Participatory Social Impact Assessment relationship (transparency and consultation).		X
2. Risk analysis: actual (past and present) impacts of the current farms, or potential impacts of the intended farm or expansion of existing farm and at least two alternatives (one of these is the "no farm or no expansion" scenario). Concepts to cover include the following:		X
(a) The type of farming, possible alternatives, and a summary of activities likely to affect biodiversity.	X	
(b) An analysis of opportunities and constraints for biodiversity (include "no net biodiversity loss" or "biodiversity restoration" alternatives).	X	
(c) Expected biophysical changes (in soil, water, air, flora and fauna) resulting from proposed or existing activities or induced by any socioeconomic changes.	X	
(d) Spatial and temporal scale of influence, identifying effects on connectivity between ecosystems, and potential cumulative effects.	X	
(e) Available information on baseline conditions and any anticipated trends in biodiversity in the absence of the proposal.	X	
(f) Likely biodiversity impacts associated with the proposal or current operations in terms of composition, structure and function of surrounding ecosystems.	X	
(g) Biodiversity services and values identified in consultation with stakeholders and anticipated magnitude, direction and timeline of changes in these (highlight any irreversible impacts).	X	
(h) Possible measures to avoid, minimize, or compensate for significant biodiversity damage or loss, making reference to any legal requirements. Information required to support decision making and summary of important gaps.	X	
(j) Proposed impact assessment methodology and timescale.		X

Table 11.2 *(Continued)*

Item	Biodiversity impact	
	Yes	No
3. Impact statement is available and contains all of the requirements listed above along with a clear indication of authors and affiliations.		X
4. Review process, reviewers (decision makers), and decisions clearly documented.		X
5. Clear understanding as to how options for mitigation and offsetting were determined and how avoidance actions were prioritized over compensation.		X
6. Names, affiliations, and experience of the reviewing specialist are documented and clear understanding of how affected groups were involved and how balanced consideration was given to conservation versus development goals in the peer review.		X
7. Clear articulation of a biodiversity management system including targets and monitoring strategies for mitigation.		X

Authors' interpretation of the utility of checklist items as they assist in preserving biodiversity.
Source: http://assets.worldwildlife.org/publications/429/files/original/ShAD_Standard_Final_Draft.pdf?1346186260

potential for fabrication of data. Thus eNGOs are to prescribe what is necessary to be examined. Because the eNGO must reduce the potential to be criticized as having been too lax by fellow eNGOs, this task of prescribing aspects of an EIA becomes exhaustive and overly complicated. In the end what is left is something more akin to an academic research project than a tool that could be used at the farm level. This was the case with regards to the development of environmental and social standards in the Shrimp Aquaculture Dialogue. The threat of extreme eNGO public criticism drove the more moderate eNGOs toward an extreme position with the result being an omnibus Biodiversity-inclusive Environmental Impact Assessment for farms seeking to adhere to these standards (http://assets.worldwildlife.org/publications/429/files/original/ShAD_Standard_Final_Draft.pdf?1346186260). It appears the Shrimp Aquaculture Dialogue did not have the capacity or potentially the breadth to identify indicators of biodiversity and thus developed a requirement that was based more on process than content (Table 11.2). Interestingly after making the shrimp standards extremely challenging to meet, the extreme eNGOs still exerted significant effort to criticize and boycott the standards.

The reliance that many eNGOs have on assessment reports such as described above to assist in formulating a biodiversity preservation approach can clutter the well-intended desire for those in the environmental community to adequately address biodiversity issues. At the core of this dilemma is the ultimate question of "what level of impact will reduce biodiversity?" It could be argued that the academic and research community has failed to apply fundamental ecological principles and methods to assess aquaculture activities. Moreover, most fundamental ecological research is conducted in areas where there is little human development, or paradoxically protected areas. There is scant evidence in the literature regarding quantitative indicators of biodiversity condition applied at the macro-level to aquaculture facilities. It is not surprising then that eNGOs default to EIAs for biodiversity and pollution issues that do not have clear and concise delineations. It is akin to the aquaculture industry's reliance on BMPs to address ecosystem impacts instead of utilizing environmental performance targets.

It is clear that aquaculture activities will have an impact on the local and global ecosystem. The degree of impact will be dependent on the size and scale of the operation, the specific site that is being exploited, the operational management, intensity of production, and species cultured. The eNGOs currently do not have a robust mechanism for determining limits to biodiversity impacts at the farm-level. This is likely the case for other industries that operate in a multi-user environment. Nevertheless there is a need for better benchmarks than EIAs or BMP adoption to describe biodiversity protection measures. The eNGOs are coming out of a place-based "protected area"-minded era, but there still remain gaps in the literature and practice of the measurable ways to describe biodiversity impacts in multi-user environments. It is likely that the way forward is to dispense with the attempts at attributing biodiversity impacts to specific aquaculture operations but rather develop reasonable thresholds (not transposing PA indices on multi-user environments) for the environment itself and work toward achieving those targets regardless of how large or small aquaculture operations are contributing to the impacts.

References

Alves, C. B. M., F. Vieira, A. L. B. Magalhães, and M. F. G. Brito. 2007. Impacts of non-native fish species in Minas Gerais, Brazil: Present situation and prospects. In T. M. Bert, editor, *Ecological and Genetic Implications of Aquaculture Activities*, pp. 291–314. Dordrecht: Springer Netherlands

Amirkolaie, A. K. 2008. Environmental impact of nutrient discharged by aquaculture waste water on the Haraz River. *Journal of Fisheries and Aquatic Science* 3:275–279.

Andrews, J. D. 1996. History of *Perkinsus marinus*, a pathogen of oysters in Chesapeake Bay 1950–1984. *Journal of Shellfish Research* 15:13–16.

Ansah, Y. B., E. A. Frimpong, and S. Amisah. 2012. Biological assessment of aquaculture effects on effluent-receiving streams in Ghana using structural and functional composition of fish and macroinvertebrate assemblages. *Environmental Management* 50:166–180.

Arthington, A. H. 1991. Ecological and genetic impacts of introduced and translocated freshwater fishes in Australia. *Canadian Journal of Fisheries and Aquatic Sciences* 48: 33–43.

Arthur, J. R. and R. P. Subasinghe. 2002. Potential adverse socio-economic and biological impacts of aquatic animal pathogens due to hatchery-based enhancement of inland open-water systems, and possibilities for their minimization. In J. R. Arthus, M. J. Phillips, R. P. Subasinghe, M. B. Reantaso, and I. H. MacRae, editors, *Primary Aquatic Animal Health Care in Rural, Small-scale, Aquaculture Development*, pp. 113–126, Fisheries Technical Paper No. 406. Rome: FAO.

Avault Jr., J. W. 1996. *Fundamentals of Aquaculture*. Baton Rouge: AVA Publishing Company, Inc.

Bain, M. B. 1993. Assessing impacts of introduced aquatic species: Grass carp in large systems. *Environmental Management* 17:211–224.

Barras, S. C. 2007. Avian predators at aquaculture facilities in the southern United States. Southern Regional Aquaculture Center Publication 400, Mississippi State University, Stoneville, Mississippi.

Bartley, D. and C. V. Casal. 1998. Impacts of introductions on the conservation and sustainable use of aquatic biodiversity. *FAO Aquaculture Newsletter* 20:15–19.

Bartley, D. M., K. Rana, and A. J. Immink. 2001. The use of inter-specific hybrids in aquaculture and fisheries. *Reviews in Fish Biology and Fisheries* 10:325–337.

Beardmore, J. A., G. C. Mair, and R. I. Lewis. 1997. Biodiversity in aquatic systems in relation to aquaculture. *Aquaculture Research* 28:829–839.

Bergh, O. 2007. The dual myths of the healthy wild fish and the unhealthy farmed fish. *Diseases of Aquatic Organisms* 75:159–164.

Beveridge, M. C. M. 2001. Aquaculture and wildlife interactions. In A. Uriarte and B. Basurco, editors, *Environmental Impact Assessment of Mediterranean Aquaculture Farms*, pp. 57–66. Zaragoza: CIHEAM.

Bondad-Reantaso, M. G., R. P. Subasinghe, J. R. Arthur, K. Ogawa, S. Chinabut, R. Adlard, Z. Tan, and M. Shariff. 2005. Disease and health management in Asian aquaculture. *Veterinary Parasitology* 132:249–272.

Boyd, C. E. 1999. *Codes of practice for responsible shrimp farming*. St. Louis: Global Aquaculture Alliance.

Boyd, C. E., A. A. McNevin, J. Clay, and H. M. Johnson. 2005. Certification issues for some common aquaculture species. *Reviews in Fisheries Science* 13:231–279.

Boyd, C. E., S. Soongsawang, E. W. Shell, and S. Fowler. 2009. Small impoundment complexes as a possible method to increase water supply in Alabama. *Proceedings 2009 Georgia Water Resources Conference*, April 27–29. Athens: University of Georgia.

Brown, C. A. and J. M. King. 1995. The effects of trout-farm effluents on benthic invertebrate community structure in rivers in the South-Western Cape, South Africa. *Southern African Journal of Aquatic Sciences* 21:3–21.

Butler, J. R. A. 2000. The economic costs of wildlife predation on livestock in Gokwe communal land, Zimbabwe. *African Journal of Ecology* 38:23–30.

Cahn, A. 1929. The effect of carp on a small lake: The carp as a dominant. *Ecology* 10:271–274.

Canonico, G. C., A. Arthington, J. K. McCrary, and M. L. Thieme. 2005. The effects of introduced tilapias on native biodiversity. *Aquatic Conservation: Marine and Freshwater Ecosystems* 15:463–483.

Carlton, J. T. and A. N. Cohen. 2003. Episodic global dispersal in shallow water marine organisms: The case history of the European shore crabs *Carcinus maenas* and *C. aestuarii*. *Journal of Biogeography* 30:1809–1820.

Chape, S., S. Blyth, L. Fish, P. Fox, and M. Spalding. 2003. 2003 United Nations List of Protected Areas. Cambridge: The World Conservation Union.

Chow, W., I. P. Murarka, and R. W. Brocksen. 1981. Entrainment and impingement in power plant cooling systems. *Journal of Water Pollution Control Federation* 53:965–973.

Courtenay Jr., W. R., H. F. Sahlman, W. W. MileyII, and D. J. Herrema. 1974. Exotic fishes in fresh and brackish waters of Florida. *Biological Conservation* 6:292–302.

Crossman, E. J. and B. C. Cudmore. 1999. Summary of fishes intentionally introduced in North America. In R. Claudi and J. H. Leach, editors, *Nonindigenous Freshwater Organisms: Vectors, Biology, and Impacts*, pp. 99–111. Boca Raton: CRC Press.

Devlin, R. H., L. F. Sundström, and W. M. Muir. 2006. Interface of biotechnology and ecology for environmental risk assessments of transgenic fish. *Trends in Biotechnology* 24:89–97.

Diana, J. S. 2009. Aquaculture production and biodiversity conservation. *BioScience* 59:27–38.

Dorrance, M. J. and L. D. Roy. 1976. Predation losses of domestic sheep in Alberta. *Journal of Range Management* 29:457–460.

Drake, L. A., M. A. Doblin, and F. C. Dobbs. 2007. Potential microbial bioinvasions via ships' ballast water, sediment, and biofilm. *Marine Pollution Bulletin* 55:333–341.

Drake, L. A., G. M. Ruiz, B. S. Galil, T. L. Mullady, D. O Friedmann, and F. C. Dobbs. 2002. Microbial ecology of ballast water during a transoceanic voyage and the effects of open-ocean exchange. *Marine Ecology Progress Series* 233:13–20.

Dunham, R. A. 1999. Utilization of transgenic fish in developing countries: Potential benefits and risks. *Journal of the World Aquaculture Society* 30:1–11.

Dunham, R. A., G. M. Umali, R. Beam, A. H. Kristanto, and M. Trask. 2008. Comparison of production traits of NWAC103 channel catfish, NWAC103 channel catfish × blue catfish hybrids, Kansas select 21 channel catfish, and blue catfish grown at commercial densities and exposed to natural bacterial epizootics. *North American Journal of Aquaculture* 70:98–106.

du Plessis, S. S. 1957. Growth and daily food intake of the white-breasted cormorant in captivity. *Ostrich* 28:197–201.

Eldridge, S. R., B. J. Shakeshaft, and T. J. Nano. 2002. The impact of wild dog control on cattle, native and introduced herbivores and introduced predators in central Australia. Unpublished report to the Bureau of Rural Sciences, Canberra, Australia.

Eng, C. T., J. N. Paw, and F. Y. Guarin. 1989. The environmental impact of aquaculture and the effects of pollution on coastal aquaculture development in Southeast Asia. *Marine Pollution Bulletin* 20:335–343.

Ewart, J. W. and S. E. Ford. 1993. History and impact of MSX and Dermo disease on oyster stocks in the northeast region. Northeast Regional Aquaculture Center (NRAC) Fact Sheet No. 200. University of Massachusetts, Dartmouth, Massachusetts.

Ford, J. S., and R. A. Myers. 2008. A global assessment of salmon aquaculture impacts on wild salmonids. *PLoS Biology* 6:33–41.

Gao, Q. F., K. L. Cheung, S. G. Cheung, and P. K. S. Shin. 2005. Effects of nutrient enrichment derived from fish farming activities on macroinvertebrate assemblages in a subtropical region of Hong Kong. *Marine Pollution Bulletin* 51:994–1002.

Garver, K. A., L. Al-Hussinee, L. M. Hawley, T. Schroeder, S. Edes, V. LePage, E. Contador, S. Russell, S. Lord, R. M. W. Stevenson, B. Souter, E. Wright, and J. S. Lumsden. 2010. Mass mortality associated with koi herpesvirus in wild common carp in Canada. *Journal of Wildlife Diseases* 46:1242–1251.

Gray, R. H., T. L. Page, D. A. Neitzel, and D. D. Dauble. 1986. Assessing population effects from entrainment of fish at a large volume water intake. *Journal of Environmental Science and Health* 21:191–209.

Greentree, C., G. Saunders, L. McLeod, and J. Hone. 2000. Lamb predation and fox control in south-eastern Australia. *Journal of Applied Ecology* 37:935–943.

Gross, M. R. 1998. One species with two biologies: Atlantic salmon (*Salmo salar*) in the wild and in aquaculture. *Canadian Journal of Fisheries and Aquatic Sciences* 55:131–144.

Guo, L. and Z. Li. 2003. Effects of nitrogen and phosphorus from fish cage-culture on the communities of a shallow lake in middle Yangtze River basin of China. *Aquaculture* 226:201–212

Hair, C., J. Bell, and P. Doherty. 2002. The use of wild-caught juveniles in coastal aquaculture and its application to coral reef fishes. In R. R. Stickney and J. P. McVey, editors, *Responsible Marine Aquaculture*, pp. 327–353. Oxon: CABI.

Hesthagen, T., and O. T. Sandlund. 2007. Non-native freshwater fishes in Norway: History, consequences and perspectives. *Journal of Fish Biology* 71:173–183.

Hygnstrom, S. E. and S. R. Craven. 1988. Electric fences and commercial repellents for reducing deer damage in cornfields. *Wildlife Society Bulletin* 16:291–296.

Islam, M. S. 2005. Nitrogen and phosphorus budget in coastal and marine cage aquaculture and impacts of effluent loading on ecosystem: Review and analysis towards model development. *Marine Pollution Bulletin* 50:48–61.

Kapuscinski, A. R. 2005. Current scientific understanding of the environmental biosafety of transgenic fish and shellfish. *Review Scientifique et Technique* 24:309–322.

Kerr, S. J., C. S. Brousseau, and M. Muschett. 2005. Invasive aquatic species in Ontario. *Fisheries* 30:21–30.

Kloskowski, J. 2005. Otter *Lutra lutra* damage at farmed fisheries in southeastern Poland, I: An interview survey. *Wildlife Biology* 11:201–206.

Kumar, A. B. 2000. Exotic fishes and freshwater fish diversity. *Zoo's Print Journal* 15:363–367.

Kutti, T., P. K. Hansen, A. Ervik, T. Hoisaeter, and P. Johannessen. 2007. Effects of organic effluents from a salmon farm on a fjord system. II. Temporal and spatial patterns in infauna community composition. *Aquaculture* 262:355–366.

Littauer, G. A., J. F. Glahn, D. S. Reinhold, and M. W. Brunson. 1997. Control of bird predation at aquaculture facilities: strategies and cost estimates. Southern Regional Aquaculture Center Publication 402, Mississippi State University, Stoneville, Mississippi.

Loch, D. D., J. L. West, and D. G. Perlmutter. 1996. The effect of trout farm effluent on the taxa richness of benthic macroinvertebrates. *Aquaculture* 147:37–55.

Lonsdale, W. M. 1994. Inviting trouble: Introduced pasture species in northern Australia. *Australian Journal of Ecology* 19:345–354.

Lowe-McConnell, R. H. 1993. Fish faunas of the African Great Lakes: origins, diversity, and vulnerability. *Conservation Biology* 7:634–643.

Margalef, R. 1958. Temporal succession and spatial heterogeneity in phytoplankton. In A. A. Buzzati-Traverso, editor, *Perspectives in Marine Biology*, pp. 323–349. Berkeley: University of California Press.

Mayhew, D. A., L. D. Jensen, D. F. Hanson, and P. H. Muessig. 2000. A comparative review of entrainment survival studies at power plants in estuarine environments. *Environmental Science and Policy* 3:S295–S301.

Minchin, D., S. Gollasch, and I. Wallentinus (editors). 2005. Vector pathways and the spread of exotic species in the sea. International Council for the Exploration of the Sea (ICES) Cooperative Research Report 271, Copenhagen, Denmark.

Moral, C. and P. F. Sale. 2011. Ongoing global biodiversity loss and the need to move beyond protected areas: A review of the technical and practical shortcomings of protected areas on land and sea. *Marine Ecology Progress Series* 434:251–266.

Nash, C. E., R. N. Iwamoto, and C. V. W. Mahnken. 2000. Aquaculture risk management and marine mammal interactions in the Pacific Northwest. *Aquaculture* 183:307–323.

Naylor, R. L., S. L. Williams, and D. R. Strong. 2001. Aquaculture—a gateway for exotic species. *Science* 294:1655–1656.

Norbury, G. and A. McGlinchy. 1996. The impact of rabbit control on predator sightings in the semi-arid high country of the South Island, New Zealand. *Wildlife Research* 23:93–97.

O'Connell-Rodwell, C. E., T. Rodwell, M. Rice, and L. A. Hart. 2000. Living with the modern conservation paradigm: Can agricultural communities co-exist with elephants? A five-year case study in East Caprivi, Namibia. *Biological Conservation* 93:381–391.

Odum, E. P. 1971. *Fundamentals of Ecology*, 3rd ed. Philadelphia: Saunders Publishing Company.

Ogada, M. O., R. Woodroffe, N. O. Oguge, and L. G. Frank. 2003. Limiting depredation by African carnivores: The role of livestock husbandry. *Conservation Biology* 17:1521–1530.

Páez-Osuna, F., S. R. Guerrero-Galván, and A. C. Ruiz-Fernández. 1998. The environmental impact of shrimp aquaculture and the coastal pollution in Mexico. *Marine Pollution Bulletin* 36:65–75.

Páez-Osuna, F., A. Gracia, F. Flores-Verdugo, L. P. Lyle-Fritch, R. Alonso-Rodríguez, A. Roque, and A. C. Ruiz-Fernández. 2003. Shrimp aquaculture development and the environment in the Gulf of California ecoregion. *Marine Pollution Bulletin* 46:806–815.

Pauly, D. and V. Christensen. 1995. Primary production required to sustain global fisheries. *Nature* 374:255–257.

Pauly, D., V. Christensen, J. Dalsgaard, R. Froese, and F. Torres, Jr. 1998. Fishing down marine food webs. *Science* 279:860–863.

Pauly, D., V. Christensen, S. Guénette, T. J. Pitcher, U. R. Sumaila, C. J. Walters, R. Watson, and D. Zeller. 2002. Towards sustainability in world fisheries. *Nature* 418:689–695.

Pillay, T. V. R. 2004. *Aquaculture and the Environment*, 2nd ed. Oxford: Blackwell Publishing.

Pimentel, D., L. Lach, R. Zuniga, and D. Morrison. 2000. Environmental and economic costs of nonindigenous species in the United States. *BioScience* 50:53–65.

Pipalova, I. 2006. A review of grass carp use for aquatic weed control and its impact on water bodies. *Journal of Aquatic Plant Management* 44:1–12.

Pistori, R. E. T., G. G. Henry-Silva, J. F. V. Biudes, and A. F. M. Camargo. 2010. Influence of aquaculture effluents on the growth of *Salvinia molesta*. *Acta Limnologica Brasiliensia* 22:179–186.

Puig, P., M. Canals, J. B. Company, J. Martin, D. Amblas, G. Lastras, A. Palanques, and A. M. Calafat. 2012. Ploughing the deep sea floor. *Nature* 489:286–289.

Rahel, F. J. 2004. Unauthorized fish introductions: Fisheries management of the people, for the people, or by the people? *American Fisheries Society Symposium* 44:431–443.

Reantaso, M. 2002. APEC, FAO, NACA, and OIE enhance capacity on risk analysis (IRA) in aquatic animal movement in Asia-Pacific region. *Aquaculture Asia* 7:4–6

Richardson, D. M., P. Pyšek, M. Rejmánek, M. G. Barbour, F. D. Panetta, and C. J. West. 2000. Naturalization and invasion of alien plants: Concepts and definitions. *Diversity and Distributions* 6:93–107.

Roberts, R. M. 2007. Genetically modified organisms for agricultural food production. In P. Weirich, editor, *Labeling Genetically Modified Food*. New York: Oxford University Press.

Ryerson, K. A. 1933. History and significance of the foreign plant introduction work of the United States Department of Agriculture. *Agricultural History* 7:110–128.

Sachs, E. S. 2012. Biotechnology and crop production. In J. S. Popp, M. D. Matlock, M. M. John, and N. P. Kemper, editors, *The Role of Biotechnology in a Sustainable Food Supply*, pp. 49–78. New York: Cambridge University Press.

Sagoff, M. 2005. Do non-native species threaten the natural environment? *Journal of Agricultural and Environmental Ethics* 18:215–236.

Schoof, R. R. and G. A. Gander. 1982. Computation of runoff reduction caused by farm ponds. *Water Resources Bulletin* 18:529–532.

Selong, J. H. and L. A. Helfrich. 1998. Impacts of trout culture effluent on water quality and biotic communities in Virginia headwater streams. *The Progressive Fish-Culturist* 60:247–262.

Senanan, W., S. Panutrakul, P. Barnette, S. Chavanich, V. Mantachitr, N. Tangkrock-Olan, and V. Viyakarn. 2009. Preliminary risk assessment of Pacific whiteleg shrimp (*P. vannamei*) introduced to Thailand for Aquaculture. *Aquaculture Asia Magazine* 14:28–32.

Sonnenholzner, S., L. Massaut, and C. E. Boyd. 2002. Ecuador study shows wild postlarvae use down. *Global Aquaculture Advocate* 5:56–57.

Sousa, O. V., A. Macrae, F. G. R. Menezes, N. C. M. Gomes, R. H. S. F. Vieira, and L. C. S. Mendonça-Hagler. 2006. The impact of shrimp farming effluent on bacterial communities in mangrove waters, Ceará, Brazil. *Marine Pollution Bulletin* 52:1725–1734.

Stephens, W. W. and J. L. Farris. 2004. Instream community assessment of aquaculture effluents. *Aquaculture* 231:149–162.

Strayer, D. L. 2010. Alien species in fresh waters: Ecological effects, interactions with other stressors, and prospects for the future. *Freshwater Biology* 55:152–174.

Strong, D. R. and R. W. Pemberton. 2000. Biological control of invading species—risk and reform. *Science* 16:288:1969–1970.

Thomas, Y., C. Courties, Y. El Helwe, A. Herbland, and H. Lemonnier. 2010. Spatial and temporal extension of eutrophication associated with shrimp farm wastewater discharges in the New Caledonia lagoon. *Marine Pollution Bulletin* 61:387–398.

Thompson, T. 2000. Intake modifications to reduce entrainment and impingement at Carolina Power & Light Company's Brunswick Steam Electric Plant, Southport, North Carolina. *Environmental Science and Policy* 3:S417–S424.

Tlusty, M. 2002. The benefits and risks of aquacultural production for the aquarium trade. *Aquaculture* 205:203–219.

Vista, A., P. Norris, F. Lupi, and R. Bernsten. 2006. Nutrient loading and efficiency of tilapia cage culture in Taal Lake, Philippines. *Philippine Agricultural Scientist* 89:48–57.

Walker, P. J. and J. R. Winton. 2010. Emerging viral diseases of fish and shrimp. *Veterinary Research* 41:51–65.

Weber, E. F. 1997. The alien flora of Europe: A taxonomic and biogeographic review. *Journal of Vegetation Science* 8:565–572.

Weber, E. and D. Gut. 2004. Assessing the risk of potentially invasive plant species in central Europe. *Journal for Nature Conservation* 12:171–179.

Yan, X., L. Zhenyu, W. P. Gregg, and L. Dianmo. 2001. Invasive species in China—an overview. *Biodiversity and Conservation* 10:1317–1341.

Yoo, K. H. and C. E. Boyd. 1994. *Hydrology and Water Supply for Aquaculture*. New York: Chapman and Hall.

Zheng, J. Y., W. Zhuang, Y. T. Yi, G. Wu, J. Gong, and H. B. Shao. 2010. Developmentally utilizing molecular biological techniques into aquaculture. *Reviews in Fisheries Science* 18:125–130.

Zhu, Z., L. He, and S. Chen. 1985. Novel gene transfer into the fertilized eggs of gold fish (*Carassius auratus* L. 1758). *Journal of Applied Ichthyology* 1:31–34.

Živić, I., Z. Marković, Z. Filpović-Rojka, and M. Živić. 2009. Influence of a trout farm on water quality and macrozoobenthos communities of the receiving stream (Trešnjica River, Serbia). *International Review of Hydrobiology* 94:673–687.

Chapter 12

Governmental regulations

Governments have developed regulations for natural resource use, methodologies for producing goods and services, and waste disposal in order to avoid conflicts, prevent overexploitation of resources, protect consumers, lessen risk of human and animal diseases, protect national interests, and avoid negative environmental impacts. Regulations may be developed at the local, state, and federal levels. Federal regulations have the benefit of imposing a uniform, minimum level of regulatory standards across an entire country. In most countries, other levels of government have the prerogative to impose more rigorous regulations. The range and rigor of regulations and degree of enforcement vary greatly among countries and even from place to place within a country.

The earliest regulations on resource use were likely related to water use rights and land tenure. As population increased and technology advanced, habitat destruction and pollution by domestic, industrial, and agricultural activities began to threaten public health and environmental stability. This led to regulations about land use and water pollution. Food production increasingly relies on use of chemicals such as fertilizers, petroleum products, and pesticides—some of which can leave residues potentially harmful to consumers. Food safety regulations were implemented as a result. Spread of animal diseases and negative effects of uncontrolled dispersal of plant and animal species brought into countries for agriculture and other purposes necessitated regulations on species introductions and movement.

All countries probably have developed a suite of resource use, environmental, and food safety regulations. Aquaculture is a latecomer to the regulatory arena. There were few regulations before the 1950s and most regulations have been imposed since the 1970s. Fact sheets were found for national aquaculture legislation in 55 countries including all major aquaculture countries (http://www.fao.org/fishery/nalo/search/en).

Aquaculture, Resource Use, and the Environment, First Edition. Claude E. Boyd and Aaron A. McNevin.
© 2015 John Wiley & Sons, Inc. Published 2015 by John Wiley & Sons, Inc.

Environmental regulations included in aquaculture legislation are varied and many, and country-level comparisons will not be provided. This chapter will only discuss the types of governmental regulations and explain why they have been imposed.

Land and water use

Land and water use has long been a source of conflict among individuals, different land and water uses, municipalities, regions, and countries. Land tenure is the relationship of individuals or groups of people to the land including the waters and forests. The rules of land tenure were developed by societies and may be legal or customary (http://www.fao.org/docrep/005/Y4307E/y4307e05.htm). A governmental entity or community may have the recognized authority to allocate and reallocate land—a power known as expropriation. In some cases, several parties may be allowed different rights to the same property, for example, one party may use the land for farming, another may acquire water from the property, while a third party may be allowed to travel across the land regularly. Several parties with the same interest may be given the collective right to use a parcel of land. Of course, land tenure disputes occur when two parties with the same interest want the exclusive use of the property or when two parties have the right to use the land and the activities of one party interferes with the use of the land by the other party.

Private control of land is common, but even in democratic and capitalist countries people do not actually own the land. They own the right to use the land which can be passed to succeeding generations, leased to others, bartered, or sold. The government can expropriate the land or certain rights to the land through eminent domain with compensation to the owner. Communal land belongs to communities where members have a right to use the holdings. Of course, community leaders may or may not have rules for using the "commons" (Hardin 1968). There may be open areas where specific rights may or may not be assigned, and no one is excluded from using the land. The state may be the owner of the land and the property rights are assigned by the appropriate authority—this is common in socialist countries.

In some countries and especially in coastal areas, zoning of activities may be imposed, and specific activities, for example, aquaculture, may be allowed only in specified areas. In addition, a few countries have made regulations on the amount of aquaculture production that can be realized in an area or the minimum allowable distance between aquaculture facilities.

Land rights usually are registered and taxes must be paid by the property owner or right holder. Thus, those conducting land-based aquaculture usually must have the recognized right to use the land and pay through taxes or other fees for the right. The major problem in aquaculture is not installation of facilities on private land, but with overlapping interest where aquaculture facilities may interfere with other parties who also have right to use the same land. Instances may occur where communal or state allocation of land to aquaculture may be affected by influence. Abuse of other parties by one or more parties may sometimes occur by the acquisition of the commons or open access areas by aquaculture projects. The development of aquaculture in coastal areas has been accused of many environmental and social ills.

Water rights usually are associated with ownership of the land above groundwater aquifers or bordering streams or other water bodies. The owner of this land is known as the riparian rights owner (Vesilind et al. 1994). Riparian owners have the right to use the water for reasonable purposes as long as they do not convey water away from the riparian property and do not interfere unduly with the water use by other riparian owners. Usually when there is competition among riparian owners for water and the concept of "making reasonable use of the water" cannot resolve the issue, the riparian owner who makes a beneficial use of the water has the greatest right.

Groundwater use often is based on the principle that the riparian owner can remove water from beneath the land for reasonable uses. Of course, groundwater use is complicated because groundwater aquifers may extend beneath the properties of several owners, but the well on one property may drawdown the water table under one or more nearby properties, diminishing the amount of water available to other owners.

In some locations, withdrawal of surface water or groundwater may require a permit and reporting of the amount of water withdrawn. There also may be limits on the amount of water that can be withdrawn or governments may impose a fee for water withdrawal. Water rights can impose limits on aquaculture. For example, in Idaho (USA), all of the water rights in the trout farming region have been assigned— either to trout producers or other users.

An interesting issue related to water rights is that in many countries, they can be separated from land ownership. A land owner may sell the water right associated with a property to another party. To illustrate, in the United States, the Texas Water Commission gave a land owner permission to install a catfish farm that relied on groundwater. The fish farmer operated the facility for a short period to establish the right to use a specific amount of water. The aquaculture farm was then closed and the land was sold but the water right was retained by the former fish farmer. He later sold the water right to a nearby municipality for a substantial sum.

In socialist countries, water rights are handled in a similar manner to land tenure— assigned by the responsible, governmental agency. Moreover, the use of coastal waters and water of large inland lakes or reservoirs may sometimes be assigned by local communities or the rights not assigned and no one is excluded from using them.

In most instances, stream water and groundwater use for aquaculture is associated with riparian ownership, land ownership, or allocations given by the government or communities. The most common cause of conflicts is when aquaculture facilities withdraw excessive water and infringe on the rights of other water users in the vicinity. Installation of cage culture or open-water culture in public waters also may lead to conflicts among water users. Governments often require permits for cage culture, and increasingly, these permits place limits on the areas where cages may be installed and the amount of production that can be realized by the permit holder.

Wetlands

A wetland is a land area that is either continuously or periodically saturated with water. The plant life in such areas must be adapted to continuous or periodic soil saturation, and it can be used to delineate wetlands from the surrounding terrestrial habitat. More precise definition of wetlands can be found in Cowardin et al. (1979); Mitsch and Gosselink (1993); and Lyon (1993). The legal definition of a wetland in the United States is "those areas that are inundated or saturated by surface or groundwater at a frequency and duration sufficient to support, and that under normal circumstances do support, a prevalence of vegetation typically adopted for life in saturated soil conditions. Wetlands generally include swamps, marshes, bogs, and similar areas" (http://water.epa.gov/type/wetlands).

Wetlands are highly productive ecosystems. They are nurseries for feeding aquatic animals and waterfowl and other bird life. They are important habitat for amphibians and some reptiles. Wetlands function as sediment traps and provide flood protection. In coastal zones, wetlands provide protection from strong waves and heavy winds of cyclones. Riparian vegetation acts as a buffer zone and filters runoff to reduce the input of suspended solids and nutrients into streams. Riparian vegetation also helps control bank erosion. Mangrove forests (Chapter 5) and other coastal wetlands provide a similar filtration and erosion-control system in the mouths and deltas of rivers. To emphasize the role of wetlands in purifying water, Mitsch and Gosselink (1993) called wetlands "the kidneys of the landscape."

Because of their important roles in ecology and biodiversity, hydrology, water purification, and source of resources for humans, the protection of wetlands has been given much emphasis. Of particular significance is the Ramsar Convention on Wetlands of International Importance. The Ramsar Convention is a treaty for national action and international cooperation to conserve wetlands and assure the wise use of their resources. The name Ramsar is from the city in Iran where the treaty was adopted in 1971 by 164 contracting countries. These countries listed a total of 2083 sites governing 197 849 428 ha of wetlands of international importance. The Ramsar Secretariat is located in Gland, Switzerland, and the Convention has five official partners: Birdlife International, International Union for Conservation of Nature, International Water Management Institute, Wetlands International, and World Wildlife Fund (http://www.ramsar.org/cda/en/ramsar-about-aboutramsar/main/ramsar/1–36%5E7687_4000_0_).

The Ramsar Convention uses a broad definition of wetlands: "lakes, rivers, swamps and marshes, wet grasslands and peat lands, oases, estuaries, deltas and tidal flats, near-shore marine areas, mangroves and coral reefs, and human-made sites such as fish ponds, rice paddies, reservoirs, and salt ponds." The Convention's mission is "the conservation and wise use of all wetlands through local and national actions and international cooperation." The Ramsar Convention promotes the use of wetlands, and defines this concept as "the maintenance of their (wetlands) ecological character, achieved through the implementation of ecosystem approaches, within the context of sustainable development."

The Ramsar Convention—like other international conventions—has lofty ideals and goals. Countries that agreed to participate made a commitment "to work toward the wise use of wetlands through national land-use planning, appropriate policies and legislation, management actions, and public education." Moreover, they agreed "to cooperate internationally concerning transboundary wetlands, shared wetland systems, shared species, and development projects that may affect wetlands." Of course, this commitment was not binding, and some countries have fulfilled their commitment better than others. Nevertheless, because of the Ramsar Convention and other organizations concerned about the disappearance and degradation of the world's wetlands, most governments have developed regulations about wetlands. These regulations and their enforcement obviously vary.

The regulations for wetlands in the United States are so many that a meaningful summary of them cannot be made in a short space. A convenient list is provided by the USGS (http://water.usgs.gov/nwsum/WSP2425/legislation.html). The United States signed the Ramsar Convention and made a commitment to no net loss of wetlands. The US Fish and Wildlife Service is responsible for the National Wetland Inventory. Several agencies have responsibilities for wetland protection but the main regulations are Section 404 of the Clean Water Act and Section 10 of the Rivers and Harbors Act (http://www.environment.fhwa.dot.gov/ecosystems/laws_rgl022.asp). The US Army Corps of Engineers has the major responsibility for issuing permits related to wetland use or modification.

The regulations allow certain uses and alterations of wetlands, and allow wetlands to be removed for approved purposes provided the removal is mitigated by creating wetlands in another area. However, careful studies must be made, permits obtained, and compliance verified.

There are incentive and disincentive measures to conserve wetlands and to improve the extent and environmental quality of these ecosystems (Copeland 2010). The USDA Natural Resource Conservation Service (USDA NRCS) is responsible for the wetland reserve program, and more than 11 000 US landowners have enrolled nearly 100 000 ha of wetlands into this program (http://www.nrcs.usda.gov/wps/portal/nrcs/main/national/programs/easements/wetlands/).

These regulations and programs appear to have been successful. Private land owners are well aware of the need to obtain permits for permissible uses and alterations of wetlands on their property. There also is much publicity by government entities, educational organizations, and eNGOs about the need to conserve wetlands. According to data provided by Dahl (2011), the rate of wetland loss in the United States averaged about 220 000 ha/year between 1600 and 2000. In the period 2004–2009, wetlands were lost at a rate of about 5040 ha/year in the United States, but because of variation in measurements, the difference in estimated wetland area in 2004 and 2009 was not statistically significant.

Bird and other predator control

Predators may do significant economic damage to aquaculture stocks, and managers attempt to reduce this damage by shooting, trapping, or poisoning them. These actions often are illegal because most countries have laws to protect many species that may prey on fish, shrimp, and other aquaculture species. For example, in the United States, the Migratory Bird Treaty Act protects all migrating birds and their feathers, nests, and eggs. One may not kill, posses, or transport a migratory bird without a special federal permit. Many of the species that are present at aquaculture facilities are migratory. It is often possible to obtain a permit for destroying migratory birds that have become a nuisance by destroying public or private property (including nursery stocks and cultured fish) or threatening public health or welfare. The permits specifies the conditions under which the birds may be controlled, the number that may be killed, and the methods that may be used for this purpose (http://www.extension.org/pages/10443/federal-laws-and-regulations/print/).

There are many procedures for frightening birds away and for preventing predator access to aquaculture facilities (Barras and Godwin 2005). Frightening techniques are initially successful but birds get used to the frightening stimulus—usually a noise, reflection, or light—and ignore it. Environmentalists encourage the nondestructive control of wildlife predation on aquaculture and there is a growing tendency to use this approach. In severe cases of predators, it may be necessary to use lethal control methods.

Importations of aquaculture species

The problems associated with importation of exotic or nonnative species are discussed in Chapter 11. Because of the adverse ecological impacts that have sometimes resulted most governments have regulations on importations of nonnative species. Some species may be banned completely, while other importations may require permits.

In the United States, imports of live fish are regulated by the USDA Animal and Plant Health Inspection Service (APHIS) (http://www.aphis.usda.gov/import_export/animals/animal_import/marine_import_fish.shtml). Most species of live finfish may be imported into the United States without import requirements from APHIS. However, eight species of fish that are susceptible to the disease spring viremia of carp now are under APHIS oversight and require permits with certain conditions. These species are common and koi carp, goldfish, grass carp, silver carp, bighead carp, crucian carp, tench, and sheatfish. Some states in the United States have additional regulations about the introduction of fish.

Many countries have stricter laws than the United States about importations of live fish and shellfish. This is especially true for some of the major shrimp aquaculture countries. However, no references could be found to the degree to which different countries enforce the regulations of importations of nonnative species.

Aquatic animal disease regulations

International trade in live eggs, fingerlings, and brood stock can result in the movement of diseases from one country to another or from one location to another in the same country (Hendrick 1996). The increase in aquaculture production has resulted in an increase in the outbreaks of traditional pathogens found in a region. Many times these diseases can be minimized to an acceptable level by good management practices to assure suitable environmental conditions in culture systems to avoid stressing the culture species. However, aquaculturists may bring in new species with a greater aquaculture potential than native species, better-performing brood stock, or fingerling and postlarvae from other regions or countries. These introductions may be contaminated with diseases not naturally present. This can lead to serious epidemics.

A good example of the spread of a new disease is the white spot virus disease of shrimp. The disease emerged in Taiwan in 1991 and 1992 and quickly spread to mainland China where it was first reported in 1992—apparently it spread to mainland China in live postlarvae for stocking in ponds. The same year the disease was reported in Japan, Korea, and India. By 1994, white spot disease reached Thailand and soon was reported in most other Asian countries with shrimp farming. In the mid-1990s, the disease was found in the United States, and during 1999 and 2000, it spread to all shrimp farming countries along the Pacific coast of the Americas. In 1999 alone, in Ecuador, the losses were estimated at $US280 million and around 150 000 jobs were lost (http://agriculture.de/acms1/conf6/ws9fish.htm).

The World Organization for Animal Health (OIE) is an intergovernmental organization responsible for improving animal health worldwide. This organization has a Fish Diseases Commission (FDC) with the primary role of developing aquatic animal health standards and regulations. The FDC has an Aquatic Animal Health Code that includes the following sections: disease diagnosis, surveillance, and notification; risk analysis; quality of aquatic animal services; recommendations for preventing and controlling diseases, importation/exportation procedures, and health certificates; veterinary public health; welfare of farmed fish; specific sections on diseases of different types of aquatic animals (http://www.oie.int/international-standard-setting/aquatic-code/access-online/).

Many countries have made regulations for importations and exportations of live eggs, fingerlings, and adults of aquaculture species based on the FDC code. However, despite these regulations new diseases continue to spread. A good example is the recent spread of the early mortality syndrome or acute hepatopancreatic necrosis syndrome of farmed shrimp that began in mainland China in 2010 and spread to Vietnam, Malaysia, and Thailand by 2011 (Lightner et al. 2012; Flegel 2012). The cause of this disease has now been identified as a strain of the bacterium *Vibrio parahaemolyticus* that is infected with a phage (Lightner et al. 2013), and it was reported unofficially in Mexico in May 2013.

Seafood regulations

The main emphasis of this chapter is environmental regulations, but brief mention of seafood regulations is necessary. The use of antibiotics, drugs, and other chemicals was discussed in Chapter 9. Some of these substances can lead to residues in aquaculture products that can be potentially harmful to consumers. Many countries have rules on maximum concentrations of various chemical residues allowable in seafood. These concentration limits are imposed on both domestic and imported seafood including aquaculture products.

The greatest concern about food safety by consumers in developed countries is usually expressed about imported products. Most developed countries have seafood inspections at the port of entry. Samples are analyzed for physical, chemical, and biological contaminants. Shipments not passing inspection are refused entry or seized and destroyed.

There are two major programs worldwide to assist producers and processors meet the food safety standards imposed on domestic and imported products. One is the HACCP program in which a company takes a preventive approach to food safety by identifying hazards that can negatively affect food safety and applying methods for eliminating the risk or reducing it to an acceptable level (http://www.22000-tools.com/what-is-haccp.html?pmc=ba).

The Codex Alimentarius Commission was established in 1961 by FAO and the WHO. The Codex Alimentarius contains many food standards and codes of practice to improve the safety, quality, and fairness of international food trade (http://www.codexalimentarius.org/about-codex/en/).

Many food producers and processors achieve certification by the ISO. The HACCP and standards and guidelines of the Codex Alimentarius are used in ISO certification of food safety.

In many countries, a country-of-origin label is required to allow consumers to distinguish between domestic and imported products. In the United States, country-of-origin labeling is required on seafood—including aquaculture products (http://www.ams.usda.gov/AMSv1.0/cool). In addition, many countries—including the United States—require method of production labeling. The label states if the product was from wild-caught or farmed aquatic animals.

Effluent discharge permits

Most countries require that discharges of wastewater be permitted. A discharge permit gives a party permission to discharge wastewater for a specific period of time—generally several years. Restrictions usually are placed on concentrations of certain water quality variables in the effluent, the volume of effluent, or both. The permits also specify the frequency of monitoring and reporting schedule for the results.

The way in which wastewater discharge permits are formulated varies greatly among countries, but the purpose is always the same—to prevent the discharge from causing water-quality deterioration in the receiving water body. In this discussion, we will describe how wastewater permitting is done in the United States as an example. The US Clean Water Act of 1965 requires a National Pollutant Discharge Elimination System (NPDES) permit for every discharge of pollutants. The NPDES does not apply to storm runoff, discharge into water treatment systems, and some other effluents. Management and enforcement of the Clean Water Act is the responsibility of the USEPA. Individual states and municipalities in the United States may establish regulations for effluents not covered at the federal level.

Stream classification

Effluents often are released into streams, and streams have different primary uses such as potable water supply, fish and wildlife habitat, recreation, irrigation, industrial water supply, and navigation. Some primary uses require better water quality than others, and streams are classified by use and assigned water quality standards based on use categories. It will be useful to explain how USEPA defines a water quality standard. Water quality standards are legally enforceable statements that describe the ambient condition that should exist in a water body. A standard provides the designated use (classification level) of a water body. The standard has water quality criteria specifying the amounts of selected pollutants that may be present in the water body without impairing the designated use. Water quality standards also include antidegradation requirements to protect existing water uses and limits on degradation of high quality waters (Gallagher and Miller 1996).

Each state in the United States was required to classify their streams, assign water quality standards for each use class, and implement the NPDES. For example, in Alabama, the Alabama Department of Environmental Management (ADEM) developed eight classification levels listed in order of decreasing water quality status: Outstanding Alabama Water, Public Water Supply, Swimming and Other Whole Body Water-Contact Sports, Shellfish Harvesting, Fish and Wildlife, Limited Warmwater Fishery, and Agricultural and Industrial Water Supply (http://www.adem.alabama.gov/alEnviroRegLaws/files/Division 6Vol1.pdf). Water quality standards also were made for most large reservoirs in Alabama. The criteria for discharge differs among the waters of Alabama, for example, the dissolved oxygen concentration should not fall below 6 mg/L in an Outstanding Alabama Water, but it may fall to 3 mg/L in a stream classified as Agricultural and Industrial Water Supply. Other water quality variables in the stream classification standards and reservoir standards vary by a similar amount but it is not feasible to list them. Moreover, the stream classification systems and associated water quality standards vary from state to state.

Wastewater discharge permits

The objective of wastewater discharge permits is threefold: (1) to avoid mortality of aquatic species or because of ecological nuisances in the area where the effluent mixes with the receiving waters; (2) to prevent excessively high concentrations of pollutants throughout the receiving water body; (3) protect human health. A small discharge of a highly concentrated effluent might cause serious water quality degradation near the outfall into the receiving water, but because of dilution, it might not have a measureable effect upon the water body as a whole. A large discharge of a relatively dilute effluent might not cause toxicity of ecological perturbations in the mixing zone near the effluent outfall, but it might over time cause a high concentration of one or more pollutants in the entire receiving water body. Moreover, wastewater might have acceptable ranges of chemical and physical water quality variables, but contain microorganisms harmful to human health.

The common types of pollution are suspended solids (turbidity) mainly from erosion; nitrogen, phosphorus, and other nutrients that cause excess algal blooms leading to eutrophication; organic matter that decomposes to cause a low dissolved oxygen concentration; toxicity from organic or inorganic chemicals; thermal pollution caused by heated effluents; pollution with potentially pathogenic organisms. Each effluent will be different with respect to kinds, intensity, and amounts (loads) of pollutants, and each receiving water body will have a different capacity to dilute and assimilate pollutants. Thus, the water quality variables that must be included in a given wastewater discharge permit and the concentrations, amounts, and discharge rates—usually referred to as effluent limitations—must be established on a case-by-case basis.

Water pollution characteristics usually are quite similar across a given industry. The USEPA makes national rules for regulating wastewater discharges on an industry-by-industry basis. The federal rule usually establishes guidelines for exempting specific types of operations within the industry from regulations, recommends treatment methods, and establishes priority pollutants and effluent limitation guidelines for them. The rule is made by the USEPA but they allow stakeholder involvement in the process. The public also has the opportunity to comment on the proposed rules. Once the rule is finalized and published in the Federal Register, it becomes law.

State water pollution control agencies in the United States must work to assure that all facilities within the various industries comply with the federal effluent rules. Individual states, however, can implement stricter effluent limitations than necessary to comply with the federal rule where they think it necessary to do so in order to prevent noncompliance of water bodies with water quality standards.

The permit holder may not be able to comply with the conditions of the permit immediately, and a schedule for compliance will be specified. In the United States, and most other countries, water pollution control agencies depend upon self-monitoring to verify whether or not compliance with a permit is achieved and maintained. The permit will specify the frequency of sampling, method of securing the samples, variables to be measured, methods of measurement, and reporting requirements. Of course, environmental agencies often make their own analyses for quality control or in cases of suspected problems with reported self-monitoring results.

Enforcement of wastewater discharge permits normally requires administrative review of the periodic reports from monitoring. In theory, the permit holder should know when noncompliance occurs, and should work to correct the situation. The administrative actions related to noncompliance usually result in a warning and possibly a compliance schedule will be mandated. If the warning is not heeded, fines may be imposed, and serious violations may lead to termination of the discharge permit, or in cases of flagrant disregard for the terms of the permit or fraudulent reporting, criminal prosecution.

Some parties who believe that their interests have been damaged by wastewater discharges—even if the discharge complies with permit conditions—may bring civil lawsuits against permit holders. It also is not uncommon for environmental groups to initiate lobbying efforts in attempts to obtain stronger water quality standards, more rigorous wastewater discharge permits, and stricter enforcement of permits. These groups also may initiate civil actions when environmental agencies do not respond to their requests.

The water quality criteria in wastewater discharge permits can be quite varied. The simplest types of permits are those that contain only criteria regarding permissible concentrations of selected water quality variables. Examples of concentration-based criteria in a wastewater discharge permit are as follows:

Variable	Permissible range
pH	6–9
Dissolved oxygen	5 mg/L or above
Biochemical oxygen demand (BOD)	30 mg/L or less
Total suspended solids (TSS)	25 mg/L or less

These effluent limitations put restrictions on how high or low concentrations of variables may be, and still protect against excessively high concentrations of pollutants in the mixing zone. However, the limits can be met by intentionally adding water to dilute the effluent without installing means for reducing the total quantities of pollutants discharged. If this practice is prohibited, then simple effluent limitations will prevent the permittee from increasing pollutant discharge above the amount that is necessary to attain the limit. As mentioned earlier, simple effluent limitations do not necessarily avoid a gradual decrease in water quality within the receiving water body as a whole.

Water quality standards also may have mass- or load-based criteria (concentration multiplied by effluent volume). The BOD limit might be given as a daily maximum discharge of 100 kg BOD/day. This means that compliance might be achieved with a high BOD concentration if the discharge is low, but compliance might not be possible with a much lower BOD but greater discharge. For example, at a discharge of 10 000 m^3/day, the maximum permissible BOD load would be 10 mg/L. At a discharge of 1000 m^3/day, the maximum allowable BOD load would be reached at a BOD concentration of 100 mg/L. It is necessary to specify maximum concentrations of pollutants in mass-based criteria to avoid an excessive concentration of one or more variables in the mixing zone where the effluent is diluted by receiving waters. The BOD criteria in a wastewater discharge permit could be as follows: a maximum BOD load of 100 kg/day with the maximum BOD concentration not to exceed 30 mg/L during any 24-hour period.

Limiting the load of an effluent is not necessarily helpful unless the capacity of the receiving water to assimilate the waste is known. Of course, the load-based criteria prevent the permittee from increasing the amount of pollution over time.

In the United States, there is a program to develop total maximum daily loads (TMDLs) for streams and other receiving waters. This involves calculating the maximum amount of a pollutant from all sources (natural or pollution) that can be allowed without causing the water body to violate its standard. In such situations, the TMDL of a pollutant must be allocated among the different sources of the pollutant. For example, suppose a stream reach has a TMDL for phosphorus of 500 kg/day, several industries want to discharge into the stream reach, and natural sources of phosphorus are 100 kg/day. The maximum amount of phosphorus that can be contained in effluents from the industries is 400 kg/day, and this amount normally would not be allowed in order to have a safety factor. If a safety factor of 2 is used, only 200 kg/day of phosphorus would be allowed. This load would then have to be allocated among the different industries.

Allotments for maximum pollutant loads in individual wastewater discharge permits allow trading in allotments among permittees If a permittee does not need the entire TMDL allocated, the excess allocation can be sold to another permittee discharging into the same water body. Concentration limits also may be imposed in TMDL-based standards to protect water quality in the mixing zone.

Limits may sometimes be put on total discharge volume. For example, a standard might restrict average daily discharge for any month to 100 000 m^3/day. Typically a maximum daily discharge also would be specified. In the above example, the maximum daily discharge might be limited to 250 000 m^3/day.

Qualitative criteria may be included in wastewater discharge permits. There might be a provision that a visible turbidity plume at the outfall is unacceptable. Other examples are to prohibit odors or foam in effluents.

Sometimes a wastewater discharge permit will specify the maximum allowable increase for one or more variables. The total suspended solids (TSS) criteria could specify that TSS concentration in effluent cannot exceed the ambient TSS concentration of intake water by more than 15 mg/L. In other cases, the criteria might give the permissible effluent concentration of a variable as an increase above the receiving water concentration. For example, the turbidity cannot be more than 20 nephlometer turbidity units greater than that of the receiving water.

The USEPA has published criteria for more than 50 pollutants to be used in water quality standards. Criteria for toxic pollutants are applied so that acutely toxic conditions do not exist in any state waters of any designated use. According to Gallagher and Miller (1996), USEPA regulations require NPDES permits to include limitations for all pollutants that may be discharged at levels that can cause the receiving water to violate its use classification standard.

It is difficult to establish concentration-based criteria for toxicants in wastewater discharge permits because the toxicity of some metals and organic chemicals vary greatly with water quality conditions. Thus toxicity-based limitations may be established by whole effluent toxicity testing. The toxicity tests are conducted by exposing certain species of aquatic organisms to the effluent in question under controlled conditions. An NPDES permit may require that toxicity testing be done to prove that the effluent is not acutely toxic to organisms in the receiving water body.

Biological criteria or biocriteria also may be included in a permit. Biocriteria may be used to supplement the traditional water quality criteria or used as an alternative where traditional criteria have not been established. Development of biocriteria requires a reference condition (minimal impact) for each use classification, measurement of community structure and function in reference to water quality to establish biocriteria, and a protocol for determining if community structure and function has been impaired.

In addition to meeting the standard for the use classification, the dilution and mixing of the discharge in the receiving water must be considered. The purpose of this activity is to prevent water quality deterioration in an area around the outfall, yet recognizing that standards may be exceeded within the mixing zone but without acutely toxic conditions existing. Also, water quality in the mixing zone must be adequate to assure the well-being of the established aquatic community in the water body and to allow designated use of the water to continue.

Table 12.1 Definition of concentrated aquatic animal production (CAAP) facilities in the United States.

Coldwater CAAP facilities

Includes ponds, raceways, or other similar structures which discharge at least 30 days per year but does not include
- Facilities which produce less than 9090 harvest weight kilograms per year.
- Facilities which feed less than 2272 kg during the month of maximum feeding.

Warmwater CAAP facilities

Includes ponds, raceways, or other similar structures which discharge at least 30 days per year but does not include
- Closed ponds which discharge only during periods of excess runoff.
- Facilities which produce less than 45 454 harvest weight kilograms per year.

Source: Modified from Federal Register (2004).

Aquaculture effluent rule

The first effort to develop a national aquaculture effluent rule was in the 1970s (USEPA 1974). However, this plan was soon discarded because of pressure applied by the aquaculture industry, and the opinion of USEPA that there were much more important industries needing effluent rules. Nevertheless, the effort resulted in the USEPA defining warmwater- and coldwater-concentrated aquatic animal production (CAAP) facilities (Table 12.1).

The development of the federal effluent rule for US aquaculture was finally initiated in the late 1990s in response to the environmental lobby. Several eNGOs had filed a complaint in federal court that the USEPA was not implementing the Clean Water Act in a timely manner. They won a consent decree in which the judge ordered USEPA to add more industries to the list of industries for which effluent rules would be developed before a specific date. Soon after this eNGOs became concerned about effects on the environment of cage culture of salmon and marine shrimp farming. This concern "spilled over" to those interested in the environmental effects of aquaculture in the United States, and the EDF published an assessment of the environmental effects of aquaculture in the United States (Goldburg and Triplett 1997). Information in this publication was used by the EDF and several other eNGOs to convince the USEPA to start an effluent rule-making activity for US aquaculture. The USEPA removed the shipping container industry from the list of industries for new effluent rules and replaced it with the aquaculture industry.

The USEPA asked the USDA to convene an Effluent Task Force to provide information on aquaculture to them and a consulting company that was hired to assist in obtaining information on aquaculture production and resulting environmental impacts. Based on inputs of the Effluent Task Force and the consulting company, USEPA made a draft rule for aquaculture (Federal Register 2002), and following review and comment on the proposal by the aquaculture industry, eNGOs, and other stakeholders, the final rule was published (Federal Register 2004).

The USEPA concluded that aquaculture was not a major contributor to the pollution of surface waters in the United States. They concluded further that there were no cost-effective means of treating most aquaculture effluents. Because of this the

Table 12.2 Basic elements of BMP plan recommended by United States Environmental Protection Agency for concentrated aquatic animal production facilities.

Management of removed solids and excess feed

- Minimize TSS in water supply.
- Minimize excess feed.
- Minimize discharge of unconsumed feed.
- Clean raceways to minimize discharge of accumulated solids at harvest.

Proper operation and maintenance

- Maintain in-flow system technology.
- Store drugs and chemicals in a manner to prevent spills.
- Disposal of biological wastes in a responsible way.
- Prevent escape of nonnative species.
- Educate workers about BMP plan.

Source: Modified from Federal Register (2004).

USEPA did not recommend effluent limitation guidelines for ponds and most other production facilities.

An NPDES permit was required for pond aquaculture operations that qualified as CAAP facilities, but instead of mandating effluent limitations for these facilities, USEPA recommended the use of aquaculture BMPs. The basic elements of the BMP plan are provided in Table 12.2.

USEPA provided effluent limitation guidelines for recirculating systems and net pen (cage) systems producing more than 45 454 kg harvest weight per year. For recirculating systems, maximum daily TSS concentration was limited to 50 mg/L but the maximum monthly average TSS concentration cannot exceed 30 mg/L. These facilities must have BMP plans (Table 12.3) for removal of solids and excess feed, and for operation and maintenance.

Many NGOs consider this rule too weak to be of much benefit, and several states actually have effluent rules stricter than the federal rule. Several of the major aquaculture states—Alabama, Mississippi, Arkansas, and Louisiana—apparently have not yet implemented the USEPA effluent rule.

It is interesting to note that several other studies have come—like USEPA—to the conclusion that aquaculture effluents are elevated in certain pollutants relative to receiving waters, but nevertheless not major polluters of surface waters (Boyd et al.

Table 12.3 Requirements for the United States National Pollutant Discharge Elimination System discharge permit for net pen and cage culture operations that qualify as concentrated aquatic animal production facilities.

- Monitor rate of feed consumption to prevent feed from falling through net.
- BMP plan
  - Minimize net fouling organisms discharged when cleaning nets.
  - Avoid discharging blood, viscera, fish carcasses, etc.
  - Prevent escape of nonnative species.
  - Avoid discharge of feed bags and other solid wastes, cleaning chemicals, and materials containing tributyl tin compounds.
- Alaskan net pen systems for native species are excluded.

Source: Modified from Federal Register (2004).

2000; Bosma et al. 2009; Anh et al. 2010; De Silva et al. 2010). For example, Anh et al. (2010) estimated that *Pangasius* culture of over 1 Mt/year in Vietnam contributes less than 1% of the TSS, nitrogen, and phosphorus loads to the Mekong River.

Other regulations

Regulations regarding resource use and operation of aquaculture facilities vary considerably from place to place and there possibly are requirements in addition to those already mentioned. Some examples of additional regulations—although not complete—are listed as follows: business licenses; aquaculture licenses; land use and other taxes; construction permits; landfill operation permits; minimum wage; sanitation requirements; fire prevention regulations; worker safety; disposal of wastes.

Conclusions

Governments have imposed numerous regulations on aquaculture. Regulations in many countries theoretically should avoid wasteful use of resources and prevent most environmental perturbations. However, some countries may not have adequate regulations, and even in countries with excellent regulations, enforcement can be inadequate. Environmentalists seem to have a distrust of environmental regulations and this has led to programs going beyond legal requirements with regard to resource conservation and environmental protection that will be discussed in Chapters 13 and 14.

The eNGO perspective

The eNGOs have a special goal of going beyond what is being accomplished by current activities of either the government or the private sector with regard to environmental conservation. Whether the *Great Smog* (1952) in London which killed 4000 people, the Torrey Canyon vessel that ran aground off the coast of Cornwall in 1967 and leaked large amounts of oil in the coastal zone, the Three Mile Island partial meltdown in the United States in 1979, the Bhopal Union Carbide disaster in India in 1984 that killed tens of thousands of people, or mercury poisoning of the Japanese community of Minamata throughout the 1960s (McCormick 1995), the momentum of the environmental movement picked up speed from the 1950s onward. Outrage over governmental complacency and disregard of environmental safeguards by industry were keys to the formation of a variety of NGOs both engaged in human health and environmental concerns. The formation of these organizations was a response to incidences rather than a preemptive desire to do away with capitalism as many industry advocates contend.

Regardless of the specific events that led to the formation of specific eNGOs, there is a common thread of distrust for actors that contend protections are in place to prevent environmental harm when these measures directly compete with capital allocations from private sector or government-backed development efforts. Although there has been a significant amount of improvement and acknowledgement of past mistakes by a variety of actors as they relate to environmental conservation in the governmental and private sectors, eNGOs still consider their role as a watchdog of these actors vital for the environment.

While aquaculture is clearly a small actor globally with regard to environmental impacts, there has been environmental degradation because of this activity. Relativity of impacts aside, the eNGOs will continue to play the watchdog role as it relates to aquaculture. The rapid increase in growth and subsequent land conversion, introductions and escapes of exotic species, chemical use, water pollution, and wild fish dependency will continue to be issues that eNGOs point to as needing greater attention. It is difficult to challenge that improvements still remain for the aquaculture industry as they remain for any other food and fiber production sector. Different sectors of the aquaculture industry have different impacts, and different countries and locales have varied levels of governance to address these issues. The eNGOs see their role as constantly providing pressure to priority aquaculture activities such that an improved environmental condition will be realized. While the most recent push has been on the private sector to adopt standards and certification, it is clear that this represents a very small fraction of the total global aquaculture production. The next frontier for eNGOs concerned with aquaculture will likely be toward better policy and stricter enforcement of environmental regulations related to domestic production in leading aquaculture-producing nations. Domestic production is after all the lion's share of aquaculture production.

References

Anh, P. T., C. Kroeze, S. R. Bush, and A. P. J. Mol. 2010. Water pollution *Pangasius* production in the Mekong Delta, Vietnam: cause and options for control. *Aquaculture Research* 42:108–128.

Barras, S. C. and K. C. Godwin. 2005. Controlling bird predation at aquaculture facilities: frightening techniques. *Southern Regional Aquaculture Center Publication 401*. Stoneville, Mississippi: Mississippi State University.

Bosma, R. H., C. T. T. Hanh, and J. Potting (editors). 2009. *Environmental Impact Assessment of the Pangasius Sector in the Mekong Delta*. Wageningen, The Netherlands: MARD/DAQ, Wageningen University.

Boyd, C. E., J. Queiroz, J. Lee, M. Rowan, G. N. Whitis, and A. Gross. 2000. Environmental assessment of channel catfish, *Ictalurus punctatus*, farming in Alabama. *Journal of the World Aquaculture Society* 31:511–544.

Copeland, C. 2010. Wetlands: an overview of issues. *Congressional Research Service Paper 37*. Lincoln, Nebraska: University of Nebraska.

Cowardin, L. M., V. Carter, F. C. Golet, and E. T. LaRoe. 1979. *Classification of Wetlands and Deepwater Habitats of the United States*. Washington, DC: Department of the Interior, US Fish and Wildlife Service.

Dahl, T. E. 2011. *Status and Trends of Wetlands in the Conterminous United States 2004 to 2009*. Washington, DC: Department of the Interior, US Fish and Wildlife Service.

De Silva, S. A., B. A. Ingram, P. T. Nguyen, T. M. Bui, G. J. Gooley, and G. M. Turchini. 2010. Estimation of nitrogen and phosphorus in effluent from the striped catfish farming sector in the Mekong Delta, Vietnam. *Ambio* 39:504–514.

Federal Register. 2002. Effluent limitation guidelines and new source performance standards for the concentrated aquatic animal production point source category: proposed rule. *Federal Register* 67(117):57872–57928. Office of the Federal Register, National Archives and Records Administration, Washington, DC.

Federal Register. 2004. Effluent limitation guidelines and new source performance standards for the concentrated aquatic animal production point source category: final rule. *Federal Register* 69(162):51892–51930. Office of the Federal Register, National Archives and Records Administration, Washington, DC.

Flegel, T. W. 2012. Historic emergence, impact, and current status of shrimp pathogens in Asia. *Journal of Invertebrate Pathology* 110:166–173.

Gallagher, L. M. and L. A. Miller. 1996. *Clean Water Handbook*. Rockville: Government Institutes, Inc.

Goldburg, R. and T. Triplett. 1997. *Murky Waters: Environmental Effects of Aquaculture in the United States*. Washington, DC: Environmental Defense Fund.

Hardin, G. 1968. Tragedy of the commons. *Science* 162:1243–1248.

Hendrick, R. P. 1996. Movements of pathogens with the international trade of live fish: problems and solutions. *Reviews of Science and Technology of the Office of International Epizoology* 15:523–531.

Lightner, D. V., R. M. Redman, C. R. Pantoja, B. I. Noble, and L. Tran. 2012. Early mortality syndrome affects shrimp in Asia. *Global Aquaculture Advocate* 15:40.

Lightner, D. V, Redman, R. M., Pantoja, C. R., Noble, B. L., Nunan, L. M., and Tran, L. 2013. Documentation of an emerging disease (early mortality syndrome) in SE Asia and Mexico. Presented at the Joint Final Technical and National Consultations on Early Mortality Syndrome (EMS) or Acute Hepatopancreatic Necrosis Syndrome (AHPNS) of Cultured Shrimp (FAO/TCP/VIE/3304), 25–27 June, Hanoi, Vietnam.

Lyon, J. G. 1993. *Practical Handbook for Wetland Identification and Delineation*. Boca Raton, Florida: CRC Press.

McCormick, J. 1995. *The Global Environmental Movement*. London: John Wiley.

Mitsch, W. J. and J. G. Gosselink. 1993. *Wetlands*. New York: Van Nostrand Reinhold.

USEPA (United States Environmental Protection Agency) 1974. *Development Document for Proposed Effluent Limitation Guidelines and New Source Performance Standards for the Fish Hatcheries and Farms*. Denver: National Field Investigations Center.

Vesilind, P. A., J. J. Peirce, and R. F. Weiner. 1994. *Environmental Engineering*. Boston, Massachusettes: Butterworth-Heinemann.

Chapter 13

Best management practices

In response to environmental concerns, the FAO included a section on responsible aquaculture in the Code of Conduct for Responsible Fisheries (FAO 1995). The aquaculture industry did not want a bad environmental image that could negatively affect sales of aquaculture products and possibly lead to stricter governmental regulations and possible bans on some kinds of aquaculture. The industry and its advocates developed codes of conduct and codes of practices based on the FAO Code of Conduct for Responsible Aquaculture. Codes of conduct made statements about commitments to environmental objectives, while the codes of practices listed procedures that should or should not be implemented at aquaculture facilities.

The first codes of practices for aquaculture apparently were made by the Irish Salmon Growers Association (1991)—before the FAO code—the Australian Prawn Farmers Association (Donovan 1997), a shrimp farmer association in Belize (Dixon 1997), and the Global Aquaculture Alliance (Boyd 1999). Many more codes soon were prepared by governmental agencies, aquaculture associations, international development organizations, and eNGOs.

Codes of practices typically consisted of 10 to 15 terse statements about minimizing specific impacts. For example, a common statement in most codes of practices is to apply feeds efficiently to avoid waste that could cause pollution. Specific instructions on how to use feed efficiently usually are not provided.

Practices that can conserve resources and lessen pollution are called best management practices or BMPs. Application of BMPs is the most common approach to reducing the negative environmental impacts of agriculture and they sometimes are required as part of the conditions for compliance with wastewater discharge permits or other environmental regulations (Chapter 12).

Some purists object to the term best management practices, because the best in BMP implies that a practice cannot be improved. The purists prefer the terms better management practices or good management practices. This concern may seem trivial, but it does emphasize the importance of revising BMPs as better ways of minimizing impacts are found.

Aquaculture, Resource Use, and the Environment, First Edition. Claude E. Boyd and Aaron A. McNevin.
© 2015 John Wiley & Sons, Inc. Published 2015 by John Wiley & Sons, Inc.

This chapter discusses the use of BMPs in agriculture and other industries, and the development of BMPs for aquaculture.

Background

Agriculture primarily causes nonpoint source pollution in runoff that enters natural waters. Runoff cannot be treated by conventional means as can point source pollution because of its large volume and diffuse nature. The usual way of contending with nonpoint source pollution is to apply BMPs to minimize or stop it at its source on the catchment. The most common types of BMPs are those used for nutrient management, integrated pest management (IPM), and controlling erosion and contamination of runoff.

In nutrient management, fertilizers are applied in a manner to avoid their removal in surface runoff or in water infiltrating through the soil (Havlin and Tisdale 2005). Fertilizer should be applied to crops in quantities not to exceed the rate that crop plants can absorb the nutrients in them (http://www.epa.gov/oecaagct/ag101/cropnutrientmgt.html). Soil tests are used to determine the exact nutrient needs of crops, and fertilizers are applied at times when nutrients are not likely to be removed in surface runoff. Manure applications to fields should be based on tests of the availability of nutrients in manures. This allows the amounts of chemical fertilizers to be reduced accordingly. Chemical fertilizers should be applied just before they are needed by the particular crop, for example, corn needs nitrogen mostly when plants are 3–4 weeks of age. Conservation tillage and erosion control also minimize loss of phosphorus absorbed on soil particles.

IPM involves adopting pest control measures that rely on a combination of biological, mechanical, cultural, and chemical techniques (Burns et al. 1987). The objective of IMP is to achieve effective pest control but to minimize the amounts of pesticides used. It involves an understanding of the pest so that a combination of control practices can be selected and applied at the correct time. Proper storage, handling, mixing, and application of pesticides also are critical aspects of IMP. By using less pesticide, IMP reduces the amount of pesticide drift and pesticide contained in surface runoff and in water infiltrating into aquifers. Weather conditions are particularly important for safe application of pesticides. Windy conditions favor aerosol drift, and rainfall can cause fluvial movement of pesticides.

Erosion and runoff control are important in preventing soil loss from agricultural areas. The result is lower turbidity and sediment loads in runoff. Erosion control also lessens the loss of nutrients and pesticides from agricultural land (Toy et al. 2002). No-till farming is a classic example of erosion control because tilled land erodes more easily than non-tilled land. There are many other BMPs for erosion control; some examples are strip-cropping, cover crops, shelter belts, conservation tillage, and continuous cropping. Grass-lined waterways trap sediment, and vegetation prevents erosion of waterways. Terraces may be installed to avoid gullies in agricultural fields, and bare areas of noncultivated land are covered with grass or other vegetation to lessen erosion.

Table 13.1 Representative rates of erosion for selected land uses.

Land use	Erosion (t/ha/year)	Relative erosion rate (forest = 1)
Forest	0.034	1
Grassland	0.34	10
Cropland	6.8	200
Harvested forest	17.0	500
Construction	68.0	2000

Source: United States Environmental Protection Agency (1979).

Erosion is a natural process and the key factor determining the shape of the earth's land surface. The more disturbance of the land caused by an activity, the greater is the resulting erosion rate (Table 13.1). Application of BMPs for reducing erosion is not confined to agriculture. They often are used in controlling erosion at construction sites, during and after logging operations, at surface mining locations, etc. Silt fences, sediment bags, geofabric reinforcement, grass strips, proper sloping of embankments, road grades and ditches, planting grass over finished earthwork, less destructive log skidding methods, and installation of settling basins are some examples of common BMPs used in construction, logging, and mining.

In many countries, agricultural agencies have programs—often with incentives—to assist farmers in developing BMPs to avoid a wide range of negative impacts. Many guidelines for making agricultural BMPs have been published in manuals or posted on websites. Development of BMPs should be based on sound scientific and engineering principles, and BMPs to lessen negative social impacts of activities on watersheds may even be included. Use of BMPs is site specific and requires information on the particular situation so that the correct BMPs can be selected and implemented in the proper way.

The USDA NRCS is the lead agency that interacts with other state and federal agencies and private land owners in the United States to conserve soil, water, and other natural resources on agricultural land.

Many rural ponds in the United States were constructed through programs of the USDA Soil Conservation Service (now NRCS) for trapping sediment, supplying water for farm use, increasing wildlife habitat, and providing sportfishing. This federal agency has traditionally provided advice on pond construction (USDA SCS 1971, 1982), and they have helped in the design of all types of farm ponds—including aquaculture ponds. There were a reported 2 100 000 farm ponds in the United States in 1980 (USDA SCS 1982) and Smith et al. (2002) suggested that the number was around 2 600 000 in 2000. The construction of ponds in rural areas is an excellent way of trapping sediment in runoff and reducing sediment loads to streams (Renwick et al. 2005). Thus ponds can be considered a component of erosion-control BMPs.

The adoption of BMPs usually is voluntary, but in some countries, programs that provide incentives to farmers may require the use of BMPs. For example, any type of landscape alteration or construction on land put into conservation programs of NRCS must be done according to approved BMPs. It is not uncommon for wastewater discharge permits to contain BMPs to provide protection against potential sources of pollution with which the permittee must comply. Secondary containment around fuel storage tanks is an example of such a BMP. If there is a spill of fuel from the main storage tank, the spilled fuel will be captured in the secondary containment around the tank to avoid soil and water pollution. Periodic inspections of facilities by the environmental agency are made to verify the application of BMPs. The adoption of BMPs also may be a way to comply with water quality criteria in wastewater discharge permits or with standards of ecolabel certification programs (Chapter 14).

Development of aquaculture BMPs

The use of BMPs in aquaculture is relatively new, but in the last few years, there has been a rapid proliferation of manuals and website postings of BMPs for many types of aquaculture. Most early efforts at BMPs were rudimentary and not well designed; they were made basically as a response to the complaints leveled at aquaculture by environmentalists. The aquaculture industry had not been portrayed as an environmental villain before, and many within the industry seemed to think that putting forth some codes of conduct or lists of BMPs would make the environmentalists back off. Obviously, the aquaculture industry knew little about the resolve of the environmental community or of their distrust of resource users, businesses, and governments.

There still is relatively little guidance for those developing aquaculture BMPs, nevertheless the substance of aquaculture BMPs have improved greatly over the past decades. The best single source of aquaculture BMPs is the book edited by Tucker and Hargreaves (2008) that contains BMPs for various major species, species groups, and production methodologies. One paper touches on the procedures for developing BMPs (Boyd et al. 2008). Boyd (2003) presented guidelines for aquaculture effluent management that gave some guidance on developing BMPs. In addition, a publication on BMPs for responsible aquaculture also provides guidance for development of better practices (http://pdacrsp.oregonstate.edu/pubs/featured_titles/boyd.pdf).

The publications mentioned provide guidance for developing BMPs for specific impacts; they do not provide information on how to identify the possible negative impacts by an industry or facility. Thus, we will discuss the process used for making BMPs for the Alabama catfish farming industry. This effort began when it was announced that the USEPA would likely initiate a rule-making procedure for US aquaculture. The Alabama Catfish Producers (ACPs) contracted with Auburn University in 1997 to make an EIA of Alabama catfish farming and to lead an effort to develop BMPs to reduce impacts identified in the EIA (Boyd et al. 2000). Procedures described by Wood (1995) were followed for conducting the EIA.

General information

An environmental survey was conducted to obtain information on the nature of catfish farming and how this industry fit into and affected the surrounding environment. This effort consisted of gathering information on the entire Alabama catfish industry to include annual production; value of production; number and sizes of farms; weather records (air temperature, water temperature, precipitation, and pan evaporation); soil characteristics and land use practices in the production area; extent of surface and groundwater resources; water bodies receiving catfish farm effluents; basic water quality of surface and groundwater in the area.

The general information was obtained from records maintained by the Alabama Field Office of USDA National Agricultural Statistics Service (NASS), county soil survey reports prepared by NRCS, county-level geology and groundwater reports of the Alabama Geological Survey, annual Alabama weather data records of the Alabama Agricultural Experiment Station (AAES) at Auburn University, the US census, and topographic maps and county-level highway maps from USGS and Alabama Department of Transportation, respectively. Information on stream water quality in the production area was provided by ADEM.

Farm survey

A sample of 25 farms was selected over the five-county area containing the majority of channel catfish production in Alabama. A satellite image of a typical farm is shown in Fig. 13.1. The farms were chosen with help of the extension aquaculturist at the Alabama Fish Farming Center to represent the ranges in farm size, topography and soils, water sources, production intensity, and management approach. These farms had a total pond water surface area for production of 2378 ha or about 25% of the total Alabama catfish industry.

An interview and farm evaluation guide sheet listing observations to be made, data to be sought, and specific questions for farmers (Table 13.2) was prepared. The

Figure 13.1 Satellite image of channel catfish farm in Alabama, USA. *Source:* Map data © 2013 Google Earth Pro.

Table 13.2 Catfish farm survey for environmental impact assessment; information requested from farmers in interviews and observations made by investigators.

General
Date/owner/location
Farm area/water surface area/watershed area
Watershed: activities/topography/soils/cover/erosion
Wells: number/size/access to salty groundwater

Fry and fingerling production
Type, number, and area of ponds
Production practices: stocking rates/feeding rates/use of fertilizers, lime, and other
   chemicals/amount and type of aeration
Harvest method
Production: harvest weight/feed conversion ratio

Food fish production
Type, number, and area of ponds
Production practices: stocking rates/feeding rates/use of fertilizers, lime, and other
   chemicals/amount and types of aeration
Harvest method
Production: harvest weight/feed conversion ratio

Condition of earthwork
Embankments: presence of cover/evidence of erosion by rainfall
Evidence of erosion by aerators
Evidence of sedimentation within ponds
Evidence of erosion on pond watersheds

Effluents
Normal time for draining (month)
Draining time to empty ponds (days)
Frequency of partial drawdown/complete draining
Months with overflow
Number of discharge points on farm
Type of discharge structure (trickle tube or pipe)
Type of receiving water body
Description of discharge (into ditch, onto land, into wetland, directly into stream, etc.
Observation on erosion at effluent outfalls
Evidence of erosion-prevention devices
Evidence of sedimentation around outfalls

Specific questions
Is water exchange used?
Could water be conserved by transfer to other ponds?
Could water be discharged through a settling basin?
Could water be applied to pasture or crops?
Is it necessary to completely drain ponds for reworking levees?
Is it necessary to remove sediment at intervals of several years?
Is sediment disposed of in repairing levees or put outside ponds?
What chemicals do you use in pond management and at what frequency and rate?
How are chemicals stored and disposed of?
What activities are conducted on pond watersheds?
Do you anticipate building more ponds in the future? How many hectares?
Do you use medicated feed?
Do you use other species with catfish?

owner or manager of each farm was personally interviewed. Investigators also toured farm facilities and surrounding land to observe infrastructure, watershed topography, water management, discharge points, and other items of interest. Where data analyses revealed inconsistencies or deficiencies in information and answers, clarification was done during a second visit and interview or by telephone conversations with farmers. Investigators also hired a light airplane and pilot and flew over the farms to observe watershed activities in addition to catfish farming.

Literature search

The Alabama catfish farming industry had been studied since its inception in the mid-1960s by researchers at Auburn University. A review of this literature provided information on sources, concentrations, and fates of nutrients, organic matter, and suspended solids in pond waters. There also were studies of the use of various chemical treatments and mechanical aeration practices for improving water quality and avoiding dissolved oxygen stress or mortality to fish. Much valuable information came from previous investigations on the quality of farm effluents during pond overflow and pond draining for harvest. The literature review extended to studies of catfish production in other states, and to information on specific topics such as effects of chemical use in ponds and role of aeration in oxidizing wastes in ponds regardless of culture species or geographical origin.

Special studies

The farm survey and literature review did not provide all information necessary. Studies were conducted to obtain further information on the use of salt (sodium chloride) in ponds to counteract nitrite toxicity (Tavares and Boyd 2003) and the fate of copper sulfate applied to ponds for control of phytoplankton responsible for off-flavor problems in fish (McNevin and Boyd 2004).

   The USEPA had announced that settling basins would be considered as a possible way of treating pond effluent. This necessitated studies of the rate of sedimentation of suspended soil particles and plankton from pond water (Ozbay and Boyd 2003), and the size of settling basins needed to remove suspended solids from overflow and draining effluents (Boyd and Queiroz 2001; Ozbay and Boyd 2004).

   It is interesting how unexpected happenings often affect the outcomes of endeavors. The USEPA had paid no attention to information provided by the aquaculture industry to show that sedimentation basins for pond overflow and draining effluent would be impractical because of the large volumes of water involved. Some USEPA scientists and their consultant accepted an invitation to visit Alabama catfish farms in 2002 before the draft aquaculture effluent rule was prepared. The night before this visit, an 18-cm rainfall event occurred in the catfish farming region of Alabama. Upon arrival at the farms, ponds were overflowing heavily and it was obvious to all from pond water levels that overflow would continue for several days. Nothing more was said about sedimentation basins during the visit, and there was no further mention by USEPA of a requirement for sedimentation basins for ponds.

Aeration has an important influence on fish well-being, but it also is critical for oxidizing organic and nitrogenous wastes; thus, the optimum aeration requirement of ponds was assessed (Boyd 2004). In addition, it was observed that a single stream received effluents of about 50% of the catfish production area in Alabama. A water quality study was conducted on this stream (Silapajarn and Boyd 2005).

Water management and effluents

Typical water management schemes of farms were ascertained, and effects of water use by farms on surface water and groundwater resources were estimated. A study by the USGS on groundwater use by catfish farming (Kidd and Lambert 1995) was particularly valuable. The amounts and timing of outflow from ponds and discharge of draining effluents were assessed by calculations based on hydrologic, weather, soil characteristics, land cover data, and responses to questions by farmers. Using the water quality characteristics of pond effluents gleaned from the literature and the calculations of effluent volume, loads of pollutants discharged into surface water by the entire industry were estimated.

The maximum expected concentrations and discharge volumes for farms allowed predictions of effects on stream water quality. In addition, the manner in which farm outfalls were installed was noted, and areas around these outfalls were observed to determine if there was evidence of erosion, sedimentation, or other ecological nuisances.

Farm operations

The main objective was to obtain information on all inputs to ponds and to ascertain how feeds, chemicals, and other substances were stored, handled, and applied. Attention also was given to feed use efficiency because feed is the major input of nutrients and organic matter (Boyd and Tucker 1998). The range in amounts of aeration applied also was ascertained in order to determine if producers applied enough aeration to maintain good water quality in ponds. The placement of aerators (Fig. 13.2) was observed, and aerator-induced erosion patterns were noted (Gross et al. 2000; Boyd 2012).

Information on application of potentially toxic chemicals, for example, herbicides to control unwanted plants around edges of ponds, algicides to control phytoplankton, and antibiotic and other medicinal compounds, also was requested from farmers. Farmers also were asked about salt applications to counteract nitrite toxicity and the use of saline groundwater in some ponds, because these two practices could lead to salination of surface waters.

Assessment of possible impacts

Results of the efforts described above led to the identification of possible, negative impacts of Alabama catfish farming. The main issue was potential pollution of

Figure 13.2 Typical placement of aerators in a channel catfish pond.

streams receiving pond discharges—especially the final draining effluent from ponds. Nitrogen, phosphorus, and organic matter concentrations in farm discharges were elevated above ambient concentrations in receiving streams. But the concentrations of these substances in pond effluents were not particularly high when compared to domestic wastewater and effluents of many other industries. The amounts of chemical substances applied to ponds were not great enough to cause excessively high concentrations in farm discharges—especially if these substances were not applied when ponds were flushing rapidly in response to heavy rainfall.

The concentrations of total suspended solids in overflow and final draining effluent were particularly high related to ambient concentrations in receiving waters. The solids resulted from dense plankton blooms, erosion of pond earthwork by aerators, and from erosion on watersheds. Cattle sometimes were held on pasture land in watersheds and allowed to enter ponds. Manure from cattle increases the organic load in ponds, and cattle cause erosion—especially if they enter ponds (Fig. 13.3). In addition, the discharge from ponds often caused erosion at outfalls into streams.

Despite the potential sources of pollution mentioned above, it was found that the stream into which about 50% of the catfish production area discharged still complied with its water quality standard (Silapajarn and Boyd 2005). There were, however, measurable increases in concentrations of nutrients, total suspended solids, biochemical oxygen demand, and chloride in the stream between a sampling station above any catfish farm outfalls and one below the production area.

Sediment was sometimes removed from ponds and erosion of improperly disposed sediment could result in highly turbid runoff entering the streams. Storage of fertilizers, fuels, liming material, salt, feed, and other products for aquaculture at farms presented opportunity for spills and associated water pollution.

Figure 13.3 Cattle in an aquaculture pond. Courtesy of David Cline.

The issue of land use conversion was considered. Catfish ponds were sited primarily in former pasture land, and wooded areas seldom had been cleared to make ponds. Farms were located on private property and rarely near neighbors. Water for supplying ponds comes mainly from surface runoff. Once many of the catfish farmers in Alabama used groundwater to avoid water level fluctuations in dry months (Boyd and Brown 1990), but because of the high cost of pumping water, this practice had been largely abandoned by 1997 and 1998 when the environmental survey was made. No evidence of conflicts among property owners and other stakeholders related to land and water use or other issues were detected. The catfish farming industry is in an economically depressed region of Alabama and the jobs and services resulting from catfish farming were welcomed.

Fish mortalities occurred in ponds, and dead fish were observed during the study floating at the edges of many ponds. Occasional mortalities are removed from ponds by various species of birds, raccoons, and other animals living in the vicinity. This prevents dead fish from accumulating in ponds in large numbers, and farmers did not remove dead fish from ponds. Massive die-offs of fish occasionally occur because of low dissolved oxygen concentration or disease. Overflow pipes in ponds were typically not screened and dead fish could be transported from ponds in overflow and enter streams. If left in ponds, fish decompose and could harm pond water quality. Decomposing fish are not sanitary in production ponds, and odor from decomposing fish is offensive.

Drafting of BMPs

Findings of the EIA revealed the most likely negative environmental impacts that might result from catfish farming in Alabama. The next step was to develop BMPs

Table 13.3 Categories for Alabama catfish farm best management practices.

1. Reducing storm runoff into ponds
2. Managing ponds to reduce effluent volume
3. Erosion control on watersheds and pond embankments
4. Pond management to minimize erosion
5. Control of erosion by effluents
6. Settling basins and wetlands
7. Feed management
8. Pond fertilization
9. Managing ponds to improve quality of overflow effluent
10. Managing ponds to improve quality of draining effluent
11. Therapeutic agents
12. Water quality enhancers
13. Fish mortality management
14. General operations and worker safety
15. Emergency response and management

for use in preventing or minimizing these impacts. The procedure began with preparation of draft BMPs by researchers at Auburn University. Draft BMPs were sent to extension aquaculturists in the state, the state engineer for NRCS, the agricultural wastewater specialist for ADEM, and a few selected aquacultural scientists and environmental specialists outside the state. Comments from these individuals were incorporated into the draft BMPs. A meeting was arranged by the ACP and the draft BMPs were presented to over 100 farmers. The farmers' comments on the BMPs were used to make another set of draft BMPs. A second meeting of farmers was convened; fewer attended this time, but their comments were used to make the final revision of the BMPs.

The initial draft included only 20 BMPs, some of which were rather broad; after incorporation of comments by specialists and farmers, the final list included 71 individual BMPs that were more highly focused. With help from ADEM and NRCS, the 71 BMPS were separated into 15 impact categories (Table 13.3). Each impact category of BMPs was presented individually according to the format as illustrated in Table 13.4.

An example of one of the shorter BMP sheets for Alabama catfish farming is provided in the following box.

Table 13.4 Format for presenting each category of Alabama catfish farm best management practices (BMPs).

1. Title—the name of the BMP category
2. Definition—the environmental, social, or safety problem being addressed
3. Explanation—why and how the problem should be solved
4. List of practices—the BMPs for category are listed
5. Implementation notes—suggestions to help the producer select and implement the BMPs
6. Selected references for further reading

Control of erosion by effluents

Definition

Water discharging from ponds can cause erosion at the pipe outlet and in the downstream conveyance ditches or water courses. Suspended soil particles can increase the possibility of turbidity and sedimentation in receiving streams. On the other hand, if pipe outlets are properly protected and ditches are designed and constructed for permissible velocities, erosion will be minimal.

Explanation

Pond water is usually discharged through a pipe extending through the dam or embankment, into ditches or directly into streams. Erosion may occur at the point of exit from the drainpipe in ditches during conveyance of effluent from ponds to streams. Therefore, methods to reduce water velocity and impact on the soil are essential to maintaining quality of effluents from catfish ponds.

Control methods

Practices

- Install structures such as riprap plunge basins to prevent drainpipe discharge from impacting and eroding soil.
- Construct discharge ditches of adequate size, with proper side slopes, and with proper vegetative or structural measures to prevent excessive water velocity and resulting erosion of bottoms and sides.
- Install riprap in bottoms of ditches in places that are susceptible to erosion.

Implementation notes

Control of erosion by effluent is a simple matter that involves reducing the energy of impact of discharge upon soil, reducing water velocity in ditches to prevent scouring, and extending drainpipes beyond critical points for erosion.

Drainpipes from ponds should be extended at least 2 m beyond toes of dams or embankments at an elevation near the ditch bottom. Also, the outlet area of the drainpipe should be protected with a riprap plunge pool (See Fig. 13.4). The stone-protected pool will prevent water from impacting on soil, and it will reduce the energy of the water to lessen the potential for erosion as the water flows away from the initial impact zone.

Where drainpipes discharge directly into streams they should extend far enough over the stream bank to prevent discharge from causing erosion (Fig. 13.5), and be located at an elevation near the normal water level of the stream. Erosion of the stream bottom by falling water can be avoided by installing riprap in the area of impact as illustrated in Fig. 13.4. Where extension of pipes into streams is not practical, riprap protection should be provided from the pipe to a stable outlet.

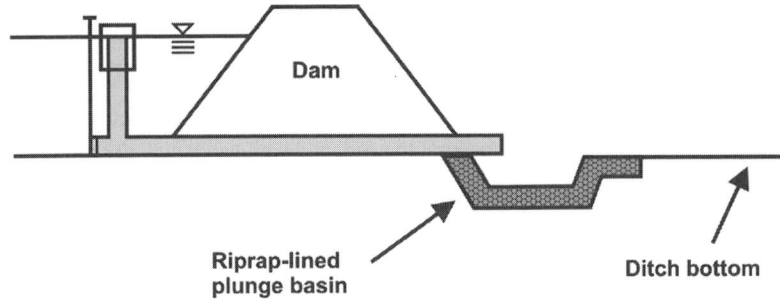

Figure 13.4 Riprap-lined plunge basin to prevent erosion by pond discharge pipe.

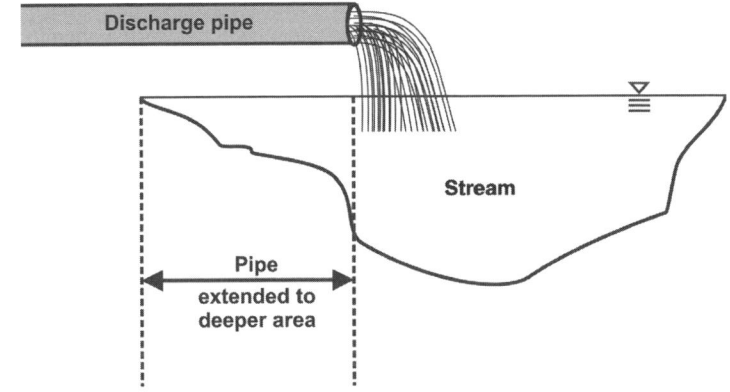

Figure 13.5 Extension of pipe to prevent erosion by discharge at the shallow edge of the stream.

Ditches for conveying water to streams should be designed according to permissible velocities for the type of soil and vegetation. Structural protection, such as riprap, may be necessary in ditch bottoms where vegetation cannot be maintained. Where permissible velocities cannot be maintained from the pond to stream, grade control structures may be necessary.

Further reading

Yoo, K. H. and C. E. Boyd. 1994. *Hydrology and Water Supply for Pond Aquaculture.* New York: Chapman and Hall.

The BMPs for two other impact categories are provided in Table 13.5. The use of BMPs is particularly important in effluent management, and the management of effluents includes BMPs for water use, feeding, water quality enhancers, and others. A discussion of practices that can be used to lessen volume and improve the quality of aquaculture effluents was included in Chapter 10 on water pollution.

The BMPs for Alabama catfish farming were combined and presented in a special AAES publication (Boyd et al. 2003). The BMPs also were posted on the NRCS website for Alabama (http://efotg.nrcs.usda.gov/ references/public/AL/INDEX.pdf).

Table 13.5 List of best management practices for two other impact categories for use on Alabama catfish farms.

Feed management

- Select high quality feeds that contain adequate, but not excessive, nitrogen and phosphorus.
- Store feed in well-ventilated, dry bins, or if bagged, in a well-ventilated, dry room. The feed should be used by the expiration date suggested by the manufacturer.
- Apply feed uniformly with a mechanical feeder.
- Do not apply more feed than fish will eat.
- Feeding rates should not exceed 30 kg/ha/day in unaerated ponds. In ponds with 5 horsepower of aeration per hectare, feeding rates usually can be increased to 100–120 kg/ha/day.
- When uneaten feed accumulates in corners of ponds, it should be manually removed.

Water quality enhancers

- Store water quality enhancers under a roof where rainfall will not wash them into surface waters.
- Copper sulfate application rate should be established by multiplying the total alkalinity concentration in the water by 0.01, for example, if the total alkalinity is 85 mg/L, the copper sulfate treatment rate is 85 mg/L × 0.01 = 0.85 mg/L.
- Sodium chloride applications should not exceed 100 mg/L.
- Lime (calcium oxide or hydroxide) applications should not exceed 100 mg/L.
- Agricultural limestone and gypsum (calcium sulfate) applications should not exceed 5000 and 2000 kg/ha, respectively.
- Calcium hypochlorite or other chlorine compounds should not be applied to catfish ponds.

Adoption and effectiveness of aquaculture BMPs

The adoption of BMPs is voluntary—except in those cases where specific BMPs may be required in wastewater discharge permits. At a particular aquaculture facility, the negative environmental impacts must be identified and the local conditions and facility management procedures assessed in order to select the appropriate suite of BMPs. The specific suite of BMPs needed to address impacts may vary considerably with site conditions, species cultured, culture method used, layout of the facility, nature of the receiving water body, and possibly other factors.

The possible relationships between cost of implementation and effect on production that may occur among BMPs are as follows:

- No implementation cost and either no effect or increased production,
- No implementation cost and decreased production,
- Implementation cost and no effect on production,
- Implementation cost and increased production,
- Implementation cost and decreased production.

There has not been adequate research to assess the cost-benefit ratio versus the resulting environmental benefits for any of the above scenarios. Thus farmers will likely adopt BMPs that do not have a cost and have either no effect or the promise for increased production. They will be skeptical of the other classes of BMPs mentioned above, and likely adopt BMPs selectively.

Table 13.6 Effects of best management practices (BMPs) on three variables in Mississippi catfish ponds.

Variable	BMP ponds	Non-BMP ponds
Overflow (cm)	52	96
Phosphorus discharged (kg/ha)	1.6	4.3
Groundwater added (cm)	18	47

Source: Modified from Tucker and Hargreaves (2006).

Putting cost aside, some BMPS also are easier to implement than others, and degree of difficulty of implementation and continued use no doubt also leads to farmers adopting BMPs selectively. No studies of the thoroughness with which BMPs were adopted on aquaculture facilities were found.

Assuming that aquaculture BMPs are devised by a procedure to assure that they are based on sound principles and implemented with respect to site characteristics, they should lessen negative impacts. However, it is rare that monitoring programs are implemented to verify the effectiveness of BMPs. The only study found on the benefits of aquaculture BMPs was by Tucker and Hargreaves (2006). They applied the following BMPs to three ponds: limiting daily feed input to 110 kg/ha/day; using a reduced-protein content feed (28% vs. 32% crude protein); maximum fish density of 18 500 fish/ha; and replacement water refilled to 15 cm below overflow structures. These ponds were compared to ponds operated without BMPs but with normal management. The BMPs resulted in a decrease in overflow, less water required for replacement of evaporation and seepage, and a smaller amount of phosphorus discharged (Table 13.6). Production did not differ between BMP and non-BMP ponds—6425 versus 6250 kg/ha, respectively. This particular suite of BMPs did not increase production cost and it did not significantly increase production. The BMPs did provide environmental benefits. More studies of this type should be conducted to provide a better rationale for adoption of aquaculture BMPs.

There is much interest in BMPs but their dissemination is not well organized. The FAO Aquaculture Service issued a call for contributions to the development of a worldwide overview and databank on codes of practice, codes of conduct, best (better) management practices, technical guidelines, etc. in aquaculture (http://www.fao.org/docrep/012/i1614e/i1614e08.pdf). Recently NACA announced a program called the ASEM (Asia–Europe Meeting) Aquaculture Platform with the objective of promoting wider adoption of BMPs for key aquaculture commodities in NACA-member countries (http://ec.europa.eu/europeaid/where/asia/regional-cooperation/support-regional-integration/asem_en.htm). The effort appears to be directed primarily at lessening negative environmental impacts as a means of providing greater opportunity for small-scale producers to participate in markets for responsible aquaculture products. The FAO and NACA efforts should do much to advance the use of BMPs.

Environmental regulations are increasingly imposed on aquaculture facilities. These regulations include standards and criteria with which the facility must comply. Various types of certification programs for voluntary participation—but with requirements for compliance with standards and criteria—are also becoming common (Chapter 14). Possibly the most important uses of BMPs in the future will be for achieving compliance with the standards and criteria of government regulations and ecolabel certification programs.

Conclusions

The use of BMPs is a common means of lessening negative environmental impacts in agriculture, forestry, mining, and other industries. It is not surprising that the aquaculture industry began to develop BMPs when confronted with complaints that unsustainable production methods were causing environmental degradation. Aquaculture BMPs should be developed according to a process that assures transparency and stakeholder involvement, but more importantly, they should be based on sound scientific and engineering principles and be technically and economically favorable for implementation.

Voluntary adoption of BMPs by producers is laudable, and environmental education to teach producers the need for operating aquaculture facilities in an environmentally responsible manner is a worthwhile endeavor. However, it is unlikely that voluntary adoption of BMPs can be relied upon to reduce the negative impacts of aquaculture. Governmental regulations are needed to assure that aquaculture facilities are operated in an environmentally responsible manner. Moreover, private certification programs are encouraged by environmentalists because these programs may have compliance requirements stricter than those of governmental regulations. Use of BMPs can be particularly helpful to producers attempting to comply with governmental or certification requirements for environmental protection.

The eNGO perspective

It is quite likely that all eNGOs supporting improvement in the aquaculture sector recognize BMPs as a tool to move toward more efficient production and lessen environmental impacts. Some eNGOs identify BMP adoption as an indicator of more responsible producers; however, most progressive producers will contend that BMPs only assist in meeting certain targets. These targets are likely more important to the eNGO community than the adoption of BMPs because BMPs are implemented at the farm scale and eNGOs are more concerned with the macrolevel environmental scale. Thus BMP development or adoption is viewed primarily as traditional "extension" support for production.

To provide some context for the above assertion, it is interesting to examine WWF's approach to environmental improvement in the aquaculture sector. Since the beginning of WWF's Aquaculture Program, the organization has acknowledged the need and utility of BMPs, but the prescribing or certain practices is viewed as potentially stifling innovation. The WWF suggests that it is important to set meaningful targets and find the means to reach or exceed the target. Take, for example, a BMP on vegetative cover of pond embankments to reduce erosion. This BMP inherently benefits a producer in reducing the costs associated with restabilizing embankments. From an environmental perspective, there can be a marked reduction in TSS concentration and associated turbidity in effluents. Arguably from an eNGO perspective, setting a discharge concentration limit on TSS concentration with adequate enforcement would provide more evidence of environment protection than the mere observation that this particular BMP had been adopted. Granted BMPs are not solely utilized to prevent or mitigate environmental impacts, but this illustrates the perspective of eNGOs on BMPs. If BMPs foster environmental improvement, eNGOs will support their adoption. However, evidence of installation of BMPs does not necessarily denote improved environmental stewardship. The inclusion of BMPs into conformance criteria for inspection standards and certification is therefore challenging to implement.

References

Boyd, C. E. 1999. *Codes of Practice for Responsible Shrimp Farming*. St. Louis, Missouri: Global Aquaculture Alliance.

Boyd, C. E. 2003. Guidelines for aquaculture effluent management at the farm level. *Aquaculture* 226:101–112.

Boyd, C. E. 2004. Overview: mechanical pond aeration. *Global Aquaculture Advocate* 7:59–60.

Boyd, C. E. 2012. Feed input versus dissolved oxygen dynamics in aquaculture systems. *INFOFISH International* 6/2012:25–29.

Boyd, C. E. and S. W. Brown. 1990. Quality of water from wells in the major catfish farming area of Alabama. In *Proceedings 50th Anniversary Symposium, Department of Fisheries and Allied Aquacultures*, pp. 195–206. Auburn University, Alabama: Alabama Agricultural Experiment Station.

Boyd, C. E. and J. Queiroz. 2001. Feasibility of retention structures, settling basins, and best management practices in effluent regulation for Alabama channel catfish farming. *Reviews in Fisheries Science* 9:43–67.

Boyd, C. E. and C. S. Tucker. 1998. *Pond Aquaculture Water Quality Management*. Boston: Kluwer Academic Publishers.

Boyd, C. E., J. Queiroz, J. Lee, M. Rowan, G. N. Whitis, and A. Gross. 2000. Environmental assessment of channel catfish, *Ictalurus punctatus*, farming in Alabama. *Journal of the World Aquaculture Society* 31:511–544.

Boyd, C. E., P. Zajicek, and J. A. Hargreaves. 2008. Development, implementation and verification of best management practices for aquaculture. In C. S. Tucker and J. A. Hargreaves, editors, *Environmental Best Management Practices for Aquaculture*. Ames, Iowa: Blackwell Publishing.

Boyd, C. E., J. Queiroz, G. N. Whitis, R. Hulcher, P. Oakes, J. Carlisle, D. Odom, Jr., M. M. Nelson, and W. G. Hemstreet. 2003. Best management practices for channel catfish farming in Alabama. Special Report No. 1 for Alabama Catfish Producers, Alabama Agricultural Experiment Station, Auburn University, Alabama.

Burns, A. J., T. H. Coaker, and P. C. Jepson. 1987. *Integrated Pest Management*. Waltham: Academic Press.

Dixon, H. 1997. Environmental code of practice for the shrimp farming industry of Belize. Unpublished manuscript.

Donovan, D. J. 1997. *Environmental Code of Practice for Australian Prawn Farmers*. East Brisbane, Australia: Kuruma Australia Pty., Ltd.

FAO (Food and Agriculture Organization). 1995. *Code of Conduct for Responsible Fisheries*. Rome: FAO.

Gross, A., C. E. Boyd, and C. W. Wood. 2000. Nitrogen transformations and balance in channel catfish ponds. *Aquacultural Engineering* 24:1–14.

Havlin, J. L. and S. L. Tisdale. 2005. *Soil Fertility and Fertilizers: An Introduction to Nutrient Management*. Upper Saddle River, New Jersey: Prentice Hall.

Irish Salmon Growers Association. 1991. *Good Farmers, Good Neighbours*. Dublin, Ireland: Irish Salmon Growers Association.

Kidd, R. E. and D. S. Lambert. 1995. Hydrogeology and ground-water quality in the Black Belt area of west-central Alabama, and estimated water use for aquaculture. *Geological Survey, Water Resources Investigations Report 94–4074*. Tuscaloosa, Alabama.

McNevin, A. and C. E. Boyd. 2004. Copper concentrations in channel catfish *Ictalurus punctatus* ponds treated with copper sulfate. *Journal of the World Aquaculture Society* 35:16–24.

Ozbay, G. and C. E. Boyd. 2003. Particle size fractions in pond effluents. *World Aquaculture* 34:56–59.

Ozbay, G. and C. E. Boyd. 2004. Treatment of channel catfish pond effluents in sedimentation basins. *World Aquaculture* 35:10–13.

Renwick, W. H., S. V. Smith, J. D. Bartley, and R. W. Buddemeier. 2005. The role of impoundments in the sediment budget of the conterminous United States. *Geomorphology* 71:99–111.

Silapajarn, O. and C. E. Boyd. 2005. Effects of channel catfish farming on water quality and flow in an Alabama stream. *Reviews in Fisheries Science* 13:109–140.

Smith, S. V., W. H. Renwick, J. D. Bartley, and R. W. Buddemeier. 2002. Distribution and significance of small, artificial water bodies across the United States landscape. *The Science of the Total Environment* 299:21–36.

Tavares, L. H. and C. E. Boyd. 2003. Possible effects of sodium chloride on quality of effluents from Alabama channel catfish ponds. *Journal of the World Aquaculture Society* 34:217–222.

Toy, T. J., G. R. Foster, and K. G. Renard. 2002. *Soil Erosion: Processes, Prediction, Measurement, and Control*. New York: John Wiley and Sons.

Tucker, C. S. and J. A. Hargreaves. 2006. Water-level management, BMPs cut water use, pond effluents. *Global Aquaculture Advocate* 9:50–51.

Tucker, C. S. and J. A. Hargreaves, editors. 2008. *Environmental Best Management Practices for Aquaculture*. Ames, Iowa: Blackwell Publishing.

USDA SCS (United States Department of Agriculture Soil Conservation Service). 1971. *Ponds for Water Supply and Recreation. Agricultural Handbook Number 387*. Washington, DC.: US Government Printing Office.

USDA SCS (United States Department of Agriculture Soil Conservation Service). 1982. *Ponds—Planning, Design, Construction. Agricultural Handbook Number 590*. Washington, DC: US Government Printing Office.

USEPA (United States Environmental Protection Agency). 1979. *Environmental Implications of Trends in Agriculture and Silviculture*. Athens: EPA

Wood, C. 1995. *Environmental Impact Assessment: A Comparative Review*. Essex: Longman Scientific and Technical.

Chapter 14

Eco-label certification

Use of BMPs throughout the supply chains of products and services coupled with environmental monitoring and effective governmental regulations would be the best scenario for environmental protection. Unfortunately such a utopia is wistful thinking, at least for the foreseeable future, and the world's ecosystems continue to be unnecessarily abused by impacts that could be avoided or greatly reduced.

The environmental ethic of the world's population however is slowly improving as people become more affluent and better educated, and this has evoked a greater level of environmental concern in many sectors of the world's economy. Some retailers have developed environmental strategies to counter negative criticisms about various products by eNGOs and to provide consumers with goods and services produced by environmentally and socially responsible methods. Producers fearing adverse effects on the marketability of their goods and services as a result of eNGO campaigns increasingly attempt to improve their environmental and social stewardship. Product safety hazards have been included—apparently for convenience—in the liturgy of environmental and social woes caused by the production of goods and services. In this chapter responsible goods and services refer to those that are safe and produced with environmental sustainability and social fairness in mind.

The developing market for responsible goods and services depends on retailers, food service companies, and restaurants being able to procure such products. Producers wanting to take advantage of this market must meet specifications required for responsibility and safety. There must be procedures to assure that products and services advertised as responsible and safe actually meet the requirements. This is not a simple task because many of the definitions, methods of identifying impacts, standards and criteria for preventing impacts, and ways of verifying that goods and services are responsible are still developing and by no means generally agreed upon.

Some major retailers have made sourcing policies for certain responsible goods and services, and they seek producers willing to comply with the policies. Some producers or producer associations claim to use environmentally and socially friendly

Aquaculture, Resource Use, and the Environment, First Edition. Claude E. Boyd and Aaron A. McNevin.
© 2015 John Wiley & Sons, Inc. Published 2015 by John Wiley & Sons, Inc.

methods resulting in responsible products. Governments may even get into the act by specifying procedures for production or for recognition of responsible products. A few eNGOs rate products based on assessments of possible environmental and social effects of production. Organizations have developed standards and criteria for responsible goods and services, and third-party auditing procedures to certify producer compliance. A chain of custody system traceability also is necessary to assure that certified goods and services are not co-mingled with ordinary products and services of the same kind.

Advertisements and labels are used to inform consumers seeking responsible products and services. The validity of the advertising varies greatly—some labels are supported only by the affirmation of a producer or vendor, while other labels denote that products have been produced in compliance with a third-party certification system.

This chapter explains the process of differentiation of products and services based on environmental and social impacts and food safety. Current programs for differentiating responsible aquaculture products also are discussed. Readers unfamiliar with the topics may find definitions of key terms in the end notes.[1]

Environmental ratings of aquaculture products

In 1999, the Monterey Bay Aquarium in Monterey, California initiated the Seafood Watch® Program, a science-based, peer reviewed, and ecosystem use-based procedure to categorize seafood items as environmental "best choices," "good alternatives," or choices to "avoid" (http://www.montereybayaquarium.org/cr/cr_seafoodwatch/sfw_aboutsfw.aspx?c=lmn). To date this aquarium has distributed over 40 million pocket guides to aid consumers in choosing safe seafood with a good environmental and social record. The Seafood Watch program was originally for wild-caught fisheries products but it has expanded to include aquaculture products. The pocket cards in the United States are prepared to represent commonly available seafood by region: Northeast, Southeast, Central, Southwest, West Coast, and Hawaii.

The aquaculture criteria used by the Seafood Watch program for rating product choices are as follows: potential of effluents from production facilities to cause water pollution; destruction of habitat to construct farms; use of potentially harmful chemicals; feed use efficiency and conservation of fish meal and oil, potential for escapes of farm-reared species; introduction of exotic species; fish health considerations; source of stocks. The ratings are reviewed by a panel of experts from academia, government, and the seafood industry, revised, and posted on the organization's website.

A number of other organizations also developed similar seafood rating programs. These include Blue Ocean Institute, New England Aquarium, Fishwise, SeaChoice, Ocean Wise, and Fish Watch. Most of these cards are quite similar if not identical.

FishChoice was formed in 2009; the stated purpose of this organization is to connect retail, restaurant, and institutional seafood buyers to suppliers of sustainable seafood products. The purpose also includes helping sustainable aquaculture operations and wild fisheries that are fished responsibly find markets for their products (http://www.fishchoice.com/content/about-fishchoice). The seafood rating programs mentioned above and two seafood certification programs, Food Alliance and MSC, are sustainable seafood partners with FishChoice.

FishChoice presents ratings given by each seafood rating program online in a format similar to the one illustrated in Table 14.1. Suppliers can submit their products for consideration for listing by FishChoice, and buyers can view sustainable products listed on the FishChoice website.

The main problem with seafood rating is that the procedure is extremely general. If a particular species is rated as a choice to "avoid," then sustainable producers of this species are lumped with unsustainable producers. The opposite also is true; by listing an aquaculture species as a "good alternative," all producers of that species are lumped together regardless of their level of environmental stewardship. It is likely that aquaculture was originally considered from a wild-capture perspective in that if overfishing a particular stock is observed, the whole fish population would require attention. The challenge with transposing this methodology on aquaculture systems is that each producer has a different environmental impact because of a specific location and management ethic. In general the rating system seems too species or species-region oriented to be of much practical value in promoting environmental stewardship in aquaculture (see Table 4.7). It is interesting to point out that Seafood Watch likely came to the same conclusion. Currently the Seafood Watch program is applying their criteria to specific farming operations in what might be considered a pseudo-certification scheme.

Sourcing policies

Major retailers, processors, and caterers often develop their own (in-house) specifications for sourcing responsible products. These organizations usually start with existing regulations and add specific requirements thought to appeal to their client bases. The aquaculture standards for in-house sourcing policies focus on product quality and food safety, but environmental and social standards typically are included.

A company with sourcing standards typically seeks producers willing to conduct aquaculture in compliance with their standards. The sourcing standards usually extend to the processing plant and require chain of custody documentation. Some companies may send inspectors to verify compliance with the standards (second-party certification), while others hire an independent third-party auditor.

Sourcing policies are the property of a private company, and as such, there is no requirement to post the standards for public comment or to follow any particular system of developing them. A list of companies that have in-house sourcing standards was not found. However, two prominent ones are Carrefour of France and Whole Foods Market in the United States (FAO 2011).

Association standards

Aquaculture producers in a region or country may band together and develop procedures for production—including environmental, social, and food safety standards.

Table 14.1 Partial example of seafood products listings by FishChoice for farm-raised whiteleg shrimp *Litopenaeus vannamei*.

Farm type	Seafood Watch	Blue Ocean Institute	New England Aquarium	FishWise	SeaChoice	Ocean Wise	Supplier location	Certifications
Farmed contained system	G	G	A	G	G	R	Texas, USA	N
Farmed contained system	G	G	A	G	G	R	Maryland, USA	N
Farmed contained system	G	G	A	G	G	R	Florida, USA	N
Farmed coastal ponds	Y	Y	—	Y	Y	NR	Texas, USA	N
Farmed contained system	G	G	A	G	G	R	Alabama, USA	N
Farmed contained system	G	G	A	G	G	R	Michigan, USA	N
Farmed contained system	Y	Y	—	Y	Y	—	Thailand	N

G = Best choice
Y = Good alternative
A = Preferred choice
R = Recommended
NR = Not recommended
N = None.

Source: Modified from the FishChoice website: http://fishchoice.com/Find-Products.aspx?q=&p=1&s=0&l-&e-Shrimp, Whiteleg.

Such organizations typically produce an eco-labeled product to facilitate consumer recognition. Two examples of such programs are the integrated management system of Salmon Chile and the Code of Good Practice for Scottish Finfish Aquaculture by the Scottish Salmon Producers' Organization (FAO 2011).

History of eco-labeling

It is not clear when efforts to differentiate products in the market began, but in the late 1940s several organizations in North America and Europe started a program to help poverty-stricken communities sell their handicrafts in upscale markets (http://fairtradeusa.org/what-is-fair-trade/history). This effort eventually became Fairtrade International, an advocate for coffee plantation workers. In 1947, ISO was founded to develop voluntary international standards for product manufacturing and services (http://www.iso.org/iso/home.html). Rachel Carson's famous book, *Silent Spring*, no doubt provided one of the strongest momentum boasts for the safe food movement in the Western World. The Label Rouge (Red Label) certification program of the French government dates back to the early 1960s. It was initiated to assure the quality of agricultural products in France (http://poultrylabelrouge.com/012_volaille_LR.php).

The origin of the environmental movement can be traced back to the 1800s or possibly earlier and became a major phenomenon after World War II, and it has grown tremendously since the 1970s. Environmentalists realized that a potential existed for using markets to promote environmental stewardship. By educating the public about the possible negative environmental effects of certain products and services, the purchase of more environmentally friendly choices could be favored. This change in sourcing patterns would cause vendors to insist on environmental stewardship from their producers and others in the supply chain. Environmentally conscious consumers also would be concerned about the safety of their products and services and of the social responsibility exhibited throughout the supply chain. Thus it was convenient to combine the three concerns in efforts to influence consumer purchasing.

Labels attesting to responsible products allow consumers to shop preferentially for these products. Proof of compliance with requirements for responsible production agreed upon in advance allows producers to market their goods and services to processors, brokers, wholesalers, retailers, restaurants, and other vendors who selectively source responsible products. By selling responsible products vendors want to be viewed as practicing environmental and social responsibility. Moreover the safety aspect associated with responsible products adds a level of protection above that afforded by government regulations with respect to possible liability that could result from an unsafe product.

Possibly the first effort approaching an eco-label program resulted from the US Energy Policy Conservation Act of 1975 that mandated fuel efficiency standards for passenger cars and light trucks. This act required manufacturers to produce vehicles that complied with the standards and to provide fuel use data for their vehicles to consumers (http://uspolitics.about.com/od/energy/i/cafe_standards.htm). The idea of energy efficiency reporting for automobiles soon spread to major appliances.

In 1992 the EU started the EU Eco-label Program for non-food and non-medical products. Today this program has issued about 1300 licenses that have resulted in around 17 000 EU Label products (http://ec.europa.eu/environment/ecolabel/facts-and-figures.html). The EUREGAP Eco-label Program began in 1997 as an initiative by retailers belonging to the Euro-Retailer Produce Working Group. The objective was to certify agricultural goods produced by good agricultural practices that also assured environmental stewardship. The name of the program was changed to GLOBALG.A.P. in 2007 (http://www.globalgap.org/uk_en/who-we-are/history).

The Forest Stewardship Council (FSC) was formed by the WWF with many other non-government organizations in 1993 following the UN Conference on Environment and Development (The Earth Summit) in Rio de Janeiro, Brazil in 1992. The FSC promotes sustainable forestry by setting standards that allow certification and labeling of forest products from socially and environmentally responsible forestry. According to the FSC website (https://ic.fsc.org/index.htm), the organization had certified 168 million ha of forests, issued 24 518 chain of custody certificates, and 1171 forest management/chain of custody certificates by the end of 2012.

The MSC was founded in 1997 by the WWF and Unilever; MSC became fully independent in 1999. The purpose of MSC is to promote sustainable exploitation of marine fisheries. To this end MSC has established standards for sustainable fishing for which compliance will lead to certification of a fishery. The MSC also has a chain of custody standard that insures that the MSC label is only put on products from certified, sustainable fisheries. The MSC reports to have certified 189 fisheries and to have 100 more in assessment as of early 2013 (http://www.msc.org/?il8nredirect=true&set_language=en).

Certification programs such as Fairtrade International, Label Rouge, and ISO have added environmental standards to their programs in recent years. Many other eco-label certification programs also have been developed. The procedures for developing standards, the rigor of standards, audit procedures, methods for selecting inspectors, requirements for compliance, and chain of custody and traceability of products from production facilities to consumers vary among programs.

Consumers have little knowledge of the veracity of different eco-label programs—they must trust the label. As a result a substantial advertising effort may be initiated to increase consumer confidence in a particular label. Although competition among vendors of products usually is good for consumers, too many eco-label programs may confuse consumers.

Principles of eco-labeling

As experience was gained in differentiating responsible products in the market, the process for doing so has been continuously revised and improved, but certain principles have been firmly adopted. Eco-labeling almost always involves the three objectives mentioned earlier—environmental responsibility, social fairness, and safety. Moreover, programs involving animal production increasingly feature improved animal welfare.

Eco-labeling programs may be developed by governments, businesses, non-profit organizations, associations, etc. Eco-label standards are developed by various methods but they should result from a transparent process conducted by stakeholders representing industry, government, academia, environmental advocacy groups, social concerns, consumers, and other appropriate groups. Ideally the standard-making body should be accredited.

Standards should cover all the main impact categories pertinent to the particular industry. Standards do not take the place of government-imposed regulations. The first standard in eco-label programs should be to require compliance with all applicable governmental regulations and other legal requirements. The eco-label program normally will require participants to go beyond legal requirements for environmental and social stewardship, product safety, and animal welfare.

Standards must be verifiable and where possible metric-based criteria should be included. Record keeping may be the primary means of proving compliance with some standards. Traceability systems are necessary to verify that products bearing a particular eco-label at the consumer level were actually from a source in compliance with the standards of the program.

Third-party auditing gives the greatest degree of confidence that a participant in an eco-labeling program is or is not in compliance with standards. In third-party verification one entity develops the standards, while an external, independent, accredited body audits the performance of participants in the program. This arrangement assures that there are no conflicts of interest among auditors who inspect facilities, entities that make the standards, and those who own or operate the facilities.

Second-party assurance that participants are in compliance with standards is given by the entity making the standards and to which participants belong. The greatest likelihood of a conflict of interest is between the certifying body and the program participant. The individual or organization providing the goods and service provides assurance of compliance with standards in first-party certification. Obviously there is no way to avoid a conflict of interest in first-party certification.

Regardless of the level of verification clear rules should be established for correcting deficiencies found during inspection, and for revoking the certification privilege for failure to correct deficiencies.

The most widely accepted guidelines for setting social and environmental standards and operating certification programs are provided by the International Social and Environmental Labeling Alliance (ISEAL Alliance). This organization has prepared codes for standard setting, impacts, and assurance (http://www.isealalliance.org/our-work/codes-of-good-practice). The standard-setting code requires justifying the need for and defining the objectives of standards, allowing wide stakeholder participation, ensuring balance through public consultations, making the standards publically available, and regularly reviewing the standards and revising them if necessary. The standards also must contribute directly to achieving the objectives, and they must be relevant to the market and build on regulations. Standards should be locally applicable but globally consistent in interpretation. Finally all standards for a particular effort should be harmonized to avoid overlap.

The impacts code of the ISEAL Alliance provides a framework for devising a monitoring and evaluation system to examine both short-term and long-term outcomes and provide a public report of the results. Basically the impacts code requires determining whether or not compliance with standards allows the participant to meet the objectives of the standards. The results of this analysis must be revealed to the public.

The assurance code of the ISEAL Alliance provides procedures for verifying compliance with standards. This includes selection of the assurance provider, development of the audit procedure, setting qualifications for training auditors, rules for correcting non-compliance, and oversight.

The ISEAL Alliance was not founded until 2002, several years after some of the efforts to develop aquaculture eco-label certification standards had already been initiated. Several other organizations provide guidelines of standard setting, and some organizations devise their own procedure for setting standards. The FAO recently presented guidelines for aquaculture eco-label certification (FAO 2011). The FAO document essentially follows the ISEAL Alliance guidelines, but it goes further by suggesting impact categories for which standards should be made.

Organic certification

Somewhat unrelated to responsible aquaculture is the market for organic foods. According to Washington and Ababouch (2011), there are about 20–25 certifying bodies for organic aquaculture products. Some of the better known are Bio-Suisse of Switzerland, the Soil Association of the United Kingdom, and Naturland of Germany. The USDA is developing organic standards for aquaculture products in the United States (http://afsic.nal.usda.gov/aquaculture-and-soilless-farming/aquaculture/organic-aquaculture).

Organic aquaculture certification programs also contain standards related to environmental protection, social responsibility, and food safety, but the main emphasis obviously is on rules for organic production. Organic aquaculture is a niche market; Szeremeta et al. (2010) reported that organic aquaculture production was about 53 000 t in 2009. Of course the water surface area devoted to organic aquaculture is likely near 200 000 ha because the organic standards typically do not allow production rates more than twice the natural production (without feeds or fertilizers).

The International Federation of Organic Agriculture Movements (IFOAM) is an entity with more than 300 member organizations—mainly in the EU—concerned with various aspects of organic food production (http://www.ifoam.org/index.html). IFOAM provides guidelines for all types of organic agriculture and attempts to harmonize standards of different organic certification programs. A manual on organic aquaculture is available (Szeremeta et al. 2010).

Aquaculture eco-label certification programs

Aquaculture labeling can be integrated into existing certifications as has been done by numerous organic organizations and GLOBALG.A.P. There are only two major eco-labels that focus specifically on aquaculture products alone—the Best Aquaculture Practices (BAP) certification formed by the Global Aquaculture Alliance (GAA) and the Aquaculture Stewardship Council (ASC) certification formed by the WWF and the Dutch Sustainable Trade Initiative (IDH). A third, Friend of the Sea (FOS) certification, includes both aquaculture and capture fisheries.

Standards for aquaculture certification have been continuously modified to include new ideas and to comply with international guidelines recommended by FAO (2011) and ISEAL Alliance. The programs tend to rely on HACCP and Codex Alimentarius for guidance with respect to food safety. In general, standards have metric-based criteria, but they also specify practices to implement and those to avoid. Certification requires information about the following: facilities; worker salaries and benefits; farm operations; sources of stock; feed and other supplies; energy use; land use; water use; chemical use; and production data. Records that allow a certified product in the market to be traced back to the farm or culture unit of origin and to identify sources of feed and seed stock and to reveal the kinds and amounts of chemicals used are necessary.

Each program was designed to reduce the negative impacts that have been categorized by the environmental community (Table 14.2). Following formation of these certification programs FAO published guidelines for aquaculture certifications, and there has been some interest in meeting their minimum substantive criteria (FAO 2011). Standards related to particular impact categories vary among programs, and for a given impact category one program may be more stringent than another. It is important to note that standards must be feasible technologically, operationally, and economically. Thus a balance must be struck—the standard must not be strict beyond practicality or so lenient as not to accomplish the purpose. Of course producers are free to choose the program they prefer or decline participation in any, but seafood buyers may require one or more of the programs as a prerequisite.

Best aquaculture practice standards

The history of the BAP program is interesting and illustrates the difficulty in developing standards and a mechanism for auditing compliance. The BAP program had its origin in a meeting of representatives of the shrimp aquaculture industry during the World Aquaculture Society annual conference in Seattle, Washington in 1997. The eNGOs had previously conducted what was called the Shrimp Tribunal at the United Nations Commission on Sustainable Development in New York in May 1996 (http://darwin.bio.uci.edu/~sustain/shrimpecos/declare2.html). They presented a list of statements to which they urged governments to agree (Table 14.2). Soon after delegates of 21 NGOs and community organizations from Latin America, North America, Europe, and Asia met in Choluteca, Honduras and made a list of declarations and demands related to the marine shrimp farming industry that was called the Choluteca Declaration (http://darwin.bio.uci.edu/~sustain/shrimpecos/declare1.html). Representatives of the shrimp farming industry already were deeply concerned about the possible effects of the Shrimp Tribunal and the Choluteca Declaration on the future of shrimp farming, and when a large eNGO turn-out with demonstrations occurred at the World Aquaculture Society meeting in Seattle, Dr. George Chamberlain, then of Ralston Purina, convened a meeting of shrimp producers, feed manufacturers, and

Table 14.2 Statements concerning unsustainable aquaculture by environmental advocates to the United Nations Commission on Sustainable Development, May 1996.

The undersigned non-governmental organizations (NGOs) urge governments to agree to:

Ensure that artisanal fisheries and dependent coastal communities, and their access to community resources, are not adversely affected by aquaculture development or operations, including extensive and semi-intensive as well as intensive aquaculture methods.

Ensure the use of environmental and social impact statements prior to aquaculture development and the regular and continuous monitoring of the environmental and social impacts of aquaculture operations.

Ensure the protection of mangrove forests, wetlands, and other ecologically sensitive areas.

Prohibit the use of toxic and bioaccumulative compounds in aquaculture operations.

Apply the precautionary approach to aquaculture development.

Prohibit the pollution of surrounding areas resulting from the excessive discharge of organic wastes.

Prohibit the development and use of genetically modified organisms.

Prohibit the use of exotic/alien species.

Prohibit the use or salinization of fresh water supplies, including groundwater, important for drinking or agriculture.

Prohibit use of feeds in aquaculture operations consisting of fish that can be consumed by people.

Prohibit the wholesale conversion of agricultural or cultivable land to aquaculture use.

Ensure that the collection of larvae does not adversely affect species biodiversity.

Ensure that abandoned or degraded aquaculture sites are ecologically rehabilitated and that the companies or industry responsible bear the costs of rehabilitation.

Ensure that aquaculture and other coastal developments are addresses in integrated coastal zone management planning which includes the meaningful participation of all coastal user groups.

Ensure the development of aquaculture in a manner which is compatible with the social, cultural, and economic interests of coastal communities, and ensure that such developments are sustainable, socially equitable, and ecologically sound.

Ensure that multi-lateral banks, bilateral aid agencies, and the UN Food and Agriculture Organization and other relevant national and international organizations or institutions do not fund or otherwise promote aquaculture development inconsistent with the above criteria.

Source: http://darwin.bio.uci.edu/~sustain/shrimpecos/declare2.html.

processors. He suggested that an industry association be formed to promote environmentally responsible aquaculture—this association became the GAA.

The GAA quickly raised funds from shrimp producers, processors, feed manufacturers, and shrimp buyers, established an office in St. Louis, Missouri, and began a trade magazine "The Global Aquaculture Advocate" to promote sustainable aquaculture production. One of the first major efforts of GAA was to develop BMPs for use by shrimp farmers. This document was reviewed by producers, scientists, and environmental specialists, but NGO representatives refused to review it. After incorporating the comments of reviewers a manual of BMPs was printed by GAA (Boyd 1999).

The producer members of GAA were requested to implement BMPs on their farms. A self-evaluation form with scoring system was developed; producers were asked to complete the evaluation annually and scores would be posted on the GAA website. The first year producers all made a grade of 95 or above (out of a possible 100) on the evaluation—it was obvious that this approach was not going to be effective.

The response also clearly demonstrates the difficulty of obtaining a realistic self-assessment and the need for third-party certification.

The next effort was to develop a certification program for shrimp farms to confirm that the participants were improving their environmental stewardship. The standards for the program were developed; the document included the reasons for the standard, the statement of conditions for compliance with the standard, guidelines for implementing the standards, and references to the sources of information used in developing the standards. The standards were based on sound scientific and engineering principles and use of the BMPs developed earlier (Boyd 1999) were a key aspect of the program and the basis for calling the standards the BAP certification standards. However, compliance with the standards required monitoring of certain environmental data and record keeping. The effort required about 3 years because of numerous reviews and revisions necessary to incorporate the resulting comments. The reviews led to the program being extended to include social responsibility, food safety, and animal welfare. In addition, some rules about certification procedures for clusters of small-scale farmers in Asia were drafted. As they had done for the BMP document, most NGO representatives refused to review the BAP standards aside from Dr. Jason Clay now with the WWF, but the program went forward without their endorsement.

The GAA also decided to develop BAP certification standards for hatcheries, feed mills, and processing plants. With these standards a certified shrimp producer conceivably could use postlarvae and feed from certified sources and sell shrimp to a certified processing plant. The GAA also decided to begin work on standards for other species and to develop an over-arching core standard in order to assure uniformity of standards while allowing species-specific standards to be added.

The Aquaculture Certification Council (ACC) was formed to certify that participating farms and other facilities were in compliance with BAP standards. The ACC held training courses to train auditors to conduct inspections of facilities according to rules and inspection documents developed by ACC. The NGOs refused to recognize ACC as independent from GAA, and they also claimed that there was opportunity for conflicts of interest among ACC, its auditors, and program participants. The BAP program was not considered to be third-party certification by the NGOs because of the way ACC was organized and operated.

The ACC was disbanded in 2011 and the GAA became the standard setting organization and standard holder for the BAP program. The GAA posted its standards and other requirements for certification on its website. The GAA selected three accredited certification bodies, Global Trust, National Sanitation Foundation (NSF) Surefish, and Société Générale de Surveillance (SGS), to audit facilities for BAP certification.

Probably one of the most important periods for the BAP program was during the collaboration with Conservation International (CI) to bridge a deal with Wal-Mart that would later require all of its suppliers to be compliant with the BAP program. Coming to an agreement to move toward MSC certification for all its capture fisheries products, Wal-Mart also set a precedent that would change the way the seafood industry approached certification. Since the Wal-Mart agreement, the GAA program has flourished and farm standards have been made for shrimp, tilapia, channel catfish, *Pangasius*, and salmon. In addition, there are standards for seafood processing plants, shrimp hatcheries, and feed mills (http://gaalliance.org/bap/standards.php). A total of 290 farms—channel catfish (2), *Pangasius* (8); tilapia (37), salmon (107), and shrimp (136)—have been awarded BAP certification as of mid-2013. The GAA website indicates that BAP certification also has been attained by 226 processing plants, 12 repacking plants, 28 shrimp hatcheries, and 18 feed mills. These facilities are located in a total of 19 countries. At the end of October 2012, the combined annual output of aquaculture products from BAP-certified processing plants exceeded 1 Mt, and annual production from BAP-certified farms reached 445 000 t (http://www.gaalliance.org/newsroom/news.php?BAP-Program_Reaches-1-Million-Metric-Ton-Milestone-87). The GAA also reported that the BAP certification program had 128 market endorsers as of July 2013, who source their seafood products from BAP-certified processing plants.

Aquaculture Dialogues and the Aquaculture Stewardship Council

The Aquaculture Dialogues were organized under the leadership of Dr. Jason Clay of the WWF who was interested in forming partnerships with producers and other major stakeholders to encourage environmental stewardship. He became involved in shrimp culture through a study of the market potential for redressing the environmental impact of world shrimp production (Clay 1996, 2004). He concluded that shrimp production was the source of many negative impacts, but the impacts of shrimp aquaculture were more avoidable than those of shrimp fishing. Boyd and Clay (1998) published an article in *Scientific American* outlining the changes needed to make shrimp farming sustainable.

As with the GAA responsible aquaculture effort, the WWF effort began with the idea of improving management practices. The Consortium on Shrimp Farming and the Environment, another of Jason Clay's ideas, was formed; it included FAO, NACA, United Nations Environmental Program (UNEP), the World Bank (WB), and the WWF. These organizations contracted with experts to make case studies to document practices used in shrimp farming worldwide (Boyd 2000). The results of the case studies were used to prepare a manual on responsible shrimp farming (FAO/NACA/UNEP/WB/WWF 2006). It was concluded that only six to eight impacts of shrimp farming caused the majority of environmental and social problems. This publication outlined eight principles for responsible shrimp farming along with some brief comments of how to achieve the goals of each principle. According to Clay (2008), the use of targeting better performance can increase the number of farmers achieving acceptable environmental performance and shrink the range in variation of performance among producers (Fig. 14.1).

The WWF concluded that if a certification program could be designed that was a bottom-up approach where stakeholders identified the main six to eight impacts and performance targets that emulated the top 15–20% of the industry, a broader shift in markets for aquaculture products could be harnessed providing an incentive to move forward. It is historically important to note that at the time of the first Aquaculture Dialogue on salmon aquaculture, the GAA had not determined it would address species other than shrimp in their certification efforts. A study of certification issues

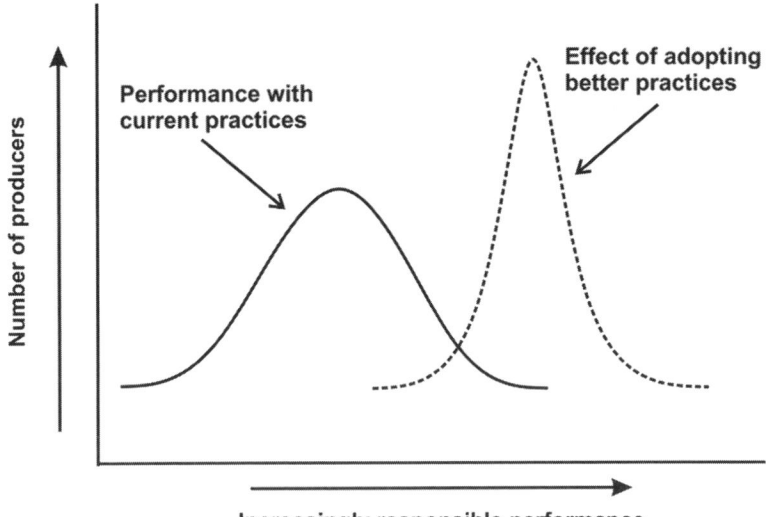

Figure 14.1 The desired effect of adopting more environmentally responsible aquaculture practices.

related to channel catfish, tilapias, rainbow trout, oysters, mussels, clams, scallops, abalone, and seaweed was conducted (Boyd et al. 2005). The WWF then initiated the Aquaculture Dialogues to develop certification standards for different species/species groups. The dialogues included farmers, retailers, NGO representatives, scientists, and other important stakeholders (http://worldwildlife.org/industries/farmed-seafood). Seven principles were addressed in the dialogues to include: legal requirements; conservation of habitat and biodiversity; conservation of water; prevention of escapes of farmed species; efficient use of feed and other resources; animal health and no unnecessary use of antibiotics and chemicals; and social responsibility. Criteria and standards—metric-based where possible—were developed for impacts identified under each principle. The Aquaculture Dialogues defined criteria for standards in an unusual way. An impact was identified and a principle was formed to describe the intention of the stakeholder group on how to minimize or eliminate the impact. Several criteria were identified under each principle as areas that would require one or more indicators for adequately addressing the impact. Indicators were characterized as the variable to be measured to determine compliance. Corresponding standards were agreed for each indicator. An example of a standard taken from the Tilapia Aquaculture Dialogue and non-aquaculture example are presented in Table 14.3. The documents containing the standards were posted on the WWF website for a minimum of two review and comment periods by interested parties and WWF posted individual responses to each public comment.

In 2010, the WWF and the IDH formed the ASC. The Aquaculture Dialogue Standards were transferred to the ASC and this entity will be responsible for the management of the standards and the ASC logo. Accreditation of third-party auditing bodies is carried out by Accreditation Services International (ASI) (Table 14.4). All companies involved in the supply of ASC-certified products must have ASC certification of the chain of custody (http://www.aquaculture-stewardship-council.com/index.html).

Table 14.3 Description of the Aquaculture Dialogue standard hierarchy.

Component	Definition	Non-aquaculture example	Aquaculture example
Impact	The problem we want to minimize	High blood pressure	Waste in effluents
Principle	The guiding principle for addressing the impact	Maintain a healthy blood pressure	Conserve water resources
Criteria	The area to focus on to address the impact	Salt consumption	Nutrient use and release
Indicator	What to measure in order to determine the extent of the impact	Milligrams of salt consumed per day	The amount of phosphorus added and released per ton of fish produced
Standard	The number and/or performance level to reach to determine if the impact is being minimized	<100 mg of salt consumed per day	Phosphorus input or utilization in tilapia aquaculture operations will not exceed 30 kg P/t fish produced and loads of phosphorus released into natural receiving waters will not exceed 22 kg P/t fish produced.

The ASC program covers all impact categories included in the BAP program—except food safety. The Aquaculture Dialogues concluded that food safety was the ultimate responsibility of the retailer or the final point in the supply chain that sold products to consumers. The ASC like the BAP emphasizes use of BMPs for achieving compliance with standards. The ASC plans to monitor the effectiveness of the certification process on achieving the expected environmental and social benefits.

It is noteworthy that ASC is independent of the WWF—the famous panda icon of WWF is not used on the certification label. Moreover, some of the other eNGOs do not support ASC. In fact, there has been particular criticism of the Shrimp Aquaculture Dialogue and resulting standards. In April 2012, an open letter of objections to the Shrimp Aquaculture Dialogue was signed by 95 environmental and social advocate organizations and 228 individuals (http://www.thebahamasweekly.com/uploads/10/OpenLetterToWWF24April2012.pdf). This illustrates the continuing rift between the aquaculture industry and its collaborators and the mainstream eNGO community. To illustrate the breadth of the chasm between the two groups, many involved in shrimp eco-label certification felt that the Shrimp Aquaculture Dialogue standards were so stringent that most shrimp farms could not achieve compliance, while the group who signed the open letter mentioned above felt that these standards were too lenient.

Friend of the sea

This organization began as the Earth Island Institute Dolphin-Safe Project. In addition to aquaculture products FOS certifies fisheries products including fish meal and

Table 14.4 Third party auditing bodies accredited by Accreditation Services International (ASI) to conduct compliance audits against Aquaculture Stewardship Council (ASC) standards and corresponding chain of custody (CoC) certification.

Company	Scope
AGRIZERT Zertifizierungsgesellschaft GmbH (AGZ)	CoC certification, worldwide.
AMITA Institute of Environmental Certification Co., Ltd. (AIEC)	CoC certification, worldwide
ARS PROBATA GmbH (ARS PROBATA)	CoC certification for ASC products for Germany, Switzerland, Austria, Poland, Lithuania, Denmark, the Netherlands, and Estonia
Bureau Veritas Certification Holding SAS (BVC)	Pangasius – Vietnam; Single Site CoC certification, worldwide
Control Union Certifications B.V. (CU)	CoC certification, worldwide
Control Union Peru SAC (CUP)	Tilapia and pangasius Single Site and CoC certification for ASC
Dansk Institut for Certificering (DIC)	CoC single site certification, worldwide
Det Norske Veritas Certification AS (DNV)	CoC certification, worldwide
Food Certification International (FCI)	CoC certification, worldwide
Institute for Marketecology (IMO)	Tilapia and pangasius Single Site and CoC certification, worldwide
Kiwa Sverige AB (KIWA)	CoC certification for ASC products in Sweden, Norway, Netherlands, Spain, Germany, and Belgium
MacAlister Elliott & Partners Ltd. (MEP)	CoC certification, worldwide
MRAG Americas, Inc (MRAG)	CoC certification, worldwide
q.inspecta GmbH (q.inspecta)	CoC certification for ASC products in Switzerland, Austria, Germany, France, Italy, Spain, Belgium, Luxemburg, Lichtenstein
SAI Global (SAI)	CoC certification, worldwide
SCS Global Services (SCS)	CoC certification, worldwide
SGS Nederland BV (SGS)	CoC certification, worldwide
TÜV Nord Cert GmbH (TN Cert)	CoC certification, worldwide

Source: http://www.accreditation-services.com/archives/standards/asc.

fish oil. It also certifies fish feed (http://www.friendofthesea.org/about-us.asp). The FOS sustainable aquaculture criteria are:

- No impact on critical habitat such as mangroves and wetlands;
- Compliance with wastewater parameters;
- Reduction of escapes of farmed animals and by-catches to a negligible level;
- No use of harmful antifoulants, genetically modified organisms, and growth hormones;
- Compliance with social accountability;
- Gradual reduction in carbon footprint.

According to Boyd and McNevin (2010), FOS certifies about 500 000 t of aquaculture products annually.

GLOBALG.A.P.

The GLOBALG.A.P. certification program for agricultural products added aquaculture to its offerings in 2005. Certification by GLOBALG.A.P. requires compliance with an All Farm Module and an Aquaculture Module. The aquaculture standards set specific criteria for site management, reproduction, chemicals, occupational health and safety, fish welfare—management and husbandry, harvesting, sampling and testing, feed management, pest control, environmental and biodiversity management, water usage and disposal, post-harvest operations, mass balance and traceability, and social issues. There also is a requirement to source compound feed from reliable suppliers certified by GLOBALG.A.P. and a GLOBALG.A.P. chain of custody standard (http://www.globalgap.org/uk_en/for-producers/aquaculture). The GLOBALG.A.P. website has a feature called My Fish into which the consumer may enter the identification number from the product package to find the identity and location of the GLOBALG.A.P.-certified farm of origin.

GLOBALG.A.P. claims that all GLOBALG.A.P.-certified products automatically undergo an environmental impact assessment as a result of tasks performed to comply with standards. The program is arranged much in the same way as the HACCP food safety inspection program with critical control points with which the participant must comply—there are 249 control points in the aquaculture module.

The program has developed standards for crustaceans (namely shrimp), molluscs, *Pangasius*, salmon, tilapia, and trout. According to GLOBALG.A.P., as of 2011, the organization had certified aquaculture farms in 20 countries and certified production for that year was over 2 Mt.

Government standards

The Label Rouge program in France has certified aquaculture products produced in the country and in other countries from which aquaculture products are imported (Mariojouls and Wessells 2002). The program focuses on food quality as already discussed, but certification of aquaculture products also requires compliance with environmental, social, and animal welfare standards.

The Thai Department of Fisheries developed a code of conduct (Thai CoC) for shrimp farming based on a modification of the BMPs developed by GAA (Boyd 1999; Tookwinas et al. 2000). This code went through several revisions by stakeholders and finally became a systematic approach intended to manage shrimp production to meet international quality standards, protect the environment, and maintain a sustainable shrimp farming industry in the country. The process for certification uses local certifiers trained by the Thai government. Producers who fail to comply with the Thai CoC can apply for the good aquaculture practices (Thai GAP) certificate. The Thai GAP program is promoted to assure a safe product, but its environmental requirements are not as rigorous as those of the Thai CoC (http://www.fao.org/fishery/legalframework/nalo_thailand/en). According to Yamprayoon and Sukhumparnich (2010), the food safety requirements in the Thai

CoC and GAP follow guidelines of the Codex Alimentarius Commission as well as those of Good Manufacturing Practices and HACCP.

Trans-national partnerships

A related effort to certification and market access is being developed by a host of organizations that primarily work with small-scale aquaculture producers to enhance their livelihoods—FAO, IDH, USAID, NACA, and others. While the specific projects of these groups vary, there is a general objective to make grassroots aquaculture producers more competitive, leading to a more successful livelihood.

Small-scale farmers sometimes are interested in entering international markets to sell their products, but more often, their entry into these markets is advocated for by processing plants seeking to increase production or by those seeking to improve the livelihood of small-scale producers. There are projects which seek to connect small-scale producers to international markets or even advertise small-scale producers and their grassroots nature as a product attribute. The ASEM is an informal process of dialogue and co-operation among the European Union member states and the European Commission with 19 Asian countries with the objective of strengthening the relationship between the two regions in a spirit of mutual respect and equal partnership. This partnership approach appears to be a result some importing countries being unsympathetic to the challenges of small-scale aquaculture producers competing with larger industrial scale aquaculture operations. This competition no doubt has led to marginalization of the small-scale sector, particularly as standards and eco-labels have proliferated and the requirements of many of the programs require a fair level of technical infrastructure and expertise. The ASEM effort addresses political, economic, and cultural issues, with the objective of strengthening the relationship between the two regions and reconciling Asian and European interests in the production and consumption of aquaculture products. Thus this effort appears to provide extension assistance to producers while communicating the particular challenges and advantages that the small-scale sector may encounter http://cordis. europa.eu/search/index.cfm?fuseaction=lib.document&DOC_LANG_ID=EN& DOC_ID=108741002&pid=3&q=55ACF4D5FF31F980798BF88BAC889B6D& type=sim.

Demand for certified products

While the motives for having certified aquaculture products are fairly clear, the interest in these products by consumers is not so obvious. Most consumers in developed countries and many in other countries likely would say that they support environmental and social stewardship. Surveys also indicate that consumers tend to believe that their purchasing choices have an effect on environmental and social outcomes (Smith 1990; Peattie 1995; Robins and Roberts 1997). Studies have noted, however, that attitudes are a poor predictor of marketplace behavior (Kraus 1995; Ajzen 2001; Vermeir and Verbeke 2006). In a survey of US and Norwegian consumers,

96% and 67%, respectively, said that they would choose certified salmon over a non-certified product at the same price. The percentage that would opt for certified salmon declined as the price differential between the two products increased (Donath et al. 2000).

As with organic products some consumers would prefer eco-labeled aquaculture products despite having to pay a greater price. At present this market is a niche market like that of organic aquaculture. Of course aquaculture certification is growing in many western, developed countries because retailers generally offer the certified products at no extra cost. In Asia, few consumers discriminate between products on environmental and social grounds and there is little interest by retailers in certified aquaculture products (Phillips et al. 2003).

Conclusions

Retailer sourcing policies and third-party certification systems—including organic certification—provide the greatest assurance that aquaculture products result from responsible production methods. There is an increase in production cost associated with implementation, maintenance, and auditing of these programs that must be borne by the producer. There usually is a price incentive to producers for participation in retailer sourcing policies and organic certification. However, these two kinds of aquaculture product differentiation programs are rather small compared to third-party certification programs such as BAP, ASC, FOS, and GLOBALG.A.P. These third-party certification programs do not provide a price incentive but producers must be certified by one of the programs to sell shrimp to certain buyers. Aquaculture certification is growing and there will likely be increased pressure on producers in the future to obtain certification from one or more entities. It is likely that retailer sourcing policies will decline in number because they can be replaced by certification programs in which most of the cost is passed down to the producer.

Small-scale producers are at a disadvantage with respect to certification unless mechanisms are provided whereby they can band together in clusters to lessen the cost of certification. There is reason to believe that certification schemes lead to further marginalization of small-scale producers (New 2003; Béné 2005; Bolton et al. 2010).

A decision to sell certified aquaculture products can be viewed as a commitment to environmental and social stewardship by retailers and restaurants. This is an inexpensive way at present of achieving such recognition if they do not have to pay more for the certified products. Environmentally conscious consumers will feel good about purchasing a certified product, and other consumers certainly will not object to certified products as long as the price is the same as for non-certified products. It is not clear what proportion of world aquaculture production could eventually be certified—presently about 5–7% is certified at the processing plant level, but maybe only 3–4% at the farm level.

Certification programs are weighted heavily toward products that are exported to developed countries. Assuming that certification leads to reduced negative environmental and social impacts—there is not actual documentation of this at present—

it cannot be expected to reduce the overall impacts of aquaculture a great deal as most aquaculture is for domestic products. Effort should be devoted to programs for improving environmental performance of aquaculture of domestic products. This effort probably will have to be driven by the voluntary use of BMPs and government regulations; it is unlikely that producers of domestic products in developing countries can afford certification programs.

The eNGO perspective

Certification is one of the most coveted tools eNGOs have to support environmental stewardship beyond the borders of protected areas. However, there is considerable discourse within the multilateral and governmental sectors and even the eNGO community on the efficacy of certification. Many eNGOs do not feel as if certification goes far enough to address environmental issues, multilateral development organizations contend certification marginalizes the poor small-holder farming community and governments see certification as either another trade impediment or "neo-colonialism." Probably the most over-arching criticism from the "political ecologists" and the like-minded is that certification is a technocentric endeavor that does not address the true problems of globalization and economic disparities. Thus techno-centric solutions to environmental degradation or poverty alleviation in the developing world are entrenched in the notion that the global capitalist approach is the appropriate model. There is likely some truth to all of these criticisms and debate will most certainly continue.

It is interesting to ponder how one might address negative impacts of any natural resource-intensive industry engaged in the globalized economy. The national government of countries would seem to be an appropriate starting point and some eNGOs are fortunate enough to have a global network which puts them in the position to lobby governments. However, the largest eNGOs are based in the developed world where most countries have a limited aquaculture industry and already have mechanisms in place to prevent negative impacts. In these regions the eNGOs can only touch the consumer community to attempt to aid their partners in other parts of the world. Considering that eNGOs do not have the manpower to reach every consumer, it is quite strategic then that they choose to alter certain markets in an attempt to address environmental issues that have become institutional priorities. The degree to which they seek to alter a form of production varies by organization and this aspect is the "Achilles heel" of the eNGOs.

One might perceive the eNGOs to be a fortified front attempting to stop environmental degradation with a common voice; however, this is not the case. In fact some of the eNGOs might be considered less aligned with each other than some of the fiercest competitors producing aquaculture product for the same market. Whether this approach is carving a specific niche for a particular eNGO for funding opportunities or potential retail partnerships (sought so fervently), the disharmony in the community shaped the landscape of certification in the aquaculture sector.

The rise of the GAA and the BAP program is probably the best perspective to take in understanding the above assertions. The eNGOs were distrustful of GAA as it was a producer-trade association and was considered to be "the fox guarding the hen house" as it relates to improving the performance on farms. The partnership between CI and Wal-Mart was the first blow to the eNGOs that sought to develop a more credible certification. Against popular opposition by several eNGOs in the United States, CI went forward with brokering the arrangement between GAA and Wal-Mart that galvanized BAP program's market penetration and jettisoned it to the dominant label in the US market.

As previously mentioned Dr. Jason Clay was the only commenter on the BAP shrimp standards. He viewed the development of a certification program as an ideal opportunity to generate forces to impart greater environmental stewardship. Following the first meetings of the Salmon Dialogue and the Bivalve Dialogue, the junior author was hired by the WWF to work in Dr. Clay's group. One of the first tasks was an attempt to collaborate with the GAA to go beyond shrimp certification and address the most economically important species. Further, the WWF sought to spin-off the ACC as an independent standards holding body to take not only the BAP shrimp standards, but also the Aquaculture Dialogue standards. At a September 2006 meeting of the WWF, GAA, ACC, and MSC the authors and representatives of those groups discussed the potential for collaboration (Chamberlain 2006). Coming with the clout of a Wal-Mart partnership, GAA was confident that they could not only develop standards for specific species aside from shrimp but do it in a timeline that they could control. The GAA saw the Aquaculture Dialogue efforts as a process that would take too long to capitalize on the market opportunities for certification. The GAA sought to control the ultimate decisions on their standards and did not desire to relinquish control to any stakeholder.

While GAA began to develop new sets of standards for a variety of species, the Aquaculture Dialogues were given a shot in the arm by the Packard Foundation with a multi-million dollar grant to ramp up activities and increase support for standards development. The prospect also arose for the Dialogue standards to be used by MSC to form a sister certification program for aquaculture products.

While GAA did not have credibility in the eNGO community, they sought on several occasions to herald their collaboration, but few eNGOs were willing to collaborate. Some eNGOs likely felt betrayed by CI's association with the Wal-Mart–GAA partnership.

It was not until a 2008 meeting of eNGOs working on seafood and funded by the Packard Foundation that the next eNGOs decided to break off and collaborate with GAA giving them more credibility. New England Aquarium which has a partnership with Darden Restaurants (a founding member of GAA because of their Red Lobster chain) and EDF which was developing plans for setting up an office in Bentonville, AR (the headquarters of Wal-Mart) were convinced that there was opportunity to improve the inner workings of GAA and ACC and to entice them into making their standards and certification more stringent. One year later, the Seafood Choices Alliance began working with GAA and in 2010, Monterey Bay Aquarium's Seafood Watch program began collaborating.

The GAA maneuvered the eNGO community with incredible sophistication, all the while continuing their push for more standards and more market share of aquaculture production, each step obtaining greater credibility from subsequent eNGO partners without relinquishing control of the BAP standards.

The lessons for the eNGO community are stark in this context. It is unknown to what degree these scenarios play out in other food production or natural resource extractive industries, but there is a need for a more united front to address the issues that eNGOs in principle appear to coalesce around.

In the end most eNGOs remained engaged in both the GAA's efforts as well as in the Aquaculture Dialogues. The MSC mulled over the prospects of absorbing the Aquaculture Dialogue standards, but finally decided against it because of the massive effort required for analyzing Wal-Mart's wild capture sources. The WWF and IDH partnered to ultimately form ASC and as the Aquaculture Dialogues wind down the hand-off of the standards to ASC continues.

Certification is ultimately a process of exclusion for those that are not certified. If all or the majority of an industry were to be certified, it would likely be charging producers to operate at the *status quo* with current product pricing. Thus certification must strike a balance between pushing the environmental performance of producers to a challenging point where they could achieve some type of price premium in the market, but not so far as to reduce volume to such a small amount that it is unrealistic for producers or buyers to participate. This is quite a balancing act and most certifications go the route of certifying as many producers as possible or limiting producers to such low numbers that the product remains isolated in a niche market with little growth potential. Thus some producers will never attain certification and some will not care to attempt it. Other producers operating ahead of the curve will likely have a relatively easy time achieving the standards. The true distinction among the main aquaculture certification programs is probably the proportion of the mainstream market to which standards are targeted. The eNGO sector would ideally be seeking to target the smallest (best) fraction of the mainstream. Because the "best" should continue to get better, one of the most important components of any certification is the mechanism for increasing the difficulty of standards as producers become more efficient—continuous improvement. The market entry and subsequent traction of a particular certification program might be more powerful a force than the efforts to continuously improve standards; after all, certification is ultimately a business endeavor.

Although the eNGOs made mistakes as did the industry throughout the upsurge of certification program development, one lesson shines through—the true properties of a sustainable aquaculture facility will likely never be embodied in a standard because these properties are intangibles. There is an ethic one cannot test for in an audit. There are numerous aquaculture operations managed by incredible individuals who take enormous strides toward environmental sustainability. These producers exemplify development and environmental stewardship and most importantly, they would be performing at this level with or without the advent of aquaculture certification.

References

Ajzen, I. 2001. Nature and operation of attitudes. *Annual Review of Psychology* 52:27–58.
Béné, C. 2005. The good, the bad and the ugly: Discourse, policy controversies and the role of science in the politics of shrimp farming development. *Development Policy Review* 23:585–614.

Bolton, B., F. Murray, J. Yound, T. Telfer, and D. C. Little. 2010. Passing the panda standard: A TAD off the mark? *Ambio* 39:2–13.

Boyd, C. E. 1999. *Codes of practice for responsible shrimp farming.* St. Louis, Missouri: Global Aquaculture Alliance.

Boyd, C. E. 2000. Case studies of world shrimp farming. *Global Aquaculture Advocate* 3:11–12.

Boyd, C. E. and J. W. Clay. 1998. Shrimp aquaculture and the environment. *Scientific American* 278:42–49.

Boyd, C. E. and A. A. McNevin. 2010. An early assessment of the effectiveness of aquaculture certification and standards. Report to RESOLVE, Washington, DC.

Boyd, C. E., A. A. McNevin, J. Clay, and H. M. Johnson. 2005. Certification issues for some common aquaculture species. *Reviews in Fisheries Science* 13:231–279.

Chamberlain, G. 2006. GAA, ACC join certification group meeting. *Global Aquaculture Advocate* 9(5):10.

Clay, J. W. 1996. *Market Potential for Redressing the Environmental Impact of Wild-Captured and Pond-Produced Shrimp.* Washington: World Wildlife Fund.

Clay, J. W. 2004. *World Agriculture and the Environment: A commodity-by Commodity Guide in Impacts and Practices.* Washington: Island Press.

Clay, J. W. 2008. The role of better management practices in environmental management. In C. S. Tucker and J. A. Hargreaves, editors, *Environmental Best Management Practices for Aquaculture*, pp. 55–72. Oxford: Blackwell Publishing.

Donath, H., C. R. Wessells, R. J. Johnson, and F. Asche. 2000. Consumer preferences for ecolabeled seafood in the United States and Norway: A comparison. Proceedings International Institute for Fisheries Economics and Trade, July 10–15, Corvallis, Oregon.

FAO (Food and Agriculture Organization of the United Nations). 2011. *Technical Guidelines on Aquaculture Certification.* Rome: FAO.

FAO/NACA/UNEP/WB/WWF. 2006. *International Principles for Responsible Shrimp Farming.* Bangkok: Network of Aquaculture Centres in Asia-Pacific.

Kraus, S. 1995. Attitudes and the prediction of behavior—a meta-analysis of the empirical literature. *Journal of Personality and Social Psychology* B21:58–75.

Mariojouls, C. and C. R. Wessells. 2002. Certification and quality signals in the aquaculture sector in France. *Marine Resource Economics* 17:175–180.

New, M. B. 2003. Responsible aquaculture: Is this a special challenge for developing countries? *World Aquaculture* 34:26–30, 60–72.

Peattie, K. 1995. *Environmental Marketing Mangement. Meeting the Green Challenge.* London: Pitman Publishing.

Phillips, B., T. Ward, and C. Chaffee. 2003. *Eco-labelling in Fisheries: What is it all about?* Oxford: Blackwell Publishing.

Robins, N. and S. Roberts (editors). 1997. *Unlocking Trade Opportunities: Changing Consumption and Production Patterns.* London: The International Institute for Environment and Development.

Smith, G. 1990. How green is my valley? *Marketing and Research Today* 18:76–82.

Szeremeta, A., L. Winkler, F. Blake, and P. Lembo (editors). 2010. *Organic Aquaculture.* Brussels: International Federation of Organic Agriculture Movements EU Group.

Tookwinas, S., S. Dirakkait, W. Prompoj, C. E. Boyd, and R. Shaw. 2000. Thailand develops code of conduct for shrimp farming. *Aquaculture Asia* V1:25–28.

Vermeir, I. and W. Verbeke. 2006. Sustainable food consumption: Exploring the consumer "attitude–behavioral intention" gap. *Journal of Agricultural and Environmental Ethics* 19:169–194.

Washington, S. and L. Ababouch. 2011. Private standards and certification in fisheries and aquaculture. FAO Fisheries and Aquaculture Technical Paper 553, FAO, Rome, Italy.

Yamprayoon, J. and K. Sukhumparnich. 2010. Thai aquaculture: achieving quality and safety through management and sustainability. *Journal of the World Aquaculture Society* 41:274–280.

Endnotes

1. The definitions below are modifications of definitions provided by FAO (2011):

   Standard setting body—An organization that sets standards and criteria needed for differentiating products or awarding product certification.

   Transparency—Implies openness, communication, documentation, and accountability in setting standards and operating product differentiation and certification schemes. In particular these schemes should be open to scrutiny by consumers, their advocates, and other interested parties.

   Standard—A statement that provides specific rules for repeated use about how a product shall be produced. Of course unless it involves governmental or international regulations, it is for voluntary adoption.

   Conflict of interest—Any situation that could make it difficult for an individual or entity to make an impartial decision related to standard setting, accreditation, auditing for compliance with standards, or maintaining the chain of custody.

   Chain of custody—A process that tracks a product from a certified production facility to the consumer to assure that it is not mixed with a non-certified product of the same kind.

   Accreditation body—An organization that has a system by which it can confirm that an entity is qualified to conduct a specific task and grants that entity accreditation.

   Certification body—An entity that has been accredited to conduct audits or oversee other bodies that carry out audits related to certification.

   Certification—A procedure by which an accredited certification body gives assurance that a product, process, or service conforms to specified requirements.

   Certification unit—The specific unit that is granted certification. This often consists of a single unit, for example, a farm or a processing plant, but it can be a group or cluster of small farms with different owners that is assessed and monitored as a whole.

Chapter 15

Some final thoughts

Species appear and disappear; fossil records show that many species of plants and animals alike have risen to dominance in ecosystems, and then—often suddenly—became extinct. It takes little imagination to conclude that humans might follow a similar pattern. Add the religious prophecies of the end of the earth, and it is easy to understand the fascination that many have for the future. Predictions about the future—as diversions and as serious efforts—are common; in particular there is a wide range of speculation about how and when the human race will end. Fortunately, most people realize that this event is not likely to happen in their lifetime or even during their children's lives; thus, planning for the future is an essential human endeavor that is done by individuals, businesses and other private organizations, and particularly governments.

There is a large consumer base for prognostications regarding every aspect of human life. Many people seek to provide information for this audience—including eNGOs, the aquaculture industry, and the authors of this book. Of course, as pointed out several times earlier, it is difficult to predict the future and none of the doomsday predictions have come to pass—at least not yet. Moreover, predictions about mundane events—such as the economy—have a rather dismal record of accuracy. But this fact does not deter the prognosticators—not even one iota!

The depletion of nonrenewable resources in the near future is more likely to occur than generally acknowledged even by many prognosticators. Known reserves of most nonrenewable resources are sufficient for only a few decades or at most a few centuries. Of particular urgency is the need to find a sustainable alternative to our almost complete dependence upon fossil fuels for energy. Moreover, renewable resources must be used more sustainably or there will be less of these essentials in the future.

Humans appeared quite recently in geological history, and for many millennia the population was small and grew slowly. The size of the human population was controlled by basically the same natural factors regulating abundances of other animals. Mankind finally learned to exert enough control over nature to assure a relatively

Aquaculture, Resource Use, and the Environment, First Edition. Claude E. Boyd and Aaron A. McNevin.
© 2015 John Wiley & Sons, Inc. Published 2015 by John Wiley & Sons, Inc.

stable food supply and to lessen mortality from common infections and diseases. The natural death rate fell well below the natural birth rate, and population growth soared. In the past two centuries, the world population increased from about 1 billion in 1800 to over 7 billion by the end of 2012. According to the UN median estimate, the population should reach about 9.3 billion by 2050, but increase more slowly to about 10 billion by 2100. Possibly the population will decline gradually afterward. The huge population requires massive amounts of primary resources: land, water, fossil fuels, agricultural crops, mineral ores, timber, etc. The per capita use of primary resources has increased more rapidly than the rate of population increase in developed countries because they have tended to become more affluent over the past two centuries. Several less developed countries—including heavily populated China and India—are now developing large middle classes and their per capita resource use is increasing because of greater affluence.

Despite there being enough remaining nonrenewable resources to meet the demand for the next 50–200 years at current use rates, they will surely become more expensive and soon run out. The fact that population growth rate is slowing and the global population is expected by some authorities to stabilize by the end of this century or soon afterward does not improve the outlook for the natural resource situation. The biggest fractions of the world's nonrenewable resources were used up in the past two centuries—much more was used than had been consumed in all of previous human history. Assuming the *status quo* it seems likely that most of the world's remaining nonrenewable resources will be depleted before the size of the world population hopefully stabilizes near the end of this century.

Renewable resource production typically requires an input of nonrenewable resources. For example, production of agricultural crops by modern farming depends on fossil fuels, phosphate fertilizers, and a few other nonrenewable resources. In addition, renewable natural resources often are overexploited, and negative environmental impacts caused by human activities can disturb natural and agricultural ecosystems causing their productivity to decline. The capture fishery illustrates this point quite nicely. The world's fishing fleets use a lot of fossil fuel in their quest for fish and other aquatic species. However, the fishing pressure exceeds the sustainable limits for many fisheries, and pollution also is having a toll on some fisheries. The capture fishery production cannot be expected to increase, and it possibly will decline.

The pessimistic viewpoint is that humans have embarked on an irreversible course to unprecedented disaster because of depletion of natural resources and disruption of ecosystems leading to a decline in essential ecological services. In other words, the world simply cannot support such a large human population, and lack of resources and ecological collapse certainly will soon cause the human race to diminish greatly in number or to disappear entirely. The pessimistic view of the future has always been an exaggeration because the critics do not understand human nature. Humans can be entirely oblivious to impending events, resistant to change, uncooperative, and totally self-centered, but they also are sometimes quite the opposite. Moreover, humankind is highly resilient and innovative; contrary to other species it has the ability to reason individually or collectively when it chooses to do so. It seems highly probable that humans will rise to the occasion by finding a new energy paradigm, recycling nonrenewable resources, using renewable resources more sustainably, and reducing the natural rate of increase so that the population peaks and eventually declines in a somewhat orderly manner to a sustainable level—let us hope for this outcome.

Humanity, of course, has had the technological capability for at least the last six decades to annihilate itself in a nuclear holocaust, but for some reason, such an event has been avoided. The senior author remembers well the graphic descriptions in news magazines of the 1950s and 1960s about the horrible consequences that would result from the impending nuclear war between the United States and the former Soviet Union. Possibly, the terrifying news pieces were responsible for avoiding a global, nuclear catastrophe. But it is more likely that the leaders of the two countries were wise enough—or at least practical enough—to conclude that the nuclear solution to their differences would be more final than desirable. The resource use and environmental issues currently faced by the world—if not solved—would not be as dramatic an end to civilization as we know it as would have been a global nuclear war, but it could have an equally negative impact on the human race.

There is also the ever-present possibility that some unexpected natural event of history-changing proportions will happen. Such events have altered both geological and human history in the past and could do so again. For example, a volcanic eruption similar or greater in magnitude to the Toba event—the tremendous volcanic explosion that formed Lake Toba in Sumatra, Indonesia about 70 million years ago—could essentially annihilate the human race. The same effect could be caused by a collision of the earth with a large asteroid, or the occurrence of a deadly pandemic disease. However, such events are extremely infrequent, and it is unlikely that one will occur before the world's nonrenewable resources are depleted, or on the other hand, before the world finds a sustainable course of resource use and environmental stewardship.

The effort to make the human population sustainable must consider all activities that use resources or cause negative environmental impacts, but the most progress can be made by focusing on the major sectors. Aquaculture is a comparatively minor consumer of most resources. Marine fish meal and oil are the only resources of which the major shares are used by aquaculture—60% and 80% of global shares, respectively. Although the eNGOs have focused intensely on this issue, the aquaculture and fish meal industries were aware of this problem even before the environmentalists. They realized that it would restrict the future growth of aquaculture, and research already has identified a variety of acceptable substitutes for marine fish meal and fish oil in aquaculture feeds and the effort continues. Aside from fish meal and oil use, there does not seem to be much basis for great concern about excessive resource use in aquaculture; it consumes less than 0.5% of annual global consumption of most resources and no more than 2% or 5% of a few others. It also produces about 6.5% of world animal protein—a seemingly good exchange with respect to resource consumption.

The negative environmental impacts of aquaculture also have been exaggerated. An example is the furor about mangrove destruction and other land use changes caused by aquaculture. However, aquaculture has been responsible for only 5–10% of loss of global mangrove area at the most and a much smaller percentage of the total loss of wetlands. The amount of agricultural land converted to aquaculture is not usually a negative impact because aquaculture is equally or more productive of food per unit area than is traditional agriculture.

Aquaculture can negatively affect biodiversity through discharge of nutrients, organic matter, suspended solids, metals, antibiotics, disinfectants and other chemicals, introductions of nonnative species or genetically modified organisms, predator control, and entrainment of organisms in pumps. However, its adverse effects on water quality and biodiversity tend to be localized and over dramatized. For example, aquaculture has been blamed for introductions of nonnative aquatic animal species—one general document by FAO states that aquaculture is the main reason for such introductions. The facts point to a much different scenario; aquatic animal introductions for sport fishing and the aquarium trade have greatly outnumbered those resulting from aquaculture of food organisms. Of course, aquaculture is the main reason for many introductions of tilapias worldwide.

Aquaculture also contributes to greenhouse gas and sulfur and nitrogen dioxide emissions. But aquaculture is responsible for a very minor share of global fuel use and therefore contributes in a proportionally minor way to atmospheric emissions.

Efforts obviously should be made to improve the efficiency of resource use and to lessen the negative environmental impacts of aquaculture. The extent of criticism directed at aquaculture on these grounds, however, seem to far outweigh the actual impacts that it causes. One only needs to observe while traveling around the countryside by car or while looking down at the landscape on a clear day from an airplane to realize that even in Asian countries with large aquaculture sectors, the imprint of aquaculture on the landscape is relatively miniscule when compared to those of traditional agriculture, expansion of heavily populated areas and associated transportation networks, and various other activities.

The attempt by some of the more radical eNGOs to change the image of aquaculture from a "blue revolution" to an "environmental pariah" that began in the late 1990s may have had the beneficial result of raising awareness within the aquaculture industry of the need for improving production practices, and in causing governments to impose more rigorous environmental regulations on aquaculture. This is not certain because there was already much effort underway within the aquaculture industry by the 1990s to improve sustainability. Aquaculture's image was not greatly tarnished by the recent assaults because it reacted proactively and improved its environmental performance. Thus the outcome possibly justifies the approach of the more radical eNGOs, but many aquaculturists would argue to the contrary.

The efforts of the more moderate eNGOs and some other organizations to promote better practices and develop eco-label certification programs for aquaculture facilities are having a positive influence on the performance of many aquaculture facilities and in raising environmental awareness within the supply chain for aquaculture products—including consumers.

The adoption of better practices also improves prospects for sustainability of aquaculture facilities. Many examples could be cited, but three will suffice. Aquaculture farms located on inferior sites are more likely to cause negative impacts and ultimately be abandoned because of poor production than do farms located on better sites, e.g., a shrimp farm located in a mangrove area versus a shrimp farm located outside the intertidal zone. Shrimp farms often share a water source and use this water body both as a water supply and an effluent recipient. Obviously, reduction in discharge from shrimp farms into their water supply will lessen the possibility of cross contamination among farms in an area—a particularly important issue in disease control. Limits on production or improvements in feed use resulting in smaller nutrient inputs to lakes or reservoirs with cage culture will avoid accelerated eutrophication that could lead to dissolved oxygen depletion and fish mortality in cages.

Better practices also can lessen resource use and improve production in many instances. Some examples are: lower FCR reduces feed input; less water use diminishes pumping requirements; adequate mechanical aeration insures plenty of dissolved oxygen and lessens fish stress; erosion control protects pond earthwork and minimizes sedimentation in ponds leading to less maintenance; good fish health management does not necessitate use of antibiotics and other chemicals. Thus adoption of better practices likely will lower production costs and improve profitability—a topic deserving further study.

Solutions to the world's resource use and environmental problems will require application of scientific and technological findings, but they also will be highly dependent upon changes in human attitude and behavior, education on conservation and the environment, legislation, planning, cooperation, etc. The rising middle class in much of the developing world will likely be the hardest to convince of the need for new paradigms of resource use and conservation. The developed world has already used up most of the world's resources and the people of developing countries with improving economies no doubt feel that they should have the right to use more resources than in the past now that they can afford to do so.

Hopefully this book will help readers put aquaculture in perspective with other aspects of global resource use, food supply, and the overall effects of human actions on the world's ecosystems. Aquaculture is a small but growing facet of world food production that should be encouraged. It is the only way of preventing a shortage of fisheries products because capture fisheries production is no longer capable of meeting the demand. As aquaculture is a relatively small sector it requires a small share of most global resources—the exception being fish meal and fish oil. It contributes in a lesser way to environmental perturbations—albeit mostly preventable ones—at the local level.

Many in the aquaculture industry may feel that the environmentalist have—to use a clique—"bigger fish to fry" than aquaculture. But the aquaculture sector should continue efforts to improve resource conservation and environmental stewardship. We must be concerned about all reasons for wasteful resource use and negative environmental impacts regardless of the size of the sector causing them. The small sectors causing local impacts combined no doubt cause rather large global effects. The small sector effects only can be reduced by addressing the problems on a sector by sector basis. We all contribute to the loaming sustainability trap, and we all should be willing or be required by governments to do our share to escape it.

The eNGO perspective

"Flush with cash and never satisfied"—this was often the characterization of the eNGOs by industry advocates in the aquaculture sector. Industry advocates saw ties to the large private foundations in Palo Alto, California, the large grants from the Global Environment Facility and million dollar partnerships with Coca Cola—this was the luxurious life up in the ivory tower. An attempt has been made to put aquaculture in perspective with the traditional agriculture sector. It is useful therefore to also put the work of the eNGOs on aquaculture in perspective with their other efforts, because those in the aquaculture industry seldom look beyond the bounds of their paradigm. While there are several eNGOs working solely on seafood-related issues, most notably the various aquaria, Seaweb, Sustainable Fisheries Partnership, Seafood Choices Alliance, Fishwise, etc., a large portion of the work of eNGOs in aquaculture is taken on by larger NGOs with a global reach. These eNGOs include WWF, Conservation International, The Nature Conservancy, Environmental Defense Fund, Friends of the Earth, Greenpeace, etc. The amount of funding allotted to aquaculture campaigns by these organizations is relatively small compared to the issues related to climate change, deforestation, water scarcity, agriculture, and marine issues. Further, much of the aquaculture work of eNGOs started at the local context with grassroots actors that saw the rapid development of the aquaculture industry and experienced some of the mistakes made by the sector. When aquaculture started to conglomerate and become part of large multinational companies such as Marine Harvest, Charoen Pokphand, Cargill, the stage was set for large international eNGOs to begin to engage in an attempt to support grassroots environmental efforts.

A junior university faculty member would scoff at the salary of the entry-level eNGO. A small university president would make more than top management personnel at most of the eNGOs. What is often misunderstood is that the eNGOs are active and prevalent because of their passion to make a change or improve the environment which for them trumps the desires for creature comforts and large incomes and retirement packages. There is an intense lobbying of the foundations for grants, just the same as lobbying is conducted to gain investors for a business. It is difficult to imagine the existence of any employee of an eNGO that did not join this community for one sole purpose—to attempt betterment of the natural environment. Their views on what they think is appropriate are based on both passion and evidence. The ability to tilt the scales on the balance of passion and evidence can be challenging for those in the aquaculture industry and academia to accept. Extreme and radical eNGOs tip the balance between passion and fact in favor of passion. These eNGOs will never consider serious negotiations with industry, and asserting that eNGOs need to be at some appropriate level of scientific sophistication to make some of the claims levied at the industry is more a misunderstanding of intent.

The impacts of aquaculture are dwarfed by the impacts of many other sectors of the global economy but it should be reiterated that many mainstream eNGOs perceive aquaculture as a way to lessen the demands on capture fisheries and a beneficial option to most fishing endeavors. Aquaculture industry advocates have noted the same attributes of aquaculture. This obviously cannot be achieved if more fish in the oceans need to be caught to satisfy future growth in aquaculture. However, the advent of more carnivorous species novel to aquaculture being produced—namely cods, groupers, and tunas would not offset but rather increase the demands on capture fisheries. Thus, while more efficient feeds and feedstuffs have been developed to lessen the burden on wild fish, the eNGOs would argue that as much as the industry attempts to increase efficiency of wild fish conversions, the greater perspective on the issue is lost if the next sectors of aquaculture to be developed are more carnivorous than the previous.

It is likely that most eNGOs see their role, in part, as trying to raise awareness about the problems caused by certain human activities; however, they would argue the more important role for them is to ward off problems that have not been realized and to act as a "watchdog." Thus, the utility of having a watchdog is not necessarily to catch every robber, rather prevent robberies from happening in the first place. The eNGOs would argue that should they not be present much more harm could be possible.

Probably more so than ever before the eNGOs are working closer with the aquaculture industry. Site visits to aquaculture facilities by eNGOs occur more regularly now than just a few years ago when many in the industry would immediately refuse these types of interactions. There is likely a growing understanding that the lion's share of aquaculture is the production of aquatic plants and molluscs. With a minor few exceptions, there is a net benefit in the production of these species on the environment. Of the 84 Mt of aquaculture product produced in the world, 42% of the production is composed of plants and molluscs. Thus, the largest portion of the global aquaculture sector does more good than harm. Another 30% of the total is comprised almost entirely of carps. In short global aquaculture is likely more sustainable as a whole than most other types of food sectors. Yet it remains the fastest growing food sector in the world which provides opportunities for large-scale expansion and mistakes at the same scale.

It is, of course, true that from a relative perspective, aquaculture is a minor user of natural resources. We hope that this book has provided the context to understand this fact. But although aquaculture uses relatively less resources than other industries, there is an inescapable fact that the cumulative use of resources by all industries is at an alarming level at present and that every sector must play a role in the minimization of natural resource dependency. Segregation of sector-specific impacts while appropriate to describe the relative impacts of different industries is important, but a consistent segregation may unduly diminish the need to come together to address issues that affect every industry and every individual. Additionally each industry sector that depends on natural resources will need to continue to advance at a more rapid pace than at present. Some sectors will move more slowly, some such as aquaculture will advance more quickly. There will always be room for improvement, and there will always be some sectors that are doing better than others. This is not an excuse to decelerate the rate of progress or become complacent. We all have great strides ahead to begin to use natural resources at a truly renewable rate. There is no doubt that the eNGOs will be present to ensure acceleration towards these ends continue.

Index

Aquaculture, Resource Use, and the Environment, First Edition. Claude E. Boyd and Aaron A. McNevin.
© 2015 John Wiley & Sons, Inc. Published 2015 by John Wiley & Sons, Inc.